Methods in Enzymology

Volume 387
LIPOSOMES
Part D

METHODS IN ENZYMOLOGY

EDITORS-IN-CHIEF

John N. Abelson   Melvin I. Simon

DIVISION OF BIOLOGY
CALIFORNIA INSTITUTE OF TECHNOLOGY
PASADENA, CALIFORNIA

FOUNDING EDITORS

Sidney P. Colowick and Nathan O. Kaplan

Methods in Enzymology

Volume 387

Liposomes

Part D

EDITED BY

Nejat Düzgüneş

DEPARTMENT OF MICROBIOLOGY
UNIVERSITY OF THE PACIFIC SCHOOL OF DENTISTRY
SAN FRANCISCO, CALIFORNIA

ELSEVIER
ACADEMIC
PRESS

AMSTERDAM • BOSTON • HEIDELBERG • LONDON
NEW YORK • OXFORD • PARIS • SAN DIEGO
SAN FRANCISCO • SINGAPORE • SYDNEY • TOKYO
Academic Press is an imprint of Elsevier

Elsevier Academic Press
525 B Street, Suite 1900, San Diego, California 92101-4495, USA
84 Theobald's Road, London WC1X 8RR, UK

This book is printed on acid-free paper. ∞

For all information on all Academic Press publications
visit our Web site at www.academicpress.com

ISBN: 0-12-182792-5

PRINTED IN THE UNITED STATES OF AMERICA
04 05 06 07 08 9 8 7 6 5 4 3 2 1

Table of Contents

Section I. Antibody- or Ligand-Targeted Liposomes

Section II. Environment-Sensitive Liposomes

Contributors to Volume 387

Article numbers are in parentheses and following the names of contributors.
Affiliations listed are current.

ZRINKA ABRAMOVIĆ (17), *J. Stephan Institute, Ljubljana, Slovenia*

ATEEQ AHMAD (14), *Department of Pharmacokinetics, Metabolism, and Bioanalytical Research and Development Division of NeoPharm Inc., Waukegan, Illinois 60085*

IMRAN AHMAD (14), *Department of Pharmacokinetics, Metabolism, and Bioanalytical Research and Development Division of NeoPharm Inc., Waukegan, Illinois 60085*

THERESA M. ALLEN (11), *Department of Pharmacology, University of Alberta, Edmonton, Alberta T6G 2H7, Canada*

YECHEZKEL BARENHOLZ (19), *Laboratory of Membrane and Liposome Research, Department of Biochemistry, Hebrew University, Hadassah Medical School, Jerusalem 91120, Israel*

JEREMY BOOMER (10), *Department of Chemistry, Purdue University, West Lafayette, Indiana 47907*

PIETER R. CULLIS (7), *Department of Biochemistry and Molecular Biology, University of British Columbia, Vancouver, British Columbia V6T 5J8, Canada; Inex Pharmaceuticals Corporation, Burnaby, British Columbia V5J 5J8, Canada*

NEJAT DÜZGÜNEŞ (8), *Department of Microbiology, University of the Pacific School of Dentistry, San Francisco, California 94115*

ROM E. ELIAZ (2), *Department of Biopharmaceutical Sciences and Pharmaceutical Chemistry, School of Pharmacy, University of California, San Francisco, San Francisco, California, 94143*

XIN GUO (9), *Departments of Pharmaceutical Chemistry and Biopharmaceutical Sciences, University of California, San Francisco, San Francisco, California 94143*

ISMAIL HAFEZ (7), *Department of Biochemistry and Molecular Biology, University of British Columbia, Vancouver, British Columbia V6T 5J8, Canada**

PAAVO HONKAKOSKI (13), *Department of Pharmaceutics, University of Kuopio, Kuopio, Finland*

ZHAOHUA HUANG (9), *Departments of Pharmaceutical Chemistry and Biopharmaceutical Sciences, University of California, San Francisco, San Francisco, California 94143*

ILPO JÄÄSKELÄINEN (13), *Department of Pharmaceutics, University of Kuopio, Kuopio, Finland*

IVAN JALŠENJAK (18), *Department of Pharmaceutics, Faculty of Pharmacy and Biochemistry, University of Zagreb, 10000 Zagreb, Croatia*

JAN A. A. M. KAMPS (16), *Department of Cell Biology, Liposome Research Section, Groningen University Institute for Drug Exploration, 9713 AV Groningen, The Netherlands*

*Current Affiliation: Zucker Lab, University of California, Berkley, California 94720

SUMSULLAH KHAN (14), *Department of Pharmacokinetics, Metabolism, and Bioanalytical Research and Development Division of NeoPharm Inc., Waukegan, Illinois 60085*

JONG-MOK KIM (10), *Department of Chemistry, Purdue University, West Lafayette, Indiana 47907*

KENJI KONO (5), *Department of Applied Materials Science, Graduate School of Engineering, Osaka Prefecture University, Sakai, Osaka 599-8531, Japan*

JULIJANA KRISTL (17), *University of Ljubljana, Faculty of Pharmacy, Ljubljana, Slovenia*

KATRIINA LAPPALAINEN (13), *Department of Dermatology, Kuopio University Hospital, Kuopio, Finland*

CAROLE LAVIGNE (12), *Insitut de Recherche Medicale Beausejour, Moncton, New Brunswick, Canada*

BERNARD LEBLEU (12), *Molecular Genetics Institute, CNRS, 34293 Montpellier, France*

ROBERT J. LEE (3), *College of Pharmacy and Comprehensive Cancer Center, The Ohio State University, Columbus, Ohio 43210*

GABRIEL LOPEZ-BERESTEIN (15), *Department of Bioimmunotherapy, Unit 422, The University of Texas MD Anderson Cancer Center, Houston, Texas 77030*

PHILIP S. LOW (3), *Department of Chemistry, Purdue University, West Lafayette, Indiana 47907*

YANTO LUNARDI-ISKANDAR (12), *Insitut de Recherche Medicale Beausejour, Moncton, New Brunswick, Canada*

NASREEN MULLAH (4), *ALZA Corporation, Mountain View, California 94043*

DAVID NEEDHAM (6), *Department of Mechanical Engineering and Materials Science, Duke University, Durham, North Carolina 27708*

SHLOMO NIR (2), *Seagram Center for Soil and Water Sciences, Hebrew University of Jerusalem, Rehovot 76100, Israel*

ŽELJKA PAVELIĆ (18), *Department of Pharmaceutics, Faculty of Pharmacy and Biochemistry, University of Zagreb, 10000 Zagreb, Croatia*

MARIA C. PEDROSO DE LIMA (8), *Department of Biochemistry, Faculty of Sciences and Technology, University of Coimbra, 3000 Coimbra, Portugal*

MASOUD QAZEN (4), *ALZA Corporation, Mountain View, California 94043*

AYELET M. SAMUNI (19), *Laboratory of Membrane and Liposome Research, Department of Biochemistry, Hebrew University, Hadassah Medical School, Jerusalem 91120, Israel*

GERRIT L. SCHERPHOF (16), *Department of Cell Biology, Liposome Research Section, Groningen University Institute for Drug Exploration, 9713 AV Groningen, The Netherlands*

ROLF SCHUBERT (18), *Department of Pharmaceutical Technology, Albert-Ludwigs-University, 79104 Freiburg, Germany*

SEAN C. SEMPLE (11), *Inex Pharmaceuticals, Burnaby, British Columbia V5J 5J8, Canada*

MARJETA ŠENTJURC (17), *J. Stephan Institute, Ljubljana, Slovenia*

SERGE SHAHINIAN (1), *Department of Biochemistry, McGill University, Montréal, Québec, Canada, H3G 1YG*

JUNHWA SHIN (10), *Department of Chemistry, Purdue University, West Lafayette, Indiana 47907*

JOHN R. SILVIUS (1), *Department of Biochemistry, McGill University, Montréal, Québec, Canada, H3G 1YG*

SÉRGIO SIMÕES (8), *Laboratory of Pharmaceutical Technology, Faculty of Pharmacy, University of Coimbra, 3000 Coimbra, Portugal*

DORIS RIEKO SIWAK (15), *Department of Bioimmunotherapy, Unit 422, The University of Texas MD Anderson Cancer Center, Houston, Texas 77030*

NATAŠA ŠKALKO-BASNET (18), *Department of Pharmaceutics, Faculty of Pharmacy and Biochemistry, University of Zagreb, 10000 Zagreb, Croatia**

VLADIMIR SLEPUSHKIN (8), *VIRxSYS Corporation, Gaithersburg, Maryland 20877*

STACY M. STEPHENSON (3), *Division of Pharmaceutics and Pharmaceutical Chemistry, College of Pharmacy, The Ohio State University, Columbus, Ohio 43210*

DARRIN D. STUART (11), *Chiron Corporation, Cancer Pharmacology, Emeryville, California 94608*

FRANCIS C. SZOKA (2, 9), *Departments of Pharmaceutical Chemistry and Biopharmaceutical Sciences, School of Pharmacy, University of California, San Francisco, San Francisco, California 94143*

TORU TAKAGISHI (5), *Department of Applied Materials Science, Graduate School of Engineering, Osaka Prefecture University, Sakai, Osaka 599-8531, Japan*

ANA MARIA TARI (15), *Department of Bioimmunotherapy, Unit 422, The University of Texas MD Anderson Cancer Center, Houston, Texas 77030*

ALAIN R. THIERRY (12), *CNRS/IGM Laboratorie Des Defenses Antitumorales, 34293 Montpellier, France*

DAVID H. THOMPSON (10), *Department of Chemistry, Purdue University, West Lafayette, Indiana 47907*

ARTO URTTI (13), *Department of Pharmaceutics, University of Kuopio, Kuopio, Finland*

SAMUEL ZALIPSKY (4), *ALZA Corporation, Mountain View, California 94043*

*Current Affiliation: The School of Pharmaceutical Sciences, Pokhara University, Pokhara, Nepal

Preface

The origins of liposome research can be traced to the contributions by Alec Bangham and colleagues in the mid 1960s. The description of lecithin dispersions as containing "spherulites composed of concentric lamellae" (A. D. Bangham and R. W. Horne, *J. Mol. Biol.* **8,** 660, 1964) was followed by the observation that "the diffusion of univalent cations and anions out of spontaneously formed liquid crystals of lecithin is remarkably similar to the diffusion of such ions across biological membranes (A. D. Bangham, M. M. Standish, and J. C. Watkins, *J. Mol. Biol.* **13,** 238, 1965). Following early studies on the biophysical characterization of multilamellar and unilamellar liposomes, investigators began to utilize liposomes as a well-defined model to understand the structure and function of biological membranes. It was also recognized by pioneers including Gregory Gregoriadis and Demetrios Papa-hadjopoulos that liposomes could be used as drug delivery vehicles. It is gratifying that their efforts and the work of those inspired by them have lead to the development of liposomal formulations of doxorubicin, daunorubicin and amphotericin B now utilized in the clinic. Other medical applications of liposomes include their use as vaccine adjuvants and gene delivery vehicles, which are being explored in the laboratory as well as in clinical trials. The field has progressed enormously in the 39 years since 1965.

This volume includes chapters on targeted and environment-sensitive liposomes, delivery of oligonucleotides, and the use of liposomes *in vivo*. I hope that these chapters will facilitate the work of graduate students, post-doctoral fellows, and established scientists entering liposome research. The next volume on "Liposomes" (Part E) will focus on the therapeutic applications of liposomes. Previous volumes in this series (367, 372, and 373) cover additional subdisciplines in liposomology.

The areas represented in this volume are by no means exhaustive. I have tried to identify the experts in each area of liposome research, particularly those who have contributed to the field over some time. It is unfortunate that I was unable to convince some prominent investigators to contribute to the volume. Some invited contributors were not able to prepare their chapters, despite generous extentions of time. In some cases I may have inadvertantly overlooked some experts in a particular area, and to these individuals I extend my apologies. Their primary contributions to the field will, nevertheless, not go unnoticed, in the citations in these volumes and in the hearts and minds of the many investigators in liposome research.

I would like to express my gratitude to all the colleagues who graciously contributed to these volumes. I would like to thank Shirley Light of Academic Press for her encouragement and support, Noelle Gracy and Cindy Minor of Elsevier for their help at the later stages of the project, and Tracy Grace and Alan Palmer of Kolam USA for their diligent attention to the printing of the book. I am especially thankful to my wife Diana Flasher for her understanding, support, and love during the seemingly never-ending editing process, and my children Avery and Maxine for their unique curiosity, creativity, cheer, and love.

Finally, I wish to dedicate this volume to the memory of my mother, the late Prof. Zeliha Düzgüneş, who was an entomologist, and the late Prof. Orhan Düzgüneş, who was a geneticist, both at the University of Ankara. They provided a loving, nurturing and peaceful environment for me to pursue my interest in physics and biology, and encouraged me to do graduate work in biophysics, even though they would not be able to see me for long periods of time. Being a parent now, I know how difficult this must have been for them.

NEJAT DÜZGÜNEŞ

METHODS IN ENZYMOLOGY

VOLUME 193. Mass Spectrometry
Edited by JAMES A. MCCLOSKEY

VOLUME 194. Guide to Yeast Genetics and Molecular Biology
Edited by CHRISTINE GUTHRIE AND GERALD R. FINK

VOLUME 195. Adenylyl Cyclase, G Proteins, and Guanylyl Cyclase
Edited by ROGER A. JOHNSON AND JACKIE D. CORBIN

VOLUME 196. Molecular Motors and the Cytoskeleton
Edited by RICHARD B. VALLEE

VOLUME 197. Phospholipases
Edited by EDWARD A. DENNIS

VOLUME 198. Peptide Growth Factors (Part C)
Edited by DAVID BARNES, J. P. MATHER, AND GORDON H. SATO

VOLUME 199. Cumulative Subject Index Volumes 168–174, 176–194

VOLUME 200. Protein Phosphorylation (Part A: Protein Kinases: Assays, Purification, Antibodies, Functional Analysis, Cloning, and Expression)
Edited by TONY HUNTER AND BARTHOLOMEW M. SEFTON

VOLUME 201. Protein Phosphorylation (Part B: Analysis of Protein Phosphorylation, Protein Kinase Inhibitors, and Protein Phosphatases)
Edited by TONY HUNTER AND BARTHOLOMEW M. SEFTON

VOLUME 202. Molecular Design and Modeling: Concepts and Applications (Part A: Proteins, Peptides, and Enzymes)
Edited by JOHN J. LANGONE

VOLUME 203. Molecular Design and Modeling: Concepts and Applications (Part B: Antibodies and Antigens, Nucleic Acids, Polysaccharides, and Drugs)
Edited by JOHN J. LANGONE

VOLUME 204. Bacterial Genetic Systems
Edited by JEFFREY H. MILLER

VOLUME 205. Metallobiochemistry (Part B: Metallothionein and Related Molecules)
Edited by JAMES F. RIORDAN AND BERT L. VALLEE

VOLUME 206. Cytochrome P450
Edited by MICHAEL R. WATERMAN AND ERIC F. JOHNSON

VOLUME 207. Ion Channels
Edited by BERNARDO RUDY AND LINDA E. IVERSON

VOLUME 208. Protein–DNA Interactions
Edited by ROBERT T. SAUER

VOLUME 209. Phospholipid Biosynthesis
Edited by EDWARD A. DENNIS AND DENNIS E. VANCE

Section I

Antibody- or Ligand-Targeted Liposomes

[1]   High-Yield Coupling of Antibody Fab' Fragments to Liposomes Containing Maleimide-Functionalized Lipids

By SERGE SHAHINIAN and JOHN R. SILVIUS

Introduction

Antibodies represent a very useful means to target liposomes to specific cell surface proteins, either to favor specific binding of the liposomes to particular cell types or to promote endocytotic uptake of the liposomes by the target cells. A variety of methods have been reported that are used to couple whole antibodies to liposomes.[1-8] However, the presence of the Fc domain in intact antibodies can, in principle, alter the targeting and/or the biological effects of liposomes coupled to intact antibodies. Antibody-derived Fab' fragments represent a useful alternative to whole antibodies for coupling to liposomes as targeting agents, particularly because the free sulfhydryl residues of Fab' fragments allow these fragments to be linked to the liposome surface via a unique and well-defined site on the protein.

To exploit optimally the advantages of Fab' fragments, it is necessary to employ strategies that permit such fragments to be prepared in good yield, and in a form that can be coupled efficiently to liposomes incorporating sulhydryl-reactive lipids. Early studies found that conventional procedures to prepare Fab' fragments often yielded only small proportions of fragments that could be coupled to liposomes incorporating maleimide-functionalized lipids. This article describes procedures by which to monitor and to optimize the production of Fab' fragments, from either mouse or rabbit IgG species, that can be coupled efficiently to liposomes. Key to this approach is a novel assay using maleimide-functionalized polyethyleneglycol-5000

[1] A. Huang, L. Huang, and S. J. Kennel, *J. Biol. Chem.* **255,** 8015 (1980).

[2] F. J. Martin and D. Papahadjopoulos, *J. Biol. Chem.* **257,** 286 (1982).

[3] R. A. Schwendener, T. Trub, H. Schott, H. Langhals, R. F. Barth, P. Groscurth, and H. Hengartner, *Biochim. Biophys. Acta* **1026,** 69 (1990).

[4] C. B. Hansen, G. Y. Kao, E. H. Moase, S. Zalipsky, and T. M. Allen, *Biochim. Biophys. Acta* **1239,** 133 (1995).

[5] K. Maruyama, T. Takizawa, T. Yuda, S. J. Kennel, L. Huang, and M. Iwatsuru, *Biochim. Biophys. Acta* **1234,** 74 (1995).

[6] S. Shahinian and J. R. Silvius, *Biochim. Biophys. Acta* **1239,** 157 (1995).

[7] S. M. Ansell, P. G. Tardi, and S. S. Buchkowsky, *Bioconj. Chem.* **7,** 490 (1996).

[8] J. A. Harding, C. M. Engbers, M. S. Newman, N. I. Goldstein, and S. Zalipsky, *Biochim. Biophys. Acta* **1327,** 181 (1997).

to derivatize Fab' fragments bearing reactive sulfhydryl groups, facilitating quantitative analysis (and optimization) of the formation of these fragments by gel electrophoresis. This article also outlines a method to couple Fab' fragments efficiently to liposomes bearing sulfhydryl-reactive maleimide groups linked to polyethyleneglycol (PEG) chains on their surface.

Preparation of Liposomes

The following protocol, based on the procedure of MacDonald *et al.*,[9] yields primarily unilamellar liposomes with a mean diameter of roughly 120 nm and includes a gel-filtration step to separate liposome-encapsulated hydrophilic solutes from unencapsulated material. Many alternative procedures (and lipid compositions) can be used to prepare liposomes and to incorporate bioactive materials into them, depending on the nature of the final liposome preparation desired and the material to be encapsulated.[10–12] Particularly interesting for some applications are "remote loading" procedures, which permit certain weakly basic (and, in principle, weakly acidic) species to be incorporated into preformed liposomes with high efficiencies.[13,14]

Reagents

Egg phosphatidylcholine (ePC) or distearoylphosphatidylcholine (DSPC) material of sufficient purity is available from several commercial sources, including Avanti Polar Lipids (Alabaster, AL) or Sigma/Aldrich (St. Louis, MO). Stock solutions of lipids (in reagent-grade methylene chloride or chloroform) should be stored at $-20°$ and (for unsaturated species) under an argon blanket. Phospholipids are typically somewhat hygroscopic, and concentrations of stock solutions are best determined by digestion and assay of released inorganic phosphorus.[15,16]

---

[9] R. C. MacDonald, R. I. MacDonald, B. P. Menco, K. Takeshita, N. K. Subbarao, and L. F. R. Hu, *Biochim. Biophys. Acta* **1061,** 297 (1991).

[10] S. Venuri and C. T. Rhodes, *Pharm. Helv. Acta* **70,** 95 (1995).

[11] D. Lichtenberg and Y. Barenholz, *Methods Biochem. Anal.* **33,** 337 (1988).

[12] A. L. Weiner, *Immunomethods* **4,** 201 (1994).

[13] P. R. Cullis, M. J. Hope, M. B. Bally, T. D. Madden, L. D. Mayer, and D. B. Fenske, *Biochim. Biophys. Acta* **1331,** 187 (1997).

[14] G. Haran, R. Cohen, L. K. Bar, and Y. Barenholz, *Biochim. Biophys. Acta* **1151,** 201 (1993).

[15] R. J. Lowry and I. J. Tinsley, *Lipids* **9,** 491 (1974).

[16] N. Düzgüneş, *Methods Enzymol.* **367,** 23 (2003).

Maleimido-PEG2000-functionalized phosphatidylethanolamine (maleimido-PEG-PE) is from Avanti Polar Lipids. If maleimido-functionalized PEG-PEs are desired with acyl chains or PEG chain length different from those available commercially, small amounts of these materials (milligrams or tens of milligrams) can be prepared as described in Shahinian and Silvius[6] from commercially available phosphatidylethanolamines and α,ω-diamino-PEGs (Sigma /Aldrich or Shearwater Corporation, Huntsville, AL).

Cholesterol: 99+ mol% pure (Calbiochem/EUD Biosciences, San Diego, CA).

Coupling buffer: 150 mM NaCl, 10 mM 3-(N-morpholino)propane-sulfonic acid (MOPS), 0.1 mM EDTA, 0.05 mM diethylenetri-aminepentaacetic acid (DTPA), pH 6.5. DTPA is toxic and should be handled with care. Immediately before use the buffer is de-aerated by bubbling with argon (or nitrogen) for 15 min and is then maintained under an argon blanket.

Buffered solution containing the material to be encapsulated, prepared to be isotonic with coupling buffer.

Procedure

Phosphatidylcholine, cholesterol, and maleimido-PEG-PE stocks (2:1 PC/cholesterol, plus 0.5–1 mol% maleimido-PEG-PE) are mixed in a glass tube, dried under a gentle stream of nitrogen or argon while warming to 40–45°, and further dried under high vacuum for 2–8 h to remove residual traces of solvent. A buffered solution containing the material to be encapsulated is added, and the sample is vortexed at a temperature above the transition temperatures of the phospholipid components (25° for egg PC, 60° for DSPC). The sample is frozen (in a dry ice/ethanol bath) and thawed three times, and is then extruded (25 passes) through a 0.1-μm pore size polycarbonate filter using a hand-held extrusion device [available from Avanti Polar Lipids or Avestin (Ottawa, Ontario, Canada)]. Dispersions containing up to at least 40 mM phospholipid can be extruded readily in this manner (more easily at 37° for DSPC-containing liposomes).

Liposomes and entrapped solutes are separated from unencapsulated material by gel filtration. Sephadex G-75 columns equilibrated with coupling buffer typically allow rapid and complete separation of liposomes from polar solutes with relatively low molecular masses (up to at least 2000 Da). Larger pore-size columns (e.g., Sepharose CL-4B) may be needed to separate liposomes from larger solutes such as proteins. Columns suitable for small-scale separations (up to 100 μl of applied sample) can be prepared by packing Sephadex in a 5 3/4 in. Pasteur pipette to a bed volume of 1 ml over a glass wool plug.

Preparation of Fab' Fragments

IgGs are converted to Fab' fragments by proteolysis to separate their F(ab')$_2$ from their Fc portions, followed by partial reduction of the isolated F(ab')$_2$ to Fab' fragments. Procedures for preparation of F(ab')$_2$ fragments, using limited digestion with pepsin, papain, bromelain or ficin, vary in nature and yield, depending on the species and subclass of antibody.

Good yields of F(ab')$_2$ can be obtained from most, although not all, subclasses of murine or rat monoclonal IgGs,[17,18] as well as from rabbit polyclonal IgG.[19] Removal of the protease after digestion can be facilitated by utilizing pepsin bound to agarose beads,[20] which is now available commercially from several sources. However, we have not found this step to be essential for the preparations described here. After digestion, F(ab')$_2$ fragments can be separated readily from Fc fragments and undigested IgG by passage over a column of protein A–Sepharose.

The following protocol, using cysteine as the reducing agent, affords good yields of Fab' fragments with reactive sulfhydryl groups from F(ab')$_2$ fragments prepared from both mouse monoclonal IgG2a and rabbit polyclonal IgG.[6] The protocol described in the following section can be used to assay, and thereby to optimize, the yields of free sulfhydryl-bearing Fab' fragments when using other antibody subclasses or reducing conditions.

Reagents

F(ab')$_2$ fragments, freed of residual intact IgG and Fc fragments, in 150 mM NaCl, 20 mM sodium phosphate, pH 7.0. The solution may be agitated gently under argon for several minutes at room temperature to reduce its oxygen content. Argon should, however, not be bubbled through protein solutions.

Cysteine, freshly dissolved to 60 mM in 200 mM Tris, pH 7.6. Prior to dissolving the cysteine the Tris buffer should be purged of air by bubbling with argon (or nitrogen) for 15 min with vigorous stirring and then maintained under an argon blanket during and after cysteine addition.

[17] E. Lamoyi, *Methods Enzymol.* **121,** 652 (1996).
[18] J. Rousseaux, R. Rousseaux-Prevost, and H. Bazin, *Methods Enzymol.* **121,** 663 (1996).
[19] K. Inouye and K. Morimoto, *J. Immunol. Methods* **171,** 239 (1994).
[20] J. H. Beeson and R. W. Wissler, *Immunochemistry* **14,** 305 (1977).

Coupling buffer: Prepared and freshly de-aerated as described in the previous section.

BioGel spin columns: Polypropylene microcentrifuge tubes (0.5 ml volume) are punctured at the bottom with an 18-gauge syringe needle, and a small glass wool plug is placed over the puncture. BioGel P-6 DG (90–180 mesh, Bio-Rad Laboratories, Hercules, CA), swollen in coupling buffer, is added until the drained bed volume is 0.4 ml. The tubes are then centrifuged in a clinical centrifuge (at 1100g), 20 μl of bovine serum albumin (BSA) (2 mg/ml in coupling buffer) is layered onto the gel bed, and the tubes are recentrifuged for 3 min and then washed three times with 20 μl coupling buffer, centrifuging after each application of buffer as just indicated.

Procedure

The F(ab′)₂ fragment (1–5 μl of 10 mg/ml, in 150 mM NaCl, 20 mM sodium phosphate, pH 7.0) is mixed in a 0.5-ml microcentrifuge tube with 5 μl of cysteine solution, plus water (if needed) to a final reaction volume of 10 μl. The tube is capped immediately under a gentle stream of argon and incubated for 15 min at 37°. The reaction is mixed with an equal volume of coupling buffer and applied immediately to a BioGel spin minicolumn packed in a 0.5-ml microcentrifuge tube (prepared as described earlier), which is placed into a 1.5-ml microcentrifuge tube containing either liposomes or maleimido-PEG, as appropriate (see following sections). The minicolumns are placed in 13 × 100-mm glass tubes and centrifuged for 3 min (1100g, room temperature). The inner tubes containing the gel are discarded, and the outer tubes are flushed rapidly with argon and capped.

The aforementioned F(ab′)₂ reduction can easily be scaled up several-fold, applying 20-μl aliquots of the diluted reaction mixture to multiple spin minicolumns. Because the Fab′ fragments are relatively stable at pH < 7.0, still larger volumes of Fab′ could be separated from the reducing agent using conventional gel-filtration columns in the cold. In the latter case, however, it is advisable to deaerate the gel and eluant solution thoroughly and to monitor routinely the quality of the eluted Fab′ fragments, as described in the following section.

Antibodies from different species, or of different subtypes, may require some variations in the aforementioned protocol to achieve optimal formation of Fab′ fragments. In such cases, we have generally found it more useful to increase the concentration of cysteine than simply to prolong the incubation time. Changing the nature of the reducing agent or (modestly) altering the pH may also be useful in such cases. In our experience, however, changing these latter variables entails a greater relative risk of overereduction of the $F(ab')_2$ fragments to separated heavy and light chains.

Monitoring the Production of Free Sulfhydryl-Containing Fab' Fragments Using Maleimido-PEG5000

Reagents

> Coupling buffer, prepared and freshly de-aerated, as described in the section "Preparation of Liposomes."
> Biogel SP-6 DG minicolumns; see previous section.
> MethoxyPEG5000-maleimide   (maleimido-PEG5000)—Shearwater Chemicals (Huntsville, AL). The reagent is dissolved in de-aerated coupling buffer shortly before use and maintained under an argon blanket thereafter.
> *N*-Ethylmaleimide (NEM)—freshly dissolved to 100 mM in coupling buffer 2× nonreducing sample buffer for SDS–PAGE: 0.1 M Tris–chloride, 2% (w/v) sodium dodecyl sulfate, 20% sucrose, pH 6.8.
> SDS–PAGE reagents—as described in Laemmli.[21]

Procedure

Twenty-microliter samples of freshly generated Fab' are centrifuged as described earlier through BioGel P-6 DG minicolumns into 80 μl of maleimido-PEG5000 (0.5 mM) in coupling buffer. After centrifugation, the tubes are capped under a gentle stream of argon and incubated at 25° for 1 h and then for a further 30 min at this temperature after the addition of NEM (10 mM final concentration). A 10-μl aliquot is mixed with 10 μl of 2× nonreducing sample buffer, heated to 100° for 5 min, and applied to a 12% discontinuous polyacrylamide gel.[21]

A typical pattern of results obtained by SDS–PAGE is shown in Fig. 1, where the positions of $F(ab')_2$, PEG5000-conjugated Fab' and unmodified Fab' fragments, as well as overreduced (heavy and light chain) fragments, are indicated. It can be seen that coupling PEG5000-maleimide to Fab' fragments shifts their mobility significantly, allowing ready discrimination of the relative proportions of maleimide-reactive (free sulfhydryl-bearing) and –nonreactive Fab'. It is evident from Fig. 1 that samples of murine $F(ab')_2$ fragments that are incubated for longer times with cysteine (10 or 30 mM) give increasing proportions of Fab'

---

[21] U. K. Laemmli, *Nature* **227,** 680 (1970).

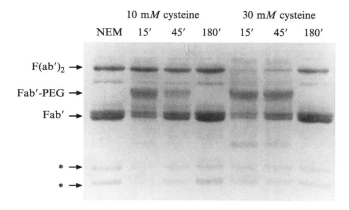

FIG. 1. SDS–PAGE analysis of cysteine reduction of the $F(ab')_2$ fragment of the murine monoclonal (IgG2) antibody TFR1. $F(ab')_2$ fragments were incubated with the indicated concentrations of cysteine (pH 7.6) for various times at 25° and then reacted successively with maleimido-PEG5000 and NEM before analyzing samples by SDS–PAGE. Bands corresponding to intact $F(ab')_2$, Fab', and maleimido-PEG-derivatized Fab', as well as products of Fab' overreduction (separated fragments of heavy and light chains; designated by asterisks), are indicated. Other conditions of sample preparation and analysis were as described in the text.

fragments that are nonreactive with PEG-maleimide. This appears to reflect the gradual formation of protein-cysteine mixed disulfides from the initially generated free protein sulhydryl groups. The reduction of individual $F(ab')_2$ disulfide linkages proceeds through the following reactions:

$$Fab' \text{-} SS \text{-} Fab' + cysteine \text{-} SH \rightarrow Fab' \text{-} SH + Fab' \text{-} SS \text{-} cysteine \quad \text{(I)}$$

$$Fab' \text{-} S \text{-} S \text{-} cysteine + cysteine \text{-} SH \rightarrow Fab' \text{-} SH + (cysteine \text{-} S)_2 \quad \text{(II)}$$

where the overall equilibrium lies far to the right of the second equation, as long as the concentration of reduced cysteine far exceeds that of both protein and $(cysteine\text{-}S)_2$. With time, however, gradual oxidation of free cysteine can shift equilibrium (II) to the left, allowing the formation of increasing amounts of Fab'-S-S-cysteine forms with blocked sulfhydryl groups. This complication could, in principle, be avoided using agents that maintain a higher reducing potential even when partly oxidized, such as dithiothreitol and tris (2-carboxyethyl) phosphine hydrochloride (TCEP). However, in our experience in using these latter agents, it is very difficult to avoid extensive overreduction of the Fab' fragments to heavy and light chains, with concomitant loss of binding activity. Scrupulous exclusion of all traces of oxygen from the cysteine-reduction mixture should likewise avoid the problem described earlier, but is difficult to achieve in practice.

Coupling of Fab' Fragments to Liposomes and Separation of Uncoupled Fragments

Reagents

Coupling buffer, prepared and freshly de-aerated as described under "Preparation of Fab' Fragments"

Biogel P-6 DG minicolumns—prepared as described under "Preparation of Fab' Fragments"

Liposomes (10 mM in coupling buffer)—prepared as described under "Preparation of Liposomes"

Cysteine, freshly prepared 10 mM solution in deaerated coupling buffer.

Procedure

Twenty-microliter samples of Fab' (diluted 1:1 with coupling biffer) are prepared and centrifuged through BioGel P-6 DG minicolumns, all as described in the section "Preparation of Fab' Fragments," into 80 μl of liposomes (10 mM) in coupling buffer. After centrifugation, the tubes are capped under a gentle stream of argon and incubated at 4° for 16 h. Residual unreacted maleimidyl groups on the liposomes are then quenched by adding cysteine (to 0.5 mM) and incubating at 25° for 15 min. Liposome-coupled and unbound Fab' fragments are separated subsequently by chromatography on Sepharose CL-4B (2.5 ml bed volume, packed in a 3-ml disposable syringe over a glass wool plug). A typical elution profile, showing well-resolved peaks for the coupled and free Fab' fragments, is shown in Fig. 2. When relatively concentrated liposome suspensions are used, elution of the liposomes from the column can be monitored by turbidity, either visually or spectrophotometrically. When desired, the efficiency of antibody–liposome coupling can be monitored by including small amounts of radioiodinated F(ab')$_2$ fragment in the reduction mixture used to generate Fab' fragments.

The efficiency of liposome–Fab' coupling achieved using the aforementioned procedure depends on both the quality of the Fab' preparation (specifically, the level of Fab' fragments bearing free sulfhydryl groups) and the nature and concentration of the maleimide-functionalized lipid incorporated in the liposomes. As illustrated in Fig. 3B, small or large

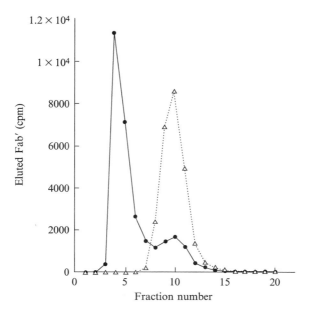

FIG. 2. Separation of liposomes coupled from free Fab' fragments on Sepharose CL-4B. Liposomes (egg phosphatidylcholine/egg phosphatidylglycerol/cholesterol, 3:1:2 molar proportions) incorporating 2.5 mol% maleimido-PEG-PE were reacted with antibody TFR1 Fab' fragments (including a small proportion of radioiodinated Fab') as described in the text and were then chromatgraphed on a column of Sepharose CL-4B (2.5 ml bed volume) (●). Fractions of 0.2 ml were collected, and samples were counted for ^{125}I-labeled Fab' radioactivity. The earlier- and later-eluting peaks correspond to liposome-coupled and -uncoupled Fab' fragments, respectively. Fab' fragments were incubated with similar liposomes omitting maleimido-PEG-PE before column separation (△).

unilamellar liposomes bearing maleimide-functionalized PEG-phosphatidylethanolamines (structure shown in Fig. 3A) give more efficient coupling of Fab' fragments than simpler maleimide-functionalized phosphatidylethanolamines, particularly when low concentrations of the maleimide-functionalized species are incorporated into the liposomal bilayer. Using as little as 0.5 mol% maleimido-PEG-PE, a high proportion of the Fab' fragments that bear free sulfhydryl groups can become liposome coupled under the conditions described earlier. An additional advantage of maleimide-functionalized PEG-PE "anchors" is that the PEG linker allows better exposure of the liposome-coupled Fab' to its target ligand, with less steric hindrance from the liposome surface and greater flexibility of attachment.

An illustration of the use of liposome-coupled Fab' fragments to promote liposome uptake by mammalian cells is shown in Fig. 4. In the experiment

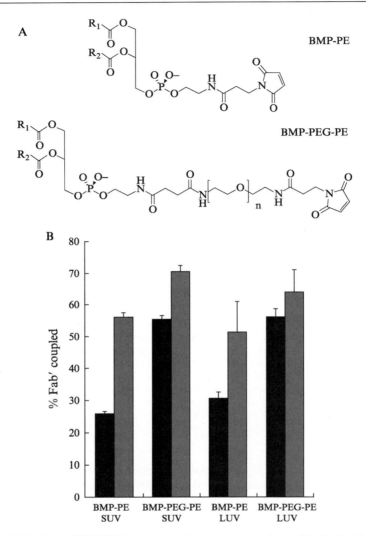

FIG. 3. Coupling of TFR1 Fab′ fragments to liposomes containing maleimide-functionalized lipids. (A) Structures of BMP- and BMP-PEG-PE. (B) Percentages of input Fab′ fragments coupled to small (SUV) or large (LUV) unilamellar liposomes prepared from 2:1:3 (molar proportions) egg phosphatidylcholine/egg phosphatidylglycerol/cholesterol and incorporating either 0.5 mol% (dark-shaded bars) or 2.5 mol% (light-shaded bars) of the indicated functionalized lipids. LUV were prepared by filter extrusion of vortexed lipid dispersions as described in the text, whereas SUV were prepared by bath-sonicating similar dispersions for 10 min. Data shown represent the average (±SD) of values determined in two to six independent experiments.

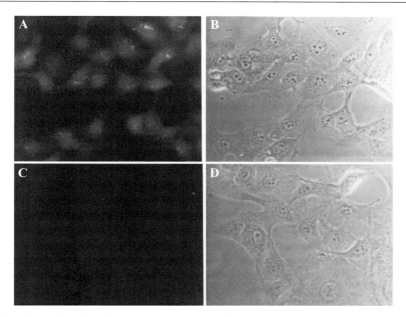

FIG. 4. Uptake of anti(transferrin receptor) Fab'-targeted liposomes by CV-1 fibroblasts. Cells were incubated for 30 min at 4° with liposomes (40 μM lipid) in Hanks' balanced salt solution/10 mM HEPES and then warmed to 37° in liposome-free media for 1 h to allow uptake of bound liposomes. [A (fluorescence) and B (phase contrast)] Cells incubated with Fab'-targeted liposomes containing 45 mM carboxyfluorescein. (C and D) Cells incubated with identical Fab'-targeted liposomes together with free Fab' fragment (1 mg/ml) to block specific binding of liposomes to transferrin receptors.

shown, CV-1 (monkey kidney fibroblast) cells were incubated with liposomes loaded with the fluorescent dye carboxyfluorescein (45 mM internal concentration) and coupled via maleimidoPEG1000-PE to Fab' fragments derived from a mouse monoclonal (IgG2) anti(transferrin receptor) antibody. Endocytosis of liposomes, and consequent exposure of the liposomes to the acidic environment of the endosomes, leads to the release of carboxyfluorescein from the liposomes to the endocytotic vesicle lumen and hence to the cytoplasm, producing a diffuse cytoplasmic fluorescence. It can be seen from Fig. 4 that the Fab'-coupled liposomes are bound and taken up efficiently by the cells (Fig. 4A), and that binding is reduced greatly when free antibody is added to block specific antibody–receptor interactions (Fig. 4C) or when the liposomes are coupled to an irrelevant antibody (not shown).

Assay of Leakage of Liposomal Contents

The procedure described in this section is useful for monitoring the potential loss of liposome-encapsulated hydrophilic solutes during antibody coupling and in subsequent manipulations. Previous studies have shown that using certain functionalized lipid anchors and liposomal lipid compositions, major losses of encapsulated solutes can occur, especially during the liposome–antibody coupling reaction.[22] The protocols and lipid compositions described in this article have been shown to give only a modest loss (<25%) of encapsulated hydrophilic contents during antibody coupling.[6] However, the following assay, based on the procedure of Allen and Cleland,[23] should be utilized to test for excessive loss of liposomal contents when new liposomal formulations (lipid compositions or types of functionalized lipids) are used.

Calcein encapsulated in liposomes at high concentrations exhibits self-quenching and hence very low fluorescence. Release of calcein from the liposomes leads to a dramatic increase in fluorescence, the magnitude of which is proportional to the extent of leakage of the dye from the vesicle. Fluorescence readings are converted easily to estimates of the extent of release of encapsulated calcein, using detergent-lysed vesicle samples to determine the fluorescence value corresponding to 100% release of vesicle contents.

Reagents

> Calcein—material of sufficient purity can be obtained from Molecular Probes (Eugene, OR) or prepared from lower-grade material by chromatography on Sephadex LH-20 Amersham Biosciences, Piscataway, NJ.[24]
>
> Liposomes, prepared as described earlier but using 100 mM calcein, pH 7.4, as the medium to be encapsulated. After formation and extrusion, the liposomes are gel filtered on Sephadex G-75 in coupling buffer.
>
> Triton X-100, 20% (w/v) in coupling buffer.

Procedure

Immediately after the calcein-loaded liposomes are eluted from the Sephadex G-75 column, the fluorescence (F) of duplicate aliquots is

[22] R. Bredehorst, F. S. Ligler, A. W. Kusterbeck, E. L. Chang, B. P. Gaber, and C.-W. Vogel, *Biochemistry* **25,** 5693 (1986).

[23] T. M. Allen and L. G. Cleland, *Biochim. Biophys. Acta* **597,** 418 (1980).

[24] E. Ralston, L. M. Hjelmeland, R. D. Klausner, J. N. Weinstein, and R. Blumenthal, *Biochim. Biophys. Acta* **649,** 133 (1981).

measured (using excitation/emission wavelengths of 490/520 nm and excitation/emission slit widths of 5 nm) after dilution into coupling buffer vs coupling buffer containing 1% (v/v) Triton X-100. Before reading their fluorescence, vesicle samples diluted into detergent-containing buffer are vortexed or bath sonicated briefly to ensure complete vesicle disruption. Appropriate blank readings (F_{blank}) are also taken using the appropriate solutions without vesicles. The normalized fluorescence (F_N) is then calculated as

$$F_N = (F - F_{blank})_{buffer}/(F - F_{blank})_{Triton} \tag{1}$$

Normalized fluorescence is measured similarly for small samples of liposomes after the antibody-coupling step (and at other stages of liposome preparation as desired). The extent of calcein release from the liposomes at any given time (t) can then be estimated using the equation

$$\% \text{ Released} = (100 \%) \frac{[F_N(t) - F_N(t = 0)]}{[1 - F_N(t = 0)]} \tag{2}$$

where the readings designated ($t = 0$) are those determined immediately after elution of the vesicles from the Sephadex G-75 column.

The aforementioned procedure provides good estimates for the extent of loss of charged or other highly polar molecules from the liposomes due to the local or global breakdown of bilayer integrity. Less polar species (e.g., small uncharged molecules) may be able to permeate across intact bilayers and hence be released from liposomes at faster rates. In such cases it may be necessary to monitor the release of such species from liposomes directly.

Conclusions

Antibody Fab' fragments can offer an attractive alternative to intact antibodies as liposomal targeting agents. The use of maleimido-PEG5000 as a reagent provides a simple means to assay specifically the formation of Fab' fragments bearing reactive sulfhydryl groups, and hence to optimize the production of such fragments from antibodies of different subtypes or from different species. This potential, and the ready availability and ease of use of diverse sulfhydryl-reactive lipid derivatives (notably including functionalized PEG-PEs), makes Fab' fragments attractive alternatives to intact antibodies as liposomal targeting agents.

Acknowledgment

This work was supported by an operating grant from the Canadian Institutes of Health Research to J.R.S.

[2] Interactions of Hyaluronan-Targeted Liposomes with Cultured Cells: Modeling of Binding and Endocytosis

By ROM E. ELIAZ, SHLOMO NIR, and FRANCIS C. SZOKA, JR.

Introduction

Hyaluronan (hyaluronic acid, HA) is a major extracellular glycosami-noglycan and is found in most, if not all, types of extracellular matrix in the mammalian body. HA is a high molecular weight glycosaminoglycan polymer (MW $= 10^6$ D) composed of the repeating disaccharide β1,3-N-acetyl glucosaminyl-β1,4-glucuronide. The presence of an extracellular matrix enriched with HA is also characteristic of the peripheral zone of in-vasive tumors. Hyaluronan influences the differentiation, migration of cells during morphogenesis, and adherence of various cell types, depending on the size and concentration of the HA and the type of cell in question.[1] These events are mediated through receptors for HA on the cell surface.[2]

CD44 is the principal cell surface receptor for HA.[3] CD44 is present at low levels on epithelial, hemopoietic, and some neuronal cells[4] and at elevat-ed levels in various carcinoma, melanoma, lymphoma, breast, colorectal, and lung tumor cells.[2–9]

In a search for selectively targeting liposomes, which have been widely used as drug carriers, we have considered attaching oligomers of hyaluro-nan to the liposome surface, with the hope that such liposomes can prefer-entially bind and be internalized by certain types of cancer cells whose surfaces include relatively large numbers of the CD44 receptors. The con-cept is that liposomes whose membranes incorporate such ligands com-posed of lipid-linked oligomers of the hyaluronan-repeating disaccharide units are likely to interact with greater avidity with cells with a high number

[1] K. Miyake, C. B. Underhill, J. Lesley, and P. W. Kincade, *J. Exp. Med.* **172,** 69 (1990).
[2] D. C. West and S. Kumar, *Exp. Cell Res.* **183,** 179 (1989).
[3] R. J. Sneath and D. C. Mangham, *Mol. Pathol.* **51,** 191 (1998).
[4] T. Hardingham and A. Fosang, *FASEB J.* **6,** 861 (1992).
[5] I. Stamenkovic, A. Aruffo, M. Amiot, and B. Seed, *EMBO J.* **10,** 943 (1991).
[6] L. H. Thomas, R. Byers, J. Vink, and I. Stamenkovic, *J. Cell Biol.* **118,** 971 (1992).
[7] M. B. Penno, J. T. August, S. B. Baylin, M. Mabry, R. I. Linnoila, V. S. Lee, D. Croteau, X. L. Yang, and C. Rosada, *Cancer Res.* **54,** 1381 (1994).
[8] T. A. Tran, B. V. Kallakury, C. E. Sheehan, and J. S. Ross, *Hum. Pathol.* **28,** 809 (1997).
[9] M. Fasano, M. T. Sabatini, R. Wieczorek, G. Sidhu, S. Goswami, and J. Jagirdar, *Cancer* **80,** 34 (1997).

of CD44 receptors on their surface than with cells with a low number of receptors. Our hypothesis was that selective targeting can be achieved by the design of a ligand with a modest affinity for the receptor, which can be incorporated in liposomes at a range of surface densities, such that the liposomes bind avidly to cells that have a high number of CD44 receptors on their surface, and bind to a low extent to cells with a low number of receptors. The ligands are mobile in a fluid bilayer and can rearrange to minimize steric constraints during their interaction with multiple receptors on the cell surface.

We present a kinetic analysis of binding and endocytosis of hyaluronan-targeted liposomes (HAL) by two types of cells: murine melanoma B16F10 cells expressing high levels of CD44 and African green monkey fibroblasts CV-1 expressing low levels of CD44.[10] We show that B16F10 cells bind avidly and internalize HAL in a temperature-dependent manner, whereas the uptake of HAL by cells expressing low levels of CD44 is minimal. Using previously described procedures,[11,12] the kinetic rate constants of ligand-conjugated liposome–cell interaction, as well as the endocytotic rate constant are obtained. Analysis of binding data includes determination of the on- and off-rate constants of liposome binding to the cells at both 4° and 37° under conditions inhibiting endocytosis. The number of receptors available for the HAL of a specific size, as well as the effect of ligand density in the liposomes on their binding avidity to the cells are also determined. The high preferential uptake of HAL by B16F10 cells turns out to be due to the large number of CD44 receptors on the surface of these cells.

Chemicals

All phospholipids [1-palmitoyl 2-oleoyl phosphatidylglycerol (POPG), palmitoyl oleoyl phosphatidylcholine (POPC), and palmitoyl oleoyl phosphatidylethanolamine (POPE)] are from Avanti Polar Lipids (Alabaster, AL) or synthesized in our laboratory. Cholesterol (Chol), bee venom, human umbilical cord hyaluronic acid, sodium cyanoborohydride, sodium fluoride, and antimycin A_2 are from Sigma Chemical Co. (St. Louis, MO). Culture medium [MEM Eagle's with Earle's BSS (EBSS) and DME H-21 (high glucose 4.5 g/liter)] is obtained from the UCSF cell culture facility.

[10] R. E. Eliaz and F. C. Szoka, Jr., *Cancer Res.* **61,** 2592 (2001).
[11] K. D. Lee, S. Nir, and D. Papahadjopoulos, *Biochemistry* **32,** 889 (1993).
[12] S. Nir, R. Peled, and K. D. Lee, *Colloids Surfaces* **A89,** 45 (1994).

Cell Culture Conditions

B16F10 murine melanoma cell line is obtained from the UCSF cell culture facility. CV-1 African green monkey kidney cells are obtained from the American Type Culture Collection (Rockville, MD). B16F10 cells are maintained in EBSS medium containing 10% fetal bovine serum (FBS), 1/100 MEM nonessential amino acids (NEAA), 1/100 sodium pyruvate (11 mg/ml stock solution), and 1/100 penicillin–streptomycin (0.1 μm sterile filtered). CV-1 cells are maintained in DME H-21 (high glucose 4.5 g/liter) medium containing 10% FBS, 1/100 MEM NEAA, 1% HEPES buffer (1 M), and 1/100 penicillin–streptomycin (0.1 μm sterile filtered). Cells are cultured with complete medium at 37° in a humidified atmosphere of 5% CO_2 in air. For all experiments, cells are harvested from subconfluent cultures using trypsin and resuspended in fresh complete medium before plating. Cells with >90% viability, as determined by trypan blue exclusion, are used.

Ligand Preparation

Bee venom hydrolysis of human umbilical cord hyaluronic acid is used to degrade high molecular weight hyaluronan into smaller fragments (two, four, six and eight saccharides). Fragments are separated on an 11 × 265-mm column of the formate form of Bio-Rad AG-3X4A ion-exchange resin (eluted with 270-ml portions of 0.015, 0.05, 0.15, 0.30, 0.50, 0.80, and 1.00 M formic acid). The phosphoethanolamine–hyaluronan conjugate (HA-PE) is prepared by reductive amination of the hyaluronan oligomers to the terminal portion of a PE lipid using sodium cyanoborohydride. The lipid derivative is purified by silicic acid column chromatography.

Liposome Preparation

Cholesterol is obtained from Sigma and recrystalized from methanol. Lipid films are prepared by drying 10 μmol of lipid including POPE-hyaluronan from solvent (butanol saturated with distilled water) under vacuum using a rotary evaporator at room temperature. Liposomes [composed of POPC:Chol:HA-POPE (60:40:3 molar ratio) or POPC:Chol: POPG (60:40:9)] are prepared by rehydrating the lipid film with 1 ml of 10 mM HEPES, 5% glucose (pH 7.4), followed by mixing on a vortex mixer for 1 min sonication for 15 min in a bath type sonicator (Laboratory Supplies Company Inc., Hicksville, NY) under argon, and extrusion through 0.2- and 0.1-μm polycarbonate membranes.[13] Liposomes are used within 1 day of preparation and stored at 4° under argon. The hydrodynamic diameter of the liposomes is determined by dynamic light scattering

(Malvern Instruments, UK). The net surface potential can be determined with a Malvern Zetasizer IV (Malvern Instruments, UK). The zeta potential of liposomes containing 3 mol% HA-PE is -9.9 mV.

Liposome Uptake Assay

Cells (4×10^5 B16F10 melanoma or CV-1 cells) are placed in each well in a 24-well plate and are grown overnight at $37°$ and 5% CO_2 in medium. The cell monolayer is rinsed with FBS-free medium, and medium containing liposomes is added. Liposomes containing trace amounts, approximately 0.01 mol percent of ^{125}I-labeled p-hydroxy-benzamidine dihexadecylphosphatidylethanolamine[14] (^{125}I-BPE), are diluted in serum-free, antibiotic-free medium and incubated with cells for 15 to 180 min at $4°$ or $37°$. At the end of the incubation, the medium is removed, and the cells are washed with three successive aliquots of 0.5 ml of ice-cold phosphate-buffered saline (PBS). The medium and washes are pooled and assayed for radioactivity. The cells can be lysed and removed from the well with 1 ml of 0.5 N NaOH. The well is then washed two more times with 1-ml PBS aliquots and the cell lysate and washes are pooled. Radioactivity associated with the cell lysate and washes is determined in a Beckman gamma scintillation spectrometer (Beckman Coulter, Inc., Fullerton, CA).

To assess the effect of hyaluronan tetrasaccharide density on liposome uptake by target cells, ^{125}I-labeled liposomes containing various amounts (0–12 mol%) of HA-PE are incubated with B16F10 cells for 0.1–5 h. The effect of hyaluronan density on cell association is repeated three more times with triplicate replications for each data point ($n = 9$). Each time course experiment is repeated two more times with triplicate wells of cells ($n = 6$). For the inhibition of endocytosis at $37°$, the cells are incubated for 15–180 min with a combination of the metabolic inhibitors, 1 mg/ml antimycin A, and 10 mM NaF.[15] The antimycin A is solubilized in ethanol and is added to the cells to a final ethanol concentration of 1%.

Analysis of Liposome Uptake

The analysis follows the procedure described previously,[11] which was extended further.[12] In the absence of endocytosis, e.g., at $4°$ or at $37°$ in the presence of inhibitors of endocytosis, the process of liposome binding

[13] F. Olson, C. A. Hunt, F. C. Szoka, W. J. Vail, and D. Papahadjopoulos, *Biochim. Biophys. Acta* **557,** 9 (1979).

[14] R. M. Abra, H. Schreier, and F. C. Szoka, *Res. Commun. Chem. Pathol. Pharmacol.* **37,** 199 (1982).

[15] J. Dijkstra, M. van Galen, and G. L. Scherphof, *Biochim. Biophys. Acta* **804,** 58 (1984).

involves association and dissociation, which are described by the rate constants C $(M^{-1} \cdot s^{-1})$ and D (s^{-1}), respectively. Another parameter required is N, the number of receptor sites per cell. If only one type of receptor site is assumed, then the measurement of liposome binding after a sufficiently long incubation time at several liposome concentrations enables the determination of N and the binding coefficient, k (M^{-1}), which is given by $k = C/D$. In Nir et al.[12] the program was extended to account for binding to two types of receptors (see also Nunes-Correia et al.[16]), but the number of parameters used was minimized, since the extent of binding of HAL to cells without specific receptors for HAL was negligibly small (e.g., binding to CV cells) in comparison to their binding to B16F10 cells. Similarly, the extent of binding of liposomes without HA was minimal. After fixing the aforementioned parameters sequentially, the kinetics of total cell association are simulated by adding the rate constant of endocytosis, ε (s^{-1}). The applicability of the model calculations is tested by the predictions for binding and total uptake kinetics for a wide range of lipid concentrations.

Interaction of HAL with Normal and Tumor Cells

Many cells express the CD44 receptor at a low level; however, CD44 expression is often increased significantly in tumors. To mimic this situation in a cell culture system the high CD44-expressing B16F10 (murine melanoma) tumor cell line is used,[17] whereas CV-1 (monkey fibroblasts) cells are used as a representative of low CD44-expressing normal cells. The tumorigenic B16F10 murine melanoma cells express CD44H, a receptor that adheres to surface-bound hyaluronan[18] and also to hyaluronan oligomers.[17] CD44 expression on B16F10 cells is almost 100 times higher than CD44 expression on CV-1 cells.[10]

Extruded unilamellar vesicles composed of POPC:Chol (6:4 mole ratio) containing various mole ratios of hyaluronan tetrasaccharide-POPE (HA-PE) are prepared. The vesicles exhibit a gaussian size distribution. The hyaluronan liposomes (HAL) with 3 mol% HA-PE have a mean diameter of 120–140 ± 40 nm with a zeta potential of −9.9 mV.

When B16F10 cells are incubated with hyaluronan liposomes, uptake of the vesicles is found to depend critically on the density of the attached hyaluronan on the liposome surface (Fig. 1). Very little uptake is observed

[16] I. Nunes-Correia, J. Ramalho-Santos, S. Nir, and M. C. Pedroso de Lima, *Biochemistry* **38**, 1095 (1999).

[17] C. Zeng, B. P. Toole, S. D. Kinney, J. W. Kuo, and I. Stamenkovic, *Int. J. Cancer* **77**, 396 (1998).

[18] A. Bartolazzi, D. Jackson, K. Bennett, A. Aruffo, R. Dickinson, J. Shields, N. Whittle, and I. Stamenkovic, *J. Cell Sci.* **108**, 1723 (1995).

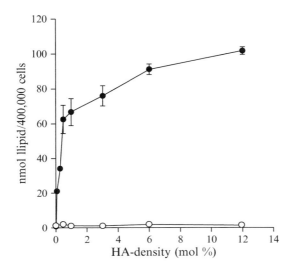

Fig. 1. Effect of hyaluronan tetrasaccharide-POPE (HA-PE) density in HAL on cell association of HAL by B16F10 and CV-1 cells. Hyaluronan-targeted liposomes (HAL) were prepared as described in the text. 0 mol% refers to POPC/Chol (6/4 mole ratio) liposomes lacking HA-PE. B16F10 (●) and CV-1 (○) cells were incubated for 3 h at 37° with liposomes (200 μM) containing various mole percents of HA-PE labeled with ^{125}I-BPE. Values presented are the mean ± SD of nine replicates.

in the absence of hyaluronan conjugation (0 mol% HA density, Fig. 1), but even at a low density (0.1–0.5 mol%) of hyaluronan, the liposomes have high affinity to the cells and show increased uptake. The extent of uptake begins to saturate when the hyaluronan density is above 3%. No significant uptake is observed when the liposomes are incubated with control CV-1 cells, regardless of the hyaluronan density.

Kinetic Parameters of Liposome Binding

In order to study the kinetics of liposome uptake by the receptor-bearing cell, B16F10 cells are incubated with three different concentrations of HAL containing 3 mol% HA-PE for various intervals. HAL uptake exhibits a slight lag phase over the first hour of incubation, is linear during the second hour of incubation, and then reaches a plateau at 3 h (Fig. 2). This decline in the rate of internalization appears to derive from saturation of the uptake system rather than depletion of liposomes, as the liposome concentration in the medium declines by only 30–37% over the course of incubation. Moreover, the same results are obtained by repeating the

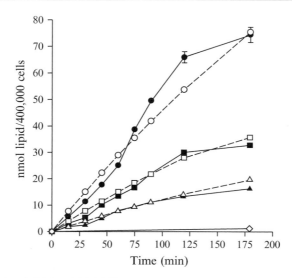

Fig. 2. Kinetics of HAL and POPG liposome association with B16F10 cells. The amount of liposomes (labeled with ^{125}I-BPE) associated with 400,000 cells is given as a function of time of incubation at 37° with varying concentrations of HAL: (●) 200 nmol/ml (200 μM), (■) 100 μM, and (▲) 50 μM. POPG liposomes (200 μM) composed of POPC/Chol/POPG (6/4/0.9 mole ratio) were incubated with the cells for 3 h (◇). Broken lines (blank symbols) show the simulated kinetic curves of liposome association obtained from calculations employing the kinetic model. Parameters and estimated uncertainties for the fit shown are delineated in Table I.

experiment with the same concentration of liposomes in half the volume of medium and using half of the number of cells. We, therefore, conclude that HAL are taken up by B16F10 cells via a saturable mechanism, followed by internalization into intracellular compartments. In contrast, little uptake is observed with POPG liposomes lacking hyaluronan that have a negative zeta potential similar to hyaluronan liposomes (Fig. 2). The POPG liposomes had a mean diameter of 127 ± 33 nm with a zeta potential of −9.8 mV. Due to the additional negative charge on the HA-PE it was necessary to include a threefold greater mole ratio of POPG than the HA-PE in the lipid composition in order to obtain a similar surface charge. After 3 h of incubation of B16F10 cells with POPG liposomes (200 nmol lipid), only 1.2 nmol lipid was associated with the cells.

To distinguish surface bound from internalized liposomes, B16F10 cells are incubated for different periods of time with three different concentrations of liposomes at either 37° in the presence of metabolic inhibitors or at 4° and then washed with cold PBS to remove unattached liposomes. The rate and extent of total association (uptake) of HAL with cells were significantly larger at 37° (Fig. 2) than at 4° (Fig. 3). The uptake at 37° was reduced dramatically by metabolic inhibitors (Fig. 4), strongly suggesting that endocytosis following binding at 37° yields the major contribution to the total uptake of the HAL. Little cell association occurs for liposomes lacking HA-PE at either temperature. Figures 3 and 4 demonstrate that during a 3-h incubation the total cell association at 37° in the presence of inhibitors was similar to the total cell association at 4°, which in both cases is due to liposomes bound to the cell surface without being endocytosed. Cells incubated with HAL at 4° or 37° in the presence of metabolic inhibitors reached a steady-state/saturation value after 90 min (Figs. 3 and 4). In contrast, at 37° the total uptake of liposomes was greater than the uptake at 4° or 37° in the presence of inhibitors and a steady-state value was not obtained until after 2 h of incubation (Fig. 2), suggesting that a substantial fraction of the cell-associated liposomes were internalized.

The binding/association of HAL to the cells at 37° exhibited a classical saturation curve characteristic of specific binding. Little binding was observed with liposomes lacking hyaluronan but containing a negatively charged phospholipid, POPG, to provide a zeta potential that was the same

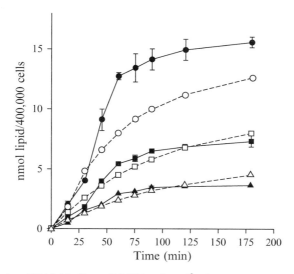

FIG. 3. Kinetics of HAL binding to B16F10 cells at 4°. The amount of liposomes bound to 400,000 cells is given as a function of time of incubation at 4° with varying concentrations of liposomes as described in Fig. 2 (same symbols). Broken lines show the simulated kinetic curves of liposome binding obtained from calculations employing the kinetic model. Parameters and estimated uncertainties for the fit shown are delineated in Table I.

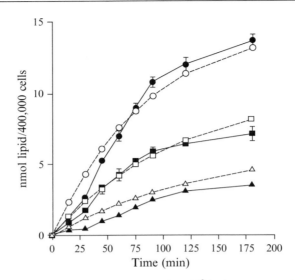

Fig. 4. Kinetics of HAL binding to B16F10 cells at 37° in the presence of inhibitors. The amount of liposomes bound to 400,000 cells is given as a function of time of incubation at 37° in the presence of metabolic inhibitors with varying concentrations of liposomes as described in Fig. 2 (same symbols). Broken lines show the simulated kinetic curves of liposome binding obtained from calculations employing the kinetic model. The kinetics of binding could be simulated and predicted using the parameters N as for 4°, $C = 1.5 \times 10^5 \ M^{-1} \cdot s^{-1}$, and $D = 7 \times 10^{-5} \ s^{-1} (R^2 = 0.95)$.

as that of HA-PE liposomes (−9.9 mV). This result suggests that the high affinity of the HAL to the B16F10 cells is not due to a nonspecific electrostatic force/attraction between the hyaluronan tetrasaccharide and the cell surface.

Analysis of Binding and Uptake

Mathematical Analysis of HAL Binding at 4°

Final extents of liposome binding to B16F10 cells are given in Fig. 5. The purpose of the calculations is to determine N, the number of receptor sites per cell for HAL, and the affinity coefficient, k. In order to avoid using an excessive number of parameters, we consider just one type of receptor sites. Figure 5 indicates that data could be fairly simulated by using $N = 3 \times 10^5$ and $k = 1.4 \times 10^9 \ M^{-1}$. The agreement between experimental and calculated values became unacceptable for N smaller than 2.5×10^5, or larger than 3.4×10^5, in which case the values of R^2 dropped below 0.7 and

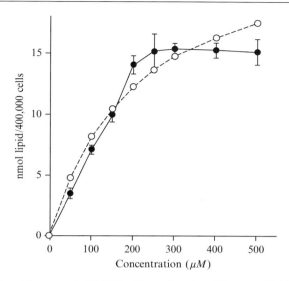

FIG. 5. Binding of liposomes to B16F10 cells as a function of liposome (lipid) concentration at 4°. The amount of liposomes containing hyaluronan (●) (HA-PE, 3 mol%) bound to B16F10 cells (nmoles total lipid bound per 400,000 cells) after 90 min of incubation at 4° was measured as a function of liposome (lipid) concentration. Values presented are the mean ± SD of six replicates. The broken line (○) shows the simulated curve of liposome binding obtained from calculations employing the kinetic model. The parameters used are given in Table I.

the deviations between experimental and calculated values increased by 50%. It cannot be excluded that the leveling off of the binding at 15 nmol, nmol, or 170,000 liposomes per cell, is close to the geometrical limit, which was not considered in the calculations.

Kinetics of liposome binding are shown in Fig. 3. By fixing the value of N, the simulation of the kinetics of binding required two additional parameters, C and D. Thus, these parameters could be fixed from two sets of data, i.e., 16 points for two lipid concentrations (50 and 200 μM lipid). Then the predictions of the model could be tested for an additional 8 points (100 μM lipid). Figure 3 demonstrates that model calculations could indeed yield reasonable predictions for the kinetics of liposome binding for the intermediate case (100 μM lipid). Because $k = C/D$, the value of k determined from these calculations is 1.7×10^9 M^{-1}, in comparison to $k = 1.4 \times 10^9$ M^{-1} in Fig. 5. This difference in k values is within the expected uncertainty. However, it might also arise from the fact that perhaps the binding results in Fig. 5 do not correspond to complete equilibration of the binding process, as the measurements were taken after 90 min of incubation. From the values of C, D, N, and the known values of liposome and cell concentrations, it was possible to estimate (according to Nir *et al.*[12]) the time required for equilibration of the binding process. The calculations yielded that about 90% of the amount of lipid bound at equilibrium could be achieved for 3 h of incubation. In fact, extending the incubation time from 1.5 to 3 h resulted in about 10% increase in the bound amounts.

Binding of HAL at 37° in the Presence of Inhibitors of Endocytosis

We assumed the same value of N at 37° as at 4°. The kinetics of binding could be simulated and predicted by using the parameters C and D (Fig. 4). Another prediction of the model calculations is illustrated in Fig. 6, which gives the binding at 90 min for lipid concentrations varying from 50 to 500 μM. Clearly, equilibration in binding was not fully attained after a 90-min incubation.

It may be noted that to a first approximation the binding results at 4° and 37° with inhibitors of endocytosis are quite similar. Such independence of the binding on temperature was observed before by Dijkstra *et al.*[15] and Lee *et al.*,[11] whereas in a similar study on the binding of influenza virus to

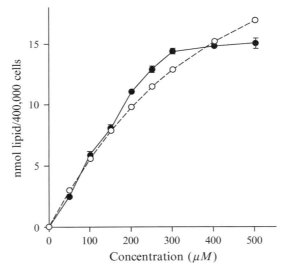

FIG. 6. Binding of liposomes to B16F10 cells as a function of liposome (lipid) concentration at 37° in the presence of metabolic inhibitors. The amount of liposomes containing hyaluronan bound to B16F10 cells (nanomoles total lipid bound per 400,000 cells) after 90 min of incubation at 37° in the presence of inhibitors was measured as a function of liposome (lipid) concentration. Values presented are the mean ± SD of six replicates. The broken line shows the predicted kinetic curve of liposome binding obtained from calculations employing the kinetic model. The parameters used are as in Fig. 4 ($R^2 = 0.93$).

TABLE I

PARAMETERS FOR THE UPTAKE OF HAL
(3 mol% HA-PE) BY B16F10 CELLS

	Temperature	
Kinetic parameters[a]	$4°$	$37°$
N	3×10^5	3×10^5
$k\ (M^{-1})$	$(1.4 - 1.7) \times 10^9$	2.1×10^9
$C\ (M^{-1} \cdot s^{-1})$	1.7×10^5	5×10^5
$D\ (s^{-1})$	10^{-4}	2.4×10^{-4}
$\varepsilon\ (s^{-1})$	—	0.002

[a] Estimated uncertainties were 10% in N, 20% in k, C, and D, and 30% in ε. The lower and higher values of k at $4°$ were obtained from binding results at 90 min (Fig. 5; $R^2 = 0.88$) and from results of incubation times between 15 and 180 min (Fig. 3; $R^2 = 0.83$), respectively. At $37°$ $R^2 = 0.95$ (Fig. 2).

cells, the fraction of virions bound declined with an increase in temperature.[16] According to the expression of C and D,[12] it may be expected in simple cases that C and D would be larger at $37°$ than at $4°$, and $k = C/D$ would decrease with an increase in temperature. A comparison of the results in Figs. 3 and 4 (see Table I) shows that C and D are slightly larger at $4°$ than at $37°$. Within the expected uncertainty, however, the corresponding values are similar. Small differences between batches can account for the observed pattern.

The deduced value of the binding coefficient, $k \approx 2 \times 10^9\ M^{-1}$, is similar to the value deduced for the binding of phosphatidylserine/phosphatidylcholine/cholesterol liposomes to murine macrophage-like cells.[11] However, the quantity, which determines the apparent affinity of binding, is $K = kN$. In the current system, we report a very large value of receptor sites. Consequently, the value of K is about $6 \times 10^{14}\ M^{-1}$, whereas in Lee et al.,[11] the largest values of K ranged from 10^{12} to $7 \times 10^{13}\ M^{-1}$, i.e., one to two orders of magnitude less. This result illustrates that the preferred binding of HAL to B16F10 cells stems from the large number of receptor sites per cell for the labeled liposomes.

Total Liposome Uptake at 37°

A comparison of the results of liposome uptake without inhibitors of endocytosis with those with inhibitors indicates severalfold more uptake in the absence of inhibitors. Ideally, the analysis should have employed the same values of N, C, and D and $(k = C/D)$ for both sets of data and just added a single parameter, ε, the rate constant of endocytosis. However, it should be recalled that liposomes incorporated by endocytosis should be first bound to the cells. Even if we ignored liposome dissociation from the cell surface and used an infinitely large value for ε, the value of $C = 1.5 \times 10^5$ $M^{-1} \cdot$ s^{-1} would underestimate the experimental results. For instance, for 50 μM lipid, and setting $D = 0$, gives 12.9% cell association at 180 min, whereas the experimental value is 32.4%. At 15 min, where endocytosis is not expected to be the dominant factor in total liposome uptake, the calculated percentage of liposome uptake is 1.3, in accordance with the experimental value in the presence of inhibitors of endocytosis, whereas the experimental value in the absence of inhibitors of endocytosis is 3.6. Accordingly, we simulated the data in Fig. 2 by using the same values of N and k as in Fig. 4, but using $C = 5 \times 10^5$ $M^{-1} \cdot$ s^{-1}, which was close to the lowest possible value that could yield a reasonable fit to data, rather than an underestimate. The overall fit to data is reasonable, despite results that the experimental percentage of lipid uptake at 180 min is higher for 200 μM lipid (37.2) than for 50 μM lipid (32.4), which can only occur if a small degree of cooperativity in binding exists. The value deduced for ε (0.002 s^{-1}) is somewhat larger than the largest value deduced in Lee et al.[11] (0.0014 s^{-1}) for liposome endocytosis by macrophage-like cells.

It may be of interest to emphasize a result mentioned previously[12,16] that for the same values of k and N, the number of liposomes bound to the plasma membranes of endocytosing cells may be significantly smaller in comparison with nonendocytosing cells. For instance, for $\varepsilon = 0.002$ s^{-1}, the calculated fraction of lipid bound (for 50 μM lipid) at 180 min is 1.4%, whereas in the absence of endocytosis the corresponding value is 9%. In this particular case, the fraction of lipid bound to the cells increases with time up to a value of 2.1% at 30 min (total association 8.2%) and from that moment it decreases slowly because the rate of liposome entry by endocytosis exceeds the rate of binding to the plasma membrane.

We have pointed out that the value of the forward rate constant of adhesion, C, which was used in simulating the total liposome uptake data at $37°$, was about three-fold larger than the value deduced from simulating the results of binding at $37°$ in the presence of inhibitors of endocytosis. The Smoluchowski–Fuchs expression of C (see also Nir et al.[12]) for the aggregation of spherical particles of the same radius is

$$C = \frac{4kT}{3\eta} \exp(-V^*/kT) 6 \times 10^{20} M^{-1} \text{ s}^{-1}$$

in which k is Boltzmann's constant, T is the absolute temperature, η is the viscosity, and V^* is the height of the potential barrier for close approach.

Extrapolation of the value of C at $4°$ to $37°$, by assuming independence of V^* of temperature, gives an increase by fivefold. The use of such a high value of C could not yield a fit to data. However, this consideration may strengthen the view that the value used for C in the absence of inhibitors is not too large, which implies that the value used in the presence of inhibitors is too low. At this stage, we can only speculate that, apart from inhibition of endocytosis, the surface of the B16F10 cells might be some-what depleted of receptors for HAL or that the structure of the receptors is modified. We favor the view that the larger C value in the absence of inhibitors, and the overall fast and extensive uptake of HAL by B16F10 cells at $37°$, is the typical behavior of this system, as two batches of liposomes and cells yielded similar results.

The Effect of HA Content in Liposomes

The increase in total lipid uptake by B16F10 cells at $37°$ as a function of HA-PE mole percent is demonstrated in Fig. 1. We assumed in the calculations that the values of N, D, and ε were independent of HA content in HAL and determined the values of C that yielded the experimental values of lipid uptake. Results in Table II demonstrate an increase of two orders of magnitude in C or k values upon an increase in HA content from 0 to 1

TABLE II
Effect of HA Content in HAL on Their
Binding[a] to B16F10 Cells at $37°$

HA content (mol%)	$C(M^{-1} \cdot s^{-1})$	$k(M^{-1})$
0	3.5×10^3	1.5×10^7
0.03	3.9×10^3	1.6×10^7
0.1	1.1×10^5	4.6×10^8
0.3	1.9×10^5	7.9×10^8
0.5	4×10^5	1.7×10^9
1	4.3×10^5	1.8×10^9
3	5×10^5	2.1×10^9
6	6.5×10^5	2.7×10^9
12	7.7×10^5	3.2×10^9

[a] See Fig. 1 for the total uptake of HAL ($200 \mu M$ lipid). Calculations were carried out by fitting exactly the results in Fig. 1. In these calculations the values of N, D, and ε were fixed at the values determined for HAL with 3 mol% HA-PE (see Table I) and the only varied parameter was C. The parameter k was given from $k = C/D$.

or 3 mol%. Remarkably, a 30-fold increase in the aforementioned values is obtained for 0.1 mol% of HA. The assumption of constant N and ε seems plausible, whereas varying D has a much smaller effect on the calculated uptake than varying C. In fact, it can be expected that D values decrease as the HA content increases, which implies that the actual increases in the affinity coefficient, k, are larger than those shown in Table II.

POPG Liposomes

We have pointed out that POPG liposomes exhibit less than 1% uptake by B16F10 cells after 3 h of incubation at 37° (Fig. 2). By employing a similar calculation as in the previous section, we could fit the uptake of POPG by fixing the values of N, ε, and D, and looking for a value of C that yielded the best fit. For the case of 200 μM lipid the fit was good by employing $C = 7000\ M^{-1} \cdot s^{-1}$, which yields $k = 2.9 \times 10^7\ M^{-1}$. These values are about 2-fold larger than the values deduced for HAL containing 0 or 0.03 mol% HA (Table II), which indicates that POPG are taken more efficiently by the cells than HAL with very low content of HA. However, these values are 70-fold lower than those obtained for HAL containing 3 mol% HA.

The Effect of Free Hyaluronan

To evaluate the role of hyaluronan in the cellular uptake of HAL, B16F10 cells are incubated with 200 μM HAL (200 nmol; 3 mol% HA-PE) in serum-free and antibiotic-free medium containing increasing concentrations of high molecular mass hyaluronan (50,000 Da). The uptake of HAL was inhibited competitively by the presence of free hyaluronan. In the absence of free hyaluronan about 75 nmol lipid was associated with the cells after a 3-h incubation. When hyaluronan was added to the medium, and the cells were first incubated with the inhibitor (free hyaluronan) for 1 h and then with the HAL for 3 more hours, 100 pM hyaluronan reduced cellular liposomes uptake by 42%. A 50% reduction of liposome uptake occurred at about 10 nmol of disaccharide equivalent. The studies of Laurent et al.[19] on the binding of HA saccharides of increasing size to liver endothelial cells demonstrate a dramatic decrease in K_d values (i.e., a dramatic increase in k values) with the size of the polymer. Thus the deduced k values are $1.3 \times 10^4, 1.7 \times 10^{10}$, and $1.1 \times 10^{11}\ M^{-1}$ for HA of 4, 2000, and 32,000 sugar residues, respectively. Hence, for hyaluronan of 50,000 Da ($\sim$270 residues), we can expect a k value of about $10^9\ M^{-1}$, which is in the range of values deduced (Tables I and II), whereas for a

[19] T. C. Laurent, J. R. Fraser, H. Pertoft, and B. Smedsrød, *Biochem. J.* **234,** 653 (1986).

disaccharide the expected value would be below 10^4 M^{-1}, i.e., five orders of magnitude less. For HAL of 200 nmol (3 mol% HA-PE) the content of HA-PE is 6 nmol. Hence, the result that hyaluronan at 10 nmol of disaccharide equivalent inhibited 50% of HAL uptake is essentially compatible with the fact that the corresponding k values are in the same range. This outcome demonstrates that coupling of the HA disaccharides to liposomes results in a significant enhancement in the binding affinity to the HA receptors, similar to that of a hyaluronan of 135 disaccharide units. Since a liposome contains about 2000 HA-PE units (composed of one complete dissacharide, while the lipid is conjugated to the second disaccharide through the carbon $\beta 3$ of N-acetyl glucosamine by reductive amination), its affinity of binding is less than that of a corresponding hyaluronan polymer, but nevertheless the HAL can compete for binding to the target cells with the free hyaluronan. Thus the HAL exhibit tight and selective binding to CD44-expressing cells. At high levels of hyaluronan (above 10 μM), HAL binding to the cells is reduced to the value observed with liposomes lacking HA-PE.

Concluding Remarks

We have cleaved enzymatically the hyaluronan polymer into smaller fragments and attached saccharides of defined chain length to the lipid. Hexameric fragments of hyaluronan have been reported to constitute the minimal sequence capable of binding to cell surface CD44 receptors.[20,21] The receptor appears to bind preferentially to a six-sugar sequence of hyaluronate and, with considerably lesser affinity, to a four-sugar sequence. However, hyaluronan tetrasaccharides at high concentration inhibit the binding of high molecular weight hyaluronan and can presumably be recognized by the receptor.[19] Thus, the low molecular weight oligomer of hyaluronic acid has a very weak affinity for the CD44 receptor, but on a liposome it may bind avidly only to cells with a high density of receptors. It was suggested[22] that hyaluronan receptors appear to act cooperatively with each other to bind high molecular weight hyaluronan. In other words, more than one receptor can interact with the same molecule of high molecular weight hyaluronan, resulting in a higher binding affinity than a single ligand–receptor interaction. This suggestion implies that large molecules of hyaluronan can interact simultaneously with more than one receptor.

[20] W. Knudson and C. B. Knudson, *J. Cell Sci.* **99,** 227 (1991).

[21] R. E. Nemec, B. P. Toole, and W. Knudson, *Biochem. Biophys. Res. Commun.* **149,** 249 (1987).

[22] C. B. Underhill and B. P. Toole, *J. Biol. Chem.* **255,** 4544 (1980).

Because the short ligand on the liposome surface has a weak affinity to a receptor, the liposome will interact weakly with dispersed or low amounts of receptors on the cell surface. Because high-affinity receptors for hyaluronic acid are enriched greatly on certain cancer cells, it was reasoned that hyaluronic acid conjugation might allow preferential targeting of liposomes to neoplastic tissues. When used as a targeting ligand on a liposome, HA could alter significantly the distribution of the targeted drug. Thus, HAL should be an optimal vehicle to target antitumor agents to tumors that overexpress the CD44 receptor.

HAL have a high affinity to B16F10 cells that express high levels of the CD44 receptor. Hyaluronan facilitates recognition of liposomes by B16F10 melanoma cells in culture, and that following cell surface binding, the liposomes are internalized into the targeted cells by hyaluronan receptor-mediated endocytosis. The binding is followed by a temperature-dependent internalization of the bound liposomes. An incubation of B16F10 cells with HAL at 4° and 37° in the presence of metabolic inhibitors for endocytosis yields severalfold less lipid association with the cells than at 37° without inhibitors, suggesting that the CD44 receptor mediates endocytosis of HAL. The B16F10 cell line has a high density of CD44, whereas the control CV-1 cells have a low CD44 density (two orders of magnitude lower than B1F10 cells), similar to the CD44 density found on many normal cells in the body. CV-1 cells show little uptake, a result indicating that HAL has high affinity to B16F10 but not to CV-1 cells. HAL binding to B16F10 cells is inhibited by hyaluronan and monoclonal antibodies directed against the CD44 receptor.[10] Thus, most of the HAL binding to B16F10 cells is apparently due to their interactions with the cell surface CD44 receptors.

This article described the determination of the parameters that affect the uptake of liposomes by cells, N, C, D, $k = C/D$, and ε. As pointed out in this article, most of the uptake (after 1 h or more) at 37° is due to endocytosis. Which of these parameters plays a unique role in the selective uptake of HAL by cells overexpressing CD44? Lee *et al.*[11] noted that the same rate of endocytosis, ε, was deduced for two types of liposomes, despite more than an order of magnitude difference in their uptake. The value of ε ($0.002\ s^{-1}$) describing the endocytosis of HAL by B16F10 cells is in the range of previously recorded values.[11] The value of D, the dissociation rate constant, has little effect on the total uptake, as long as it is significantly smaller than ε, which is observed in our case ($D = 2.4 \times 10^{-4}\ s^{-1}$ at 37°). Certainly, in binding, where k ($=C/D$) and N determine the outcome, the parameter D plays an important role. Thus, the high selectivity of uptake of HAL by B16F10 cells is due to the relatively high values of C and N. If we assume that the CD44 receptors can also bind POPG liposomes, which carry a similar amount of negative surface charge density as HAL, then the two orders of magnitude difference in their uptake in favor of HAL must be due to the difference in C values, i.e., in the affinity coefficient, which depends strongly on the surface charge density of HA in the liposome (see Table II). The value of C ($5 \times 10^5\ M^{-1} \cdot s^{-1}$), however, is not extensively large in comparison with other systems and is several orders of magnitude below the diffusion-controlled limit. Similarly, the value deduced for k ($2.1 \times 10^9\ M^{-1}$) is not extremely large in comparison with a host of other systems. Hence, a unique feature of our system is the large number of receptors per cell ($N = 3 \times 10^5$). Our calculations indicate that the total uptake is roughly proportional to N. This is in accord with the fact that the uptake of HAL containing 3 mol% HA-PE by CV-1 cells, which is two orders of magnitude less than the uptake by B16F10 cells, corresponds to the ratio of CD44 expression on these two cells.

Acknowledgments

This work was supported by Cancer Research Coordinating Committee Award No. 2-519850, the State of California Tobacco-Related Disease Research Program 8IT-0138, and NIH Grant GM61851-01A1.

[3] Folate Receptor-Mediated Targeting of Liposomal Drugs to Cancer Cells

By Stacy M. Stephenson, Philip S. Low, and Robert J. Lee

Introduction

De novo nucleotide synthesis and one-carbon metabolism require the vitamin folic acid (MW 441, Fig. 1), which cannot be synthesized by mammalian cells. Therefore, cellular uptake of folates is of prime importance to the survival of proliferating cells. Cellular acquisition of folates is mediated by both the reduced folate carrier (RFC)[1] and the membrane folate receptor (FR).[2] The RFC exhibits substrate preference for reduced folates, such as 5-methyltetrahydrofolate, and has a K_m in the micromolar range.[1] In contrast, FRs show a greater affinity for folic acid than reduced folates with K_ds in the nanomolar range.[2] While FR expression is normally low in adult

---

[1] F. M. Sirotnak and B. Tolner, *Annu. Rev. Nutr.* **19**, 91 (1999).
[2] A. C. Antony, *Annu. Rev. Nutr.* **16**, 501 (1996).

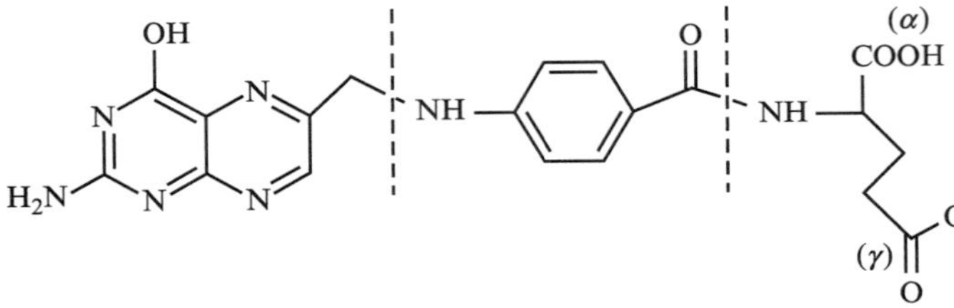

FIG. 1. Structure of folic acid.

tissues, many tumors exhibit increased FR levels.[3,4] This differential expression renders FR potentially useful for targeting therapeutic and imaging agents to cancer cells.

Folate receptors are 38-kDa glycoproteins that exist in three isoforms: two membrane glycosyl phosphatidylinositol (GPI)-anchored (α and β) forms and one soluble (γ) form secreted mainly by hematopoietic cells.[2] The α and β isoforms exhibit similar binding affinities for folic acid ($K_d \sim 0.1$–1 nM), but differ in their affinities for the stereoisomers of other folates, with FR-α having a greater affinity for the physiologic (6S) iso-mers.[5] Northern blot analysis and immunoassays demonstrate that consti-tutive expression of FRs is low in normal adult tissue.[3] Reverse transcription-polymerase chain reaction (RT-PCR) analysis has shown that in most tissues where FR is still expressed, the FR-β isoform predomi-nates.[4] However, in the placenta and kidney proximal tubules, FR-α appears to be more abundant.[4] Immunohistochemical staining and im-munoassays of many carcinomas and malignant cell lines reveal FR overex-pression.[4,6] FR-α is overexpressed consistently in nonmucinous ovarian carcinomas and amplified frequently in several other epithelial lineage tu-mors, including endometrial, lung, colorectal, breast, renal cell carcinomas, and brain metastases.[6] FR-β, however, is found to be amplified in both chronic and myelogenous leukemias.[7] Regardless of the tissue of origin, FR-α is typically the only isoform detected in tissue culture.[4]

Folate uptake by FRs has been proposed by some to follow a unique path-way called potocytosis.[8,9] In their view, potocytosis begins with the clustering

[3] S. D. Weitman, R. H. Lark, L. R. Coney, D. W. Fort, V. Frasca, V. R. Zurawski, and B. A. Kamen, *Cancer Res.* **52,** 3396 (1992).

[4] J. F. Ross, P. Chaudhuri, and M. Ratnam, *Cancer* **73,** 2432 (1994).

[5] X. Wang, F. Shen, J. H. Freisheim, L. E. Gentry, and M. Ratnam, *Biochem. Pharmacol.* **44,** 1898 (1992).

[6] P. Garin-Chesa, I. Campbell, P. E. Saigo, J. L. Lewis, L. J. Old, and W. J. Rettig, *Am. J. Pathol.* **142,** 557 (1993).

[7] J. F. Ross, H. Wang, F. G. Behm, P. Mathew, M. Wu, R. Booth, and M. Ratnam, *Cancer* **85,** 348 (1999).

of FRs in small membrane invaginations called caveolae, which are coated with the protein caveolin rather than clathrin.[8,9] FR binding then initiates the formation of plasma membrane-tethered vesicles containing the receptor–folate complex.[8,9] Folate is transported to the cytoplasm after a pH decrease causes it to dissociate from the receptor.[8,9] Unoccupied folate receptors are then returned to the plasma membrane when the caveolae reopen.[8,9]

In another model, FRs are internalized at coated pits by classical receptor-mediated endocytosis.[10–12] Following collection in primary endosomes, folate–colloidal gold conjugates are seen in such intracellular compartments as multivesicular bodies and late endosomes, and eventually substantial numbers of folate conjugates can even be observed in the cytoplasm.[13] The endosomal compartment in FR-mediated cellular internalization has been shown to be highly acidified.[14] Although both endocytotic routes could be operational, more definitive studies may be required before the correct endocytotic pathway can be defined.

The minimal constitutive expression of FR in normal tissues and its overexpression in many types of cancer render it a promising target for tumor-specific drug delivery. Folate receptor targeting can be accomplished with either monoclonal antibodies (MAbs) or folate itself.[15] Folate, as a low molecular weight agent possessing high FR affinity, has several advantages as a targeting ligand. It is a readily available chemical that exhibits superior functional stability during exposure to adverse storage conditions, organic solvents, and repeated freezing and thawing. Conjugate production with folate is also reproducible and inexpensive, as the necessary conjugation chemistry is straightforward and well defined. Folate lacks the immunogenicity associated with MAbs and their fragments, which could prevent repeated use of Mabs in a clinical setting. Finally, FR-targeted materials can continuously accumulate intracellularly due to receptor recycling. With these advantages in mind, folate-mediated targeting has been applied successfully to the selective delivery of proteins,[16–19] liposomes,[20–22] γ imaging agents,[23–26] chemotherapy agents,[27] and gene transfer vectors to cancer cells.[28]

[8] R. G. Anderson, *Science* **255,** 410 (1992).

[9] K. G. Rothberg, Y. Ying, J. F. Kolhouse, B. A. Kamen, and R. Anderson, *J. Cell Biol.* **110,** 637 (1990).

[10] M. Wu, J. Fan, W. Gunning, and M. Ratnam, *J. Membr. Biol.* **159,** 137 (1997).

[11] S. Rijnboutt, G. Jansen, G. Posthuma, J. B. Hines, J. H. Schornagel, and G. J. Strous, *J. Cell Biol.* **132,** 35 (1996).

[12] S. Mayor and F. R. Maxfield, *Mol. Biol. Cell* **6,** 929 (1995).

[13] J. J. Turek, C. P. Leamon, and P. S. Low, *J. Cell Sci.* **106,** 423 (1993).

[14] R. J. Lee, S. Wang, and P. S. Low, *Biochim. Biophys. Acta* **1312,** 237 (1996).

[15] J. Sudimack and R. J. Lee, *Adv. Drug Delivery Rev.* **41,** 147 (2000).

The use of liposomes for targeted drug delivery offers several advantages over direct conjugation of a targeting ligand to the therapeutic agent. First, the availability of functional groups for direct ligand conjugation to a drug molecule may be limited, rendering the conjugation chemistry problematic. Second, biological activity can be compromised once folate is attached, requiring the additional construction of a cleavable linker to enable drug release following endocytosis. Moreover, multiple drug molecules can be delivered upon internalization of a single liposome, whereas only single drug molecules are generally delivered following the uptake of directly conjugated agents. Thus, targeted liposomal formulations may be preferred over directly conjugated agents, provided that liposome penetration of the tumor mass is not limiting.

Steric barriers on most cell surfaces prevent liposomes containing folate linked directly to lipid head groups from being recognized by the FR. However, efficient FR binding and endocytosis of folate-derivatized liposomes can be achieved by separating folate from the liposomal surface via a polyethylene glycol (PEG) spacer.[20,21,29,30] Such folate-tethered lipids can be synthesized by a number of schemes and are commonly referred to as folate-PEG-lipids.

Two strategies can be employed to generate FR-targeted liposomes. The first method involves folate conjugation to surface functional groups on preformed liposomes.[20] The second technique entails the synthesis of lipid molecules linked to folate via a PEG spacer and subsequent incorporation of

---

[16] C. P. Leamon and P. S. Low, *Proc. Natl. Acad. Sci. USA* **88**, 5572 (1991).

[17] C. P. Leamon and P. S. Low, *J. Biol. Chem.* **267**, 966 (1992).

[18] C. P. Leamon and P. S. Low, *J. Biochem.* **291**, 855 (1993).

[19] C. P. Leamon, I. Pastan, and P. S. Low, *J. Biol. Chem.* **25**, 847 (1993).

[20] R. J. Lee and P. S. Low, *J. Biol. Chem.* **269**, 3198 (1994).

[21] R. J. Lee and P. S. Low, *Biochim. Biophys. Acta* **1233**, 134 (1994).

[22] S. Wang, R. J. Lee, G. Cauchon, D. G. Gorenstein, and P. S. Low, *Proc. Natl. Acad. Sci. USA* **92**(13), 318 (1995).

[23] S. Wang, R. J. Lee, C. J. Mathias, M. A. Green, and P. S. Low, *Bioconjug. Chem.* **7**, 56 (1996).

[24] S. Wang, J. Luo, D. A. Lantrip, D. J. Waters, C. J. Mathias, M. A. Green, P. L. Fuchs, and P. S. Low, *Bioconjug. Chem.* **8**, 673 (1997).

[25] C. J. Mathias, S. Wang, D. J. Waters, J. J. Turek, P. S. Low, and M. A. Green, *J. Nuclear Med.* **39**(9), 1579 (1998).

[26] C. J. Mathias, S. Wang, R. J. Lee, D. J. Waters, P. S. Low, and M. A. Green, *J. Nuclear Med.* **37**, 1003 (1996).

[27] C. A. Ladino, R. Chari, L. A. Bourret, N. L. Kedersha, and V. S. Goldmacher, *Int. J. Cancer* **73**, 859 (1997).

[28] R. J. Lee and L. Huang, *J. Biol. Chem.* **271**, 8481 (1996).

[29] A. Gabizon, A. T. Horowitz, D. Goren, D. Tzemach, F. Mandelbaum-Shavit, M. M. Qazen, and S. Zalipsky, *Bioconjug. Chem.* **10**, 289 (1999).

[30] R. J. Lee and P. S. Low, *J. Liposome Res.* **7**, 455 (1997).

the folate-PEG-lipid into the bilayer during liposome preparation.[21] While both strategies have been implemented successfully, formation of liposomes with a previously synthesized conjugate avoids many problems. Because the final steps of liposome assembly often require an aqueous medium, and conjugation reactions with preformed liposomes cannot be controlled precisely, the first method may result in hydrolytic by-products and unreacted head groups on both sides of the lipid bilayer. Lipid-derived by-products such as maleimide-derivatized lipids can potentially alter the surface properties of the liposomes.[20] In addition, the therapeutic cargo within the liposome can, in some cases, interfere with the ligand-coupling reactions, requiring customization of conjugation protocols for each material to be encapsulated.[20]

Even though the formation of liposomes with folate-conjugated lipids avoids these problems,[21] the technique possesses its own drawbacks. The directionality of conjugated lipid incorporation cannot be controlled, leading to the formation of liposomes with targeting ligands on both the inner and the outer leaflets.[30] However, as long as excessive targeting ligand is not employed in the liposome formulation, this consequence should not affect significantly the drug-carrying capacity of the particle. Data obtained in studies of FR-expressing tumor cells show that as little as 0.1 mol% of the conjugated lipid can be optimal for efficient targeting.[21,30] When a targeting ligand, such as an antibody, is sensitive to organic solvents or sonication procedures, the synthesis of the conjugated lipid must be performed in aqueous medium, leaving detergent dialysis, which usually results in poor drug incorporation, as the only viable method for liposome formation. Fortunately, folic acid is stable in organic solvents so the purification of folate-conjugated lipid and the preparation of FR-targeted liposomes are straightforward and may follow a method that is optimized for drug loading.[30]

Synthesis of NHS Ester of Folic Acid[31]

Materials

> Folic acid dihydrate
> N-Hydroxysuccinimide (NHS)
> Dimethyl sulfoxide (DMSO), predried using a molecular sieve
> Dicyclohexylcarbodiimide (DCC)
> Triethylamine [$N^+(Et)_3$]
> Glass wool
> Diethylether
> Anhydrous ether

[31] R. J. Lee and P. S. Low, "Methods in Molecular Medicine," Vol. 25, p. 69. Humana Press, Totowa, NJ, 2000.

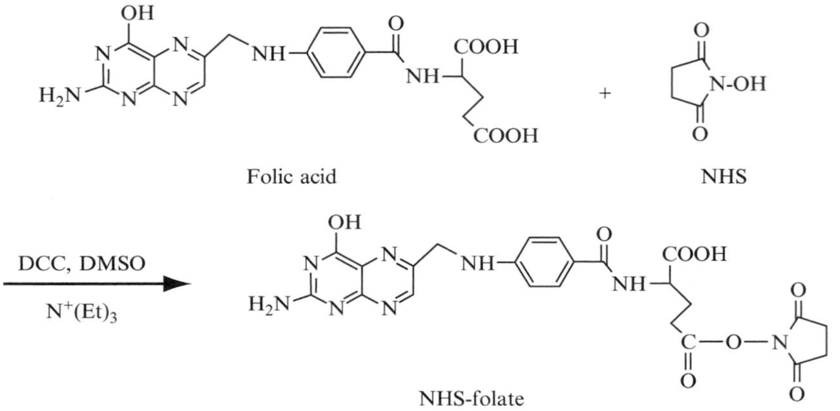

FIG. 2. Synthesis of NHS-folate.

Folic acid, shown in Fig. 1, contains two carboxyl groups that can be used for conjugation to lipids or PEG. Modification of the γ carboxyl group allows conjugation without a significant loss of binding affinity for the folate receptor.[29] Conversion of folic acid to NHS-folate, illustrated in Fig. 2, preferentially activates the γ carboxyl group for later coupling to nucleophilic groups such as primary amines on polymers and lipids.

To synthesize NHS-folate,[31] 5 g of folic acid is dissolved in 100 ml of DMSO plus 2.5 ml of triethylamine. A 1.1 molar excess of NHS (2.6 g) and DCC (4.7 g) is added. The mixture is stirred overnight at room temperature in the dark. The by-product, dicyclohexylurea, is removed by filtration through glass wool. NHS-folate, which is in the filtrate, can be precipitated with diethylether and stored as a yellow powder after several washes with anhydrous ether and desiccation. NHS-folate may also be redissolved in an appropriate volume of anhydrous DMSO for later manipulations.

Preparation of FR-Targeted Liposomes by Folate Derivatization of Preformed Vesicles[20]

Materials

Egg phosphatidylethanolamine (PE)
Phosphatidylcholine (PC)
Maleiimidocaproic acid (MC)
Dicyclohexylcarbodiimide
Traut's reagent (2-iminothiolane)

Polyoxyethylene-*bis*-amine (PEG-*bis*-amine) MW 3350

Polycarbonate filter (100 nm pore size)

Liposome extruder (either the hand-held LiposoFast extruder from Avestin, Inc. (Ottawa, Canada) or the Lipex extruder from Northern-lipids, Inc. (Vancouver, Canada)

Sepharose CL-4B size exclusion column (10 ml size)

Phosphate-buffered saline (PBS; 140 mM NaCl, 20 mM Na$_2$HPO$_4$, adjusted to pH 7.4)

Phosphate-EDTA buffer (1 mM EDTA, 100 mM phosphate, pH 8.0)

Bicarbonate buffer (1 M bicarbonate/carbonate, pH 11.0)

NHS-folate (synthesized as described earlier)

Triethylamine [N$^+$(Et)$_3$]

Preparation of Maleiimide-Activated Liposomes

N-Maleiimidocaproyl-phosphatidylethanolamine(MC-PE) is prepared by activating maleiimidocaproic acid (28.4 mg in CH$_3$Cl) with an equimolar quantity of DCC (27.7 mg) at room temperature for 60 min, followed by the addition of egg PE (100 mg in 10 ml CH$_3$Cl with 20 μl triethylamine) and incubation at room temperature for 2 h as shown in Fig. 3. For liposome preparation, lipid films are formed by dissolving a 4:1 molar ratio of PC and MC-PE in Ch$_3$Cl. The solution is dried in a rotary evaporater until a thin film is formed and is then dried further under vacuum for 2 h. Encapsulation of the desired agent can be achieved by rehydrating the lipid film in a solution containing the agent of interest. For this purpose, the lipid film is resuspended by vortexing in the presence of an aqueous solution of the desired therapeutic agent and subjecting the resulting suspension to 5 to 10 cycles of freezing and thawing, followed by extrusion

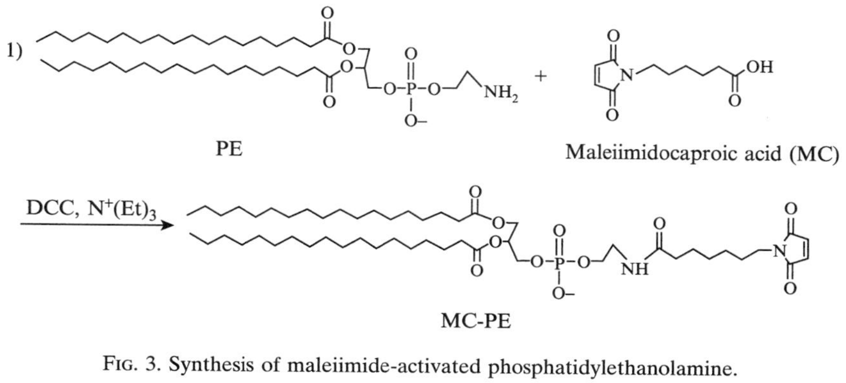

FIG. 3. Synthesis of maleiimide-activated phosphatidylethanolamine.

through a 100-nm pore size polycarbonate filter using a lipid extruder. Liposomes are then purified using a Sepharose CL-4B size exclusion column equilibrated in PBS. Liposome size can be determined by photon correlation spectroscopy (also known as dynamic or quasi-elastic light scattering) or negative-stain electron microscopy.[20]

Amino-PEGylation of Preformed Liposomes

Amino-PEG-SH is prepared by combining molar equivalents of Traut's reagent (8.2 mg in 200 μl H$_2$O) and PEG-*bis*-amine (200 mg in 1 ml phosphate-EDTA buffer) at room temperature for 60 min.

Liposomes (50 mg lipid) containing the activated MC-PE are then reacted for 60 min at room temperature with excess amino-PEG-SH (100 mg) in PBS, as shown in Fig. 4. Unreacted material is separated on the same Sepharose CL-4B column. The reacted liposomes now possess a free amino group at the distal end of the PEG, which is available for subsequent folate conjugation.[20]

Conjugation of Amino-PEGylated Liposomes to Folic Acid

Previously prepared NHS-folate (20 mg) is dissolved in 50 μl of DMSO. The pH of the liposome suspension is adjusted to 11.0 using bicarbonate buffer. The NHS-folate DMSO solution is added slowly to 2 ml of the liposome suspension while stirring. The folate conjugation reaction is

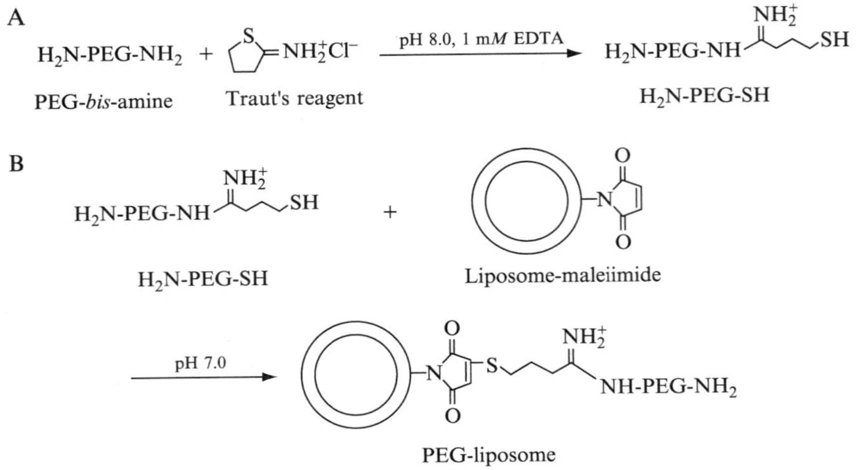

FIG. 4. Amino-PEGylation of liposomes.

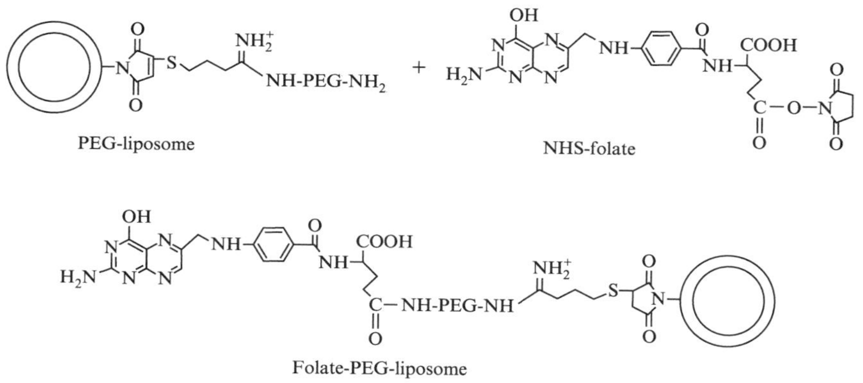

FIG. 5. Conjugation of liposomes with folic acid.

shown in Fig. 5. The reaction is allowed to proceed for 15 min at room temperature. The mixture is centrifuged briefly and the supernatant is passed down a Sepharose CL-4B column. Folate-PEG-liposomes will elute in the void fraction.[20] Liposomes may be sterilized by passing through a sterile 0.22-μm cellulose acetate filter and stored at 4°.

Preparation of FR-Targeted Liposomes Using Presynthesized Lipophilic Folate Derivatives[21,29,32]

Materials

    Distearoylphosphatidylethanolamine (DSPE)
    Distearoylphosphatidylcholine (DSPC)
    Dicyclohexylcarbodiimide (DCC)
    Methoxy-PEG2000-DSPE (PEG-DSPE)
    Cholesterol
    Cholesteryl chloroformate
    PEG-*bis*-amine MW 3350
    DMSO predried using a molecular sieve
    NHS-folate
    Succinic anhydride
    NH_4HCO_3 buffer, pH 8.0
    Diethylaminoethyl(DEAE)-trisacryl anion–exchange column
    Ninhydrin

[32] W. Guo, T. Lee, J. Sudimack, and R. J. Lee, *J. Liposome Res.* **10,** 179 (2000).

Sephadex G-25 column

Polycarbonate filter (100-nm pore size)

Liposome extruder (either the hand-held LiposoFast extruder from Avestin, Inc. or the Lipex extruder from Northernlipids, Inc.)

Phosphate-buffered saline

Pyridine

Amino-PEG-DSPE (available with various PEG lengths from Avanti Polar Lipids, Alabaster, AL)

Spectra/Por CE dialysis tubing (MW cutoff 300,000)

Synthesis of Folate-PEG-amine[32]

The method described earlier for the activation of folic acid to NHS-folate is used for this synthesis. Folate-PEG-amine is synthesized by reacting PEG-*bis*-amine (MW 3350, 500 mg in 2 ml DMSO) with a 1.1 M excess (88.3 mg) of NHS-folate, as shown in Fig. 6. The reaction mixture is incubated overnight at room temperature in the dark. A ninhydrin assay can be used to follow progression of the reaction, with conversion of approximately 50% of the initial amines indicating a complete reaction. The DMSO and low molecular weight by-products are removed by gel filtration on a Sephadex G-25 column equilibrated in deionized water. Because the folate-PEG-amine product will contain small amounts of folate-PEG-folate and PEG-*bis*-amine, folate-PEG-amine is purified on a (DEAE)-trisacryl anion–exchange column eluted with a NH_4HCO_3 gradient. Folate-PEG-amine will elute at approximately 20 mM NH_4HCO_3. The

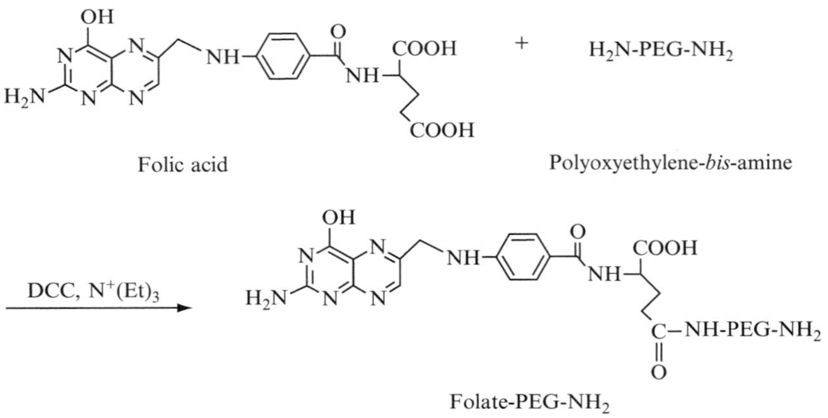

Fig. 6. Synthesis of folate-PEG-amine.

folate concentration is determined by measuring the absorbance at 363 nm using a molar extinction coefficient of 6500 M^{-1} cm^{-1}. The free amine concentration is assessed with the ninhydrin assay. The folate-to-free amine ratio should be approximately 1:1. Folate-PEG-amine can be lyophilized and stored at $-20°$.[32]

Synthesis of Folate-PEG-DSPE[21]

N-Succinyl-DSPE is synthesized by combining 100 mg DSPE in 5 ml chloroform, 10 μl pyridine, and a 1.1 molar excess (14.7 mg) of succinic anhydride, as shown in Fig. 7. The mixture is incubated overnight at room temperature, and then the N-succinyl-DSPE is precipitated with cold acetone. After verifying the presence of N-succinyl-DSPE by thin-layer chromatography (TLC), the lipid is redissolved in chloroform. The carboxyl group of N-succinyl-DSPE is activated by reacting with 1 molar equivalent of DCC for 4 h at room temperature. An equimolar quantity of the previously synthesized folate-PEG-amine, dissolved in chloroform is added and the mixture is allowed to react overnight at room temperature, as illustrated in Fig. 7. The chloroform is removed by drying under vacuum and the lipid pellet is washed twice with cold acetone. The folate-PEG-DSPE is redissolved in chloroform and stored at $-20°$. The purity of the folate-PEG-DSPE can be verified using TLC and the folate content can be determined using its extinction at 363 nm.[21] On a silica gel GF (75:36:6 chloroform/methanol/water), unreacted folate-PEG-amine will have an R_f of 0.76, whereas folate-PEG-DSPE will have an R_f of 0.57.

An alternative method of synthesizing folate-PEG-DSPE has been described by Gabizon et al.[29] where amino-PEG-DSPE is reacted with activated folate. Folic acid (100 mg; 0.244 millimoles) is dissolved in 4 ml DMSO. Four hundred milligrams (0.14 mmol) of amino-PEG-DSPE, 2 ml of pyridine, and 130 mg (0.63 mmol) of DCC are added to the folate-DMSO solution, and the mixture is incubated at room temperature for 4 h. Pyridine is removed by rotary evaporation and redissolved in 50 ml of water. The solution is centrifuged briefly to remove trace insolubles and the supernatant is dialyzed in Spectra/Por CE tubing against saline (50 mM, 2×2000 ml) and water (3×2000 ml). The dialyzate is then lyophilized. Product purity can be verified using TLC as described previously.[29]

Synthesis of Folate-PEG-Cholesterol[32]

Folate-PEG-amine is redissolved in chloroform and combined with a 1.1 molar excess of cholesteryl chloroformate, as shown in Fig. 8. The mixture is reacted overnight at room temperature and the reaction progress is assessed by following the disappearance of free amines using a ninhydrin

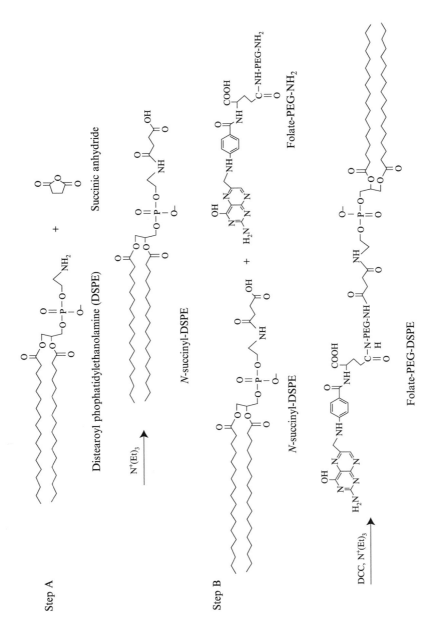

FIG. 7. Synthesis of folate-PEG-phosphatidylethanolamine.

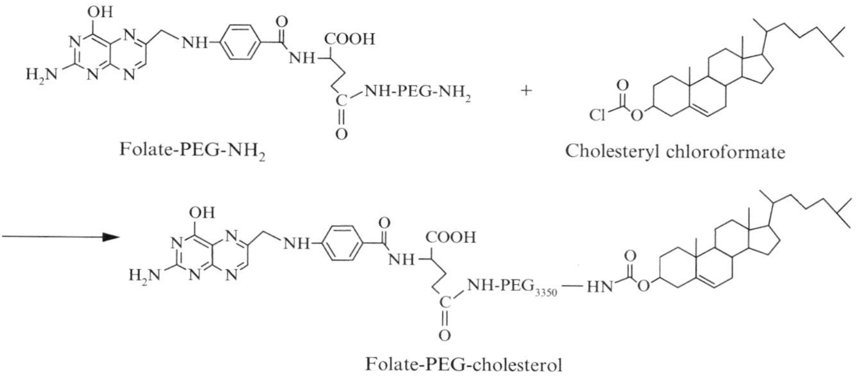

FIG. 8. Synthesis of folate-PEG-cholesterol.

assay. The folate-PEG-cholesterol is dried under vacuum, washed twice with ether to remove any residual cholesteryl chloroformate, and redissolved in chloroform. The product purity is assessed using silica gel thin-layer chromatography. In a CH_2Cl_2/methanol solvent (7/3, v/v), folate-PEG-cholesterol will have an R_f of approximately 0.86. The folate content is determined using its UV extinction at 363 nm and the compound is stored at $-20°$.[32]

Preparation of FR-Targeted Liposomes[21,29,32]

Because folate is incorporated into the liposomes as a presynthesized lipophilic derivative, no special provisions are needed for the preparation of these liposomes. Generally, a lipophilic folate derivative, folate-PEG-DSPE or folate-PEG-Chol, is added as a minor lipid component at 0.1 to 1 mol%. While increasing ligand density can sometimes enhance FR binding, excessive folate derivatization of the liposome surface may lead to increased nonspecific cell uptake or even to reduced target cell internalization (personal observation).

Applications of FR-Targeted Liposomes in Drug Delivery

Effect of PEG on FR Targeting in Cultured Cells[20]

The importance of a PEG spacer in FR targeting has been established by Lee and Low[20] in cultured KB cells (human epidermal carcinoma cell line overexpressing FR) using both untargeted and FR-targeted liposomes containing the fluorescent probe calcein. Liposomes with folate tethered to the liposome surface by a 250-Å PEG spacer showed 37-fold greater uptake than untargeted liposomes or targeted liposomes incorporating a short spacer or no spacer. Inhibition of this uptake by free folate or antibodies to the FR supported the concept that FR-targeted liposomes are internalized via a receptor-mediated endocytotic pathway. Fluorescence microscopy also supported this theory by demonstrating the saturation of cellular internalization within 4 h.[20] While investigating the effect of peptide-mediated liposomal release, Vogel et al.[33] demonstrated a similar increase in cellular uptake of the fluorophore propidium iodide with FR targeting compared to untargeted liposomes.

Gabizon et al.[29] confirmed the importance of a PEG spacer for FR-mediated liposome uptake in cultures with KB and M109 (murine lung carcinoma) cells. Folate was conjugated to liposomes with either a 2000- or a 3350-Da PEG spacer. Because a PEG coating can extend the circulation life of liposomes, approximately 4 mol% of 2000 Da PEG was included as a coating with the aforementioned folate-conjugated liposomes.[29] Although the presence of either PEG spacer increased liposomal uptake by 26-fold over untargeted liposomes, the targeted PEG coating reduced substantially the uptake for all liposomes.[29] Placement of the folate ligand on a longer PEG spacer (e.g., 3500 Da PEG) than the PEG used for steric stabilization (i.e., 2000 Da PEG), however, restored most of the FR targeting.[29] Consistent with this observation, Lee and Low[21] saw no decrease in doxorubicin delivery upon addition of a 4 mol% 2000-Da PEG coating to folate-targeted liposomes as long as the folate-tethered lipid contained a 3350-Da PEG spacer. However, liposomes containing a 4 mol% 3350-Da PEG coating were found to be readily targetable with 0.1% folate-PEG-(3350 Da) – PE (unpublished observations).

FR-Targeted Liposomal Chemotherapeutics In Vitro[21]

Targeted delivery of chemotherapeutic agents could limit unwanted side effects, such as myelosuppression, gastrointestinal toxicity, and cardiotoxicity. Therefore, several studies have investigated targeted liposome encapsulation of doxorubicin, a commonly used chemotherapeutic agent. Lee and Low[21] evaluated the cellular uptake and cytotoxicity of free and liposomal doxorubicin using flow cytometry and the MTT assay, respectively. Targeted liposomes incorporating folate-PEG-DSPE had 45-fold higher uptake and 86-fold greater tumor cell cytotoxicity than untargeted liposomes.[21] Targeted liposomal uptake and cytotoxicity, which could be

[33] K. M. Vogel, S. Wang, R. J. Lee, J. A. Chmielewski, and P. S. Low, *J. Am. Chem. Soc.* **118**, 1581 (1996).

inhibited by 1 mM free folic acid,[21] was also 1.6 and 2.7 times, respectively, higher than free doxorubicin.[21] While the difference in potency from free doxorubicin is not large, the potential reduction in normal tissue toxicity may warrant further development of this formulation.

Guo et al.[32] also demonstrated improved cytotoxicity using liposomes targeted with folate-PEG-cholesterol over untargeted liposomes and free doxorubicin. In similar studies with M109 cells, however, Goren et al.[34] reported only a 10-fold reduction in the IC$_{50}$ when comparing targeted and untargeted doxorubicin liposomes. Furthermore, they observed no difference in cytotoxicity between targeted liposomes and free doxorubicin. Despite equivalent IC$_{50}$ values, however, intracellular accumulation of targeted liposomal doxorubicin was shown to be double that of free doxorubicin after 4 h of incubation. Cell fractionation studies reiterated this finding and demonstrated a greater fraction of total doxorubicin residing in the nucleus for targeted liposomal doxorubicin than free doxorubicin.[34] In studies with folate-targeted liposomal AraC, Rui et al.[35] found a ~6000-fold enhancement in cytotoxicity with AraC entrapped in FR-targeted pH-sensitive liposomes (IC$_{50}$ 490 nM) compared to free AraC (IC$_{50}$ 2.8 mM).[35] pH-sensitive liposomes undergo rapid destabilization in response to the acidic pH in the endosomal compartment, thereby facilitating the efficient cytosolic release of membrane-impermeable agents.[36,37] The large magnitude of AraC delivery enhancement, however, may derive in part from the poor membrane permeability of the free AraC, a disadvantage not shared by free doxorubicin.

FR-Targeted Liposomes In Vivo [32,34]

Folate-derivatized liposomes have been evaluated in a syngeneic tumor model using both local and intravenous routes of administration.[34] After exposure to 10 μM free doxorubicin, untargeted liposomal doxorubicin, or targeted liposomal doxorubicin, an M109 cell suspension was injected into the footpad of BALB/c mice, and tumor growth was tracked as a function of time. Cells exposed to targeted liposomal doxorubicin showed significantly less tumor growth and lower final tumor weight than untreated cells or cells exposed to free or untargeted liposomal doxorubicin.[34]

[34] D. Goren, A. T. Horowitz, D. Tzemach, M. Tarshish, S. Zalipsky, and A. Gabizon, Clin. Cancer Res. 6, 1949 (2000).
[35] Y. Rui, S. Wang, P. S. Low, and D. H. Thompson, J. Am. Chem. Soc. 120, 11213 (1998).
[36] N. Düzgüneş, R. M. Straubinger, P. A. Baldwin, and D. Papahadjopoulos, in "Membrane Fusion" (J. Wilschut and D. Hoekstra, eds.), p. 713. Dekker, New York, 1991.
[37] R. M. Straubinger, Methods Enzymol. 221, 361 (1993).

It was, therefore, concluded that the efficacy of targeted liposomal doxorubicin is not compromised in an *in vivo* setting.[34]

Guo *et al.*[32] investigated the biodistribution of folate-targeted liposomes in C57BL/6 mice bearing subcutaneous 24JK-FBP cell-derived tumors. For this purpose, mice were injected with either folate-targeted or untargeted liposomes containing a radiolabeled lipid.[34] Tissues were harvested 24 h later and analyzed for radiolabel. The use of folate targeting did not affect significantly the biodistribution patterns, with liver and spleen uptake accounting for most of the tissue accumulation.[34]

In fact, both targeted and untargeted liposomes exhibited essentially indistinguishable tumor uptake of 6.5 and 7.1% injected dose per gram tissue, respectively.[34] It is likely that the enhanced permeability of the tumor vasculature and the consequent passive tumor accumulation, as documented previously for other liposomal formulations,[38–40] dominate the biodistribution of targeted liposomes. Attachment of a targeting ligand appears to assist in the uptake/endocytosis of the liposomal cargo following extravasation into the tumor mass.

FR-Targeted Liposomes for Delivery of Antisense Oligonucleotides[22]

Liposomes incorporating 5'-fluorescein-labeled oligonucleotides have been targeted to KB cells *in vitro* using folate-PEG-PE.[22] Folate-targeted liposomes demonstrated an increase in cellular internalization of 9- and 16-fold over untargeted liposomes and free oligonucleotide, respectively.[22] Incorporation of antisense oligonucleotides directed at the epidermal growth factor receptor (EGFR) resulted in substantial inhibition of cell growth for 4 days following incubation with the targeted liposomes.[22] Because incubation with untargeted and empty liposomes did not result in growth inhibition, the effect is likely due to the delivery of intact antisense oligonucleotides rather than any intrinsic cytotoxicity of the liposomes.[22]

The targeting of antisense oligonucleotides was expanded to include LPDII particles, which consist of liposome-entrapped, polycation-complexed oligonucleotides. Li and Huang[41] compared the transfection of KB cells with free oligonucleotides, untargeted liposomes, targeted liposomes, and targeted LPDII particles. The importance of folate targeting

---

[38] K. Hong, D. B. Kirpotin, J. W. Park, Y. Shao, R. Shalaby, G. Colbern, C. C. Benz, and D. Papahadjopoulos, *Ann. N. Y. Acad. Sci.* **886,** 293 (1999).

[39] N. Z. Wu, D. Da, T. R. Rudoll, D. Needham, A. R. Wharton, and M. W. Dewhirst, *Cancer Res.* **53,** 3765 (1993).

[40] S. K. Huang, F. J. Martin, G. Jay, J. Vogel, D. Papahadjopoulos, and D. S. Friend, *Am. J. Pathol.* **143,** 10 (1993).

[41] S. Li and L. Huang, *J. Liposome Res.* **7,** 63 (1997).

was again demonstrated by showing that both targeted liposomes and targeted LPDII vectors display increased uptake compared to untargeted liposomes and free oligonucleotides.[41] While FR-targeted liposome and LPDII uptake were found to be similar, LPDII was more effective in inhibiting cell growth.[41] Incorporation of the fusogenic lipid dioleoyl phosphatidylethanolamine (DOPE) in the LPDII formulation was likely responsible for improved endosomal disruption and the consequent cytosolic oligonucleotide delivery.[41]

FR-Targeted Liposomes for Gene Delivery[28,42,43]

While not as efficient at gene delivery as many viral vectors, nonviral systems such as lipidic vectors are less immunogenic, less restricted by DNA size, and produced more easily in clinically relevant quantities.[38] Nonviral vector systems include (i) DNA condensed with cationic polymers (polyplexes), (ii) DNA complexed to liposomes (lipoplexes), and (iii) DNA complexed to both cationic polymers and lipids (lipopolyplexes). Lipopolyplexes can incorporate either cationic liposomes (LPDI) or anionic liposomes (LPDII). FR targeting has been evaluated in LPDII vectors[28] as well as in lipoplexes.[43]

In LPDII formulations,[28] the overall charge of the vector depends on the ratio of lipid to DNA (L/D ratio), where low ratios yield vectors with an overall positive charge and high ratios result in vectors with a negative charge. Lee and Huang[28] investigated the FR-targeted transfection of KB cells *in vitro* with LPDII particles carrying plasmids containing the reporter gene luciferase. When LPDII vectors were positively charged, i.e., at a low L/D ratio, transfection efficiency was independent of folate targeting, indicating the predominance of nonspecific electrostatic interaction in mediating cell uptake.[28] In contrast, when LPDII vectors were negatively charged, i.e., at a high L/D ratio, gene expression required a targeting component, presumably to overcome electrostatic repulsion from the negatively charged cell surface.[28] While these vectors with higher L/D ratios were able to provide receptor-mediated transfection, they also exhibited greater cytotoxicity.[28] LPDII vectors with high L/D ratios also required the presence of a fusogenic lipid component such as DOPE for efficient transfection.[28]

Targeted liposomal gene delivery in the presence of serum has also been shown to be highly dependent on vector formulation. Formulations that have yielded promising results to date are composed primarily of a DNA core condensed with either cationic dendrimer or polylysine

[42] R. J. Lee and L. Huang, *Crit. Rev. Ther. Drug Carr. Syst.* **14,** 173 (1997).
[43] K. F. Pirollo, L. Xu, and E. H. Chang, *Curr. Opin. Mol. Ther.* **2,** 168 (2000).

surrounded by a lipidic bilayer containing various amounts of cationic lipids, fusogenic lipids, and folate-PEG-PE lipids. When administered intravenously to tumor-bearing mice, tumor-specific gene expression can be observed readily, with minimal or undetectable transgene expression in most normal tissues[43] (personal observations).

Concluding Remarks

In conclusion, due to the advantages of folate as a tumor-targeting ligand and a relatively simple method for the incorporation of folate into liposomes, FR-targeted liposomes are emerging as a versatile vehicle of tumor-selective drug delivery with excellent potential for clinical translation. Efforts are currently underway to evaluate the potential utility of FR-targeted liposomes in the therapy of tumors with amplified FR expression.

Acknowledgment

The authors recognize Michael A. Gosselin for his help in the manuscript preparation.

[4] Preparation of Poly(ethylene Glycol)-Grafted Liposomes with Ligands at the Extremities of Polymer Chains

By Samuel Zalipsky, Nasreen Mullah, and Masoud Qazen

Introduction

Until approximately a decade ago, most liposomal drug delivery applications were hampered by short blood lifetimes of conventional lipid vesicles and their preferential accumulation in organs of the reticuloendothelial (RES) system.[1,2] In an effort to address these problems, various liposomal surface modifiers were tested, including dicarboxylic acids, synthetic and natural polymers, and oligocaccharides (for reviews, see Refs. 2–4). Among these surface-decorating molecules, poly(ethylene glycol)

[1] D. D. Lasic, *Sci. Med.* **3**, 2 (1996).

[2] M. C. Woodle, *in* "Poly(ethylene glycol) Chemistry and Biological Applications" (J. M. Harris and S. Zalipsky, eds.), Vol. 680, p. 60. Washington, DC, 1997.

[3] T. Sato and J. Sunamoto, *Prog. Lipid Res.* **31**, 345 (1992).

[4] T. M. Allen, *Adv. Drug Delivery Rev.* **13**, 285 (1994).

(PEG) is by far the most popular one. PEG-derived lipopolymers that are utilized to make PEG-grafted liposomes are compatible with a great variety of lipid formulations.[5] The preparations of PEG-grafted lipid vesicles exhibiting optimal circulation longevity and much reduced uptake by the RES organs usually contain 3–7 mol% of methoxypolyethylene glycol (average molecular weight 2000 or 5000 Da)-distearoylphosphatiylethanolamine (mPEG-DSPE) in addition to various amounts of lecithin and cholesterol.[2,4,6]

The water-solvated, conformationally flexible, highly mobile PEG chains camouflage the surface of liposomes and thus minimize the nonspecific interactions with various plasma components. Consequently, liposomes with a covalent PEG "corona" have extended half-lives in the bloodstream (terminal $t_{1/2} \geq 48$ h in humans) and drastically reduced liver uptake.[6–8] As a result, the PEG-grafted vesicles with an average diameter of 100–200 nm are subject to an enhanced permeability and retention effect,[9] by which they tend to accumulate in sites of inflammation and tumor growth.[6,7,10] Such liposomes were named STEALTH (registered trademark of ALZA Corp.) for their ability to avoid rapid uptake by the RES. They have often been referred to in the literature as sterically stabilized liposomes on the basis of a proposed mechanism in which protein adsorption is inhibited by a steric barrier created by the surface-grafted polymer chains.[6]

Ligand-mediated targeting has been an elusive goal of liposomal drug delivery for a long time. Because of the fast clearance and low systemic exposure of conventional liposomes, their ligand-mediated targeting did not achieve any measurable success. Extended blood circulation of PEG-grafted liposomes opened up new opportunities in this area. Furthermore, PEG–liposomes can be utilized as platforms on which biologically active ligands are presented in a multivalent array. Low molecular weight ligands, which in their free form are cleared from systemic circulation rapidly, might have an opportunity to exhibit their biological activity due to their presentation on such long-circulating liposomal platforms. All

[5] M. C. Woodle, K. K. Matthay, M. S. Newman, J. E. Hidayat, L. R. Collins, C. Redemann, F. J. Martin, and D. Papahadjopoulos, *Biochim. Biophys. Acta* **1105,** 193 (1992).
[6] D. Lasic and F. Martin, eds., "Stealth Liposomes." CRC Press, Boca Raton, FL, 1995.
[7] A. Gabizon, R. Catane, B. Uziely, B. Kaufman, T. Safra, R. Cohen, F. Martin, A. Huang, and Y. Barenholz, *Cancer Res.* **54,** 987 (1994).
[8] D. Papahadjopoulos, T. M. Allen, A. Gabizon, E. Mayhew, K. Matthay, S. K. Huang, K.-D. Lee, M. C. Woodle, D. D. Lasic, C. Redemann, and F. J. Martin, *Proc. Natl. Acad. Sci. USA* **88,** 11460 (1991).
[9] H. Maeda, J. Wu, T. Sawa, Y. Matsumura, and K. Hori, *J. Control. Rel.* **65,** 271 (2000).
[10] P. Laverman, S. Zalipsky, W. J. G. Oyen, E. T. M. Dams, G. Storm, N. Mullah, F. H. M. Corstens, and O. C. Boerman, *J. Nuclear Med.* **41,** 912 (2000).

these applications require preparation of PEG–liposomes decorated with various ligands. A variety of biologically relevant ligands could be used for these purposes. These include immunoglobulins[11–16] and their fragments,[17–19] other proteins,[20,21] high[3] and low[22–25] molecular weight saccharides, peptides,[23,26,27] vitamins,[10,28–31] and even aptamers.[32]

This article summarizes the various approaches to the assembly of the ligand conjugates for preparation of ligand-bearing PEG-grafted liposomes. Since our entry into this field, we have strongly advocated positioning of the

[11] K. Maruyama, *Biosci. Rep.* **22,** 251 (2002).
[12] S. Zalipsky, C. B. Hansen, D. E. Lopes de Menezes, and T. M. Allen, *J. Control. Rel.* **39,** 153 (1996).
[13] K. Maruyama, T. Takizawa, T. Yuda, S. J. Kennel, L. Huang, and M. Iwatsuru, *Biochim. Biophys. Acta* **1234,** 74 (1995).
[14] A. L. Klibanov, K. Maruyama, A. M. Beckerleg, V. P. Torchilin, and L. Huang, *Biochim. Biophys. Acta* **1062,** 142 (1991).
[15] D. E. Lopez de Menezes, L. M. Pilarski, and T. M. Allen, *Cancer Res.* **58,** 3320 (1998).
[16] D. Goren, A. T. Horowitz, S. Zalipsky, M. C. Woodle, Y. Yarden, and A. Gabizon, *Br. J. Cancer* **74,** 1749 (1996).
[17] D. Kirpotin, J. W. Park, K. Hong, S. Zalipsky, W.-L. Li, P. Carter, C. C. Benz, and D. Papahadjopoulos, *Biochemistry* **36,** 66 (1997).
[18] K. Maruyama, N. Takahashi, T. Tagawa, K. Nagaike, and M. Iwatsuru, *FEBS Lett.* **413,** 177 (1997).
[19] E. Mastrobattista, P. Schoen, J. Wilschut, D. J. A. Crommelin, and G. Storm, *FEBS Lett.* **509,** 71 (2001).
[20] G. Blume, G. Cevc, M. D. J. A. Crommelin, I. A. J. M. Bakker-Woudenberg, C. Kluft, and G. Storm, *Biochim. Biophys. Acta* **1149,** 180 (1993).
[21] H. Iinuma, K. Maruyama, K. Okinaga, K. Sasaki, T. Sekine, O. Ishida, N. Ogiwara, K. Johkura, and Y. Yohemura, *Int. J. Cancer* **99,** 130 (2002).
[22] S. A. DeFrees, L. Phillips, L. Guo, and S. Zalipsky, *J. Am. Chem. Soc.* **118,** 6101 (1996).
[23] S. Zalipsky, N. Mullah, J. A. Harding, J. Gittelman, L. Guo, and S. A. DeFrees, *Bioconj. Chem.* **8,** 111 (1997).
[24] K. Shimada, J. A. A. M. Kamps, J. Regts, K. Ikeda, T. Shiozawa, S. Hirota, and G. L. Scherphof, *Biochim. Biophys. Acta* **1326,** 329 (1997).
[25] S. Zalipsky, N. Mullah, A. Dibble, and T. Flaherty, *Chem. Commun.* 653 (1999).
[26] S. Zalipsky, B. Puntambekar, P. Bolikas, C. M. Engbers, and M. C. Woodle, *Bioconj. Chem.* **6,** 705 (1995).
[27] V. P. Torchilin, R. Rammonohan, V. Weissig, and T. S. Levchenco, *Proc. Natl. Acad. Sci. USA* **98,** 8786 (2001).
[28] R. J. Lee and P. S. Low, *J. Biol. Chem.* **269,** 3198 (1994).
[29] J. Y. Wong, T. L. Kuhl, J. N. Israelashvili, N. Mullah, and S. Zalipsky, *Science* **275,** 820 (1996).
[30] A. Gabizon, A. T. Horowitz, D. Goren, D. Tzemach, F. Mendelbaum-Shavit, M. M. Qazen, and S. Zalipsky, *Bioconj. Chem.* **10,** 289 (1999).
[31] W. M. Li, L. M. Mayer, and M. B. Bally, *J. Pharmacol. Exp. Ther.* **300,** 976 (2002).
[32] M. C. Willis, B. Collins, T. Zhang, L. S. Green, D. P. Sebesta, C. Bell, E. Kellogg, S. C. Gill, A. Magallanez, S. Knauer, R. A. Bendele, P. S. Gill, and N. Janjie, *Bioconj. Chem.* **9,** 573 (1998).

ligands at the extremities of the PEG grafts, an approach that ultimately proved as the most promising one for the multivalent presentation of ligands[22] and for the maximal expansion of their receptor interaction range.[29] Therefore, this article focuses on the liposomes containing such constructs. Accordingly, herein we dedicated ample space to the descriptions of preparation of various ligand–PEG–lipid constructs, as well as to the synthesis of their precursors, lipopolymers equipped with conjugation-prone reactive end groups.

Experimental Procedures

Examples of Synthesis of Various Heterobifunctional X-PEG-OH Derivatives

Preparation of Pyridyldithiopropionyl (PDP)-PEG-OH.[33] Succinimidyl-3-(2-pyridyldithiopropionate (SPDP, 100 mg, 0.32 mmol) and α-amino-ω-hydroxy-PEG[34] (0.55 g, 0.275 mmol) are reacted in acetonitrile (2 ml) at 25° for 4 h. After completion of the reaction [thin-layer chromatography (TLC) using $CHCl_3/CH_3OH$, 8:2 (v/v) shows disappearance of H_2N-PEG-OH with appearance of a less polar, UV-absorbing new material ($R_f = 0.80$)], the solution is concentrated by rotary evaporation, and ethyl ether (5 ml) is added. After overnight storage at 4° the white solid is collected and dried *in vacuo* over P_2O_5. Yield: 0.5 g (90%). ^{1}H-NMR (DMSO-d6): δ 2.5 (CH_2S, overlap w/DMSO), 3.01 (t, J = 7 Hz, $CH_2C=O$, 2H), 3.20 (m, CH_2NH, 2H), 3.51 (s, PEG, $\approx$180 H), 4.52 (t, HO-PEG, 1H), 7.24 (m, pyr, 1H), 7.76 (m, pyr, 1H), 7.82 (m, pyr, 1H), 7.99 (br m, NH, 1H), 8.45 (m, pyr, 1H) ppm. The same procedure is used to prepare maleimidopropionyl (MP)-PEG-OH from *N*-succinimidyl-3-(*N*-maleimido)propionate and H_2N-PEG-OH.[17]

Preparation of tert-Butyloxycarbonyl (Boc)-HN-PEG-OH.[35] A solution of α-amino-ω-hydroxy-PEG[34] (2.0 g, 1.0 mmol) in dioxane (8 ml) is treated with Boc_2O (0.436 g, 2.0 mmol) overnight. The solution is concentrated under reduced pressure and then treated with ethyl ether (40 ml) to facilitate precipitation of the product. The product is collected and dried *in vacuo* over P_2O_5. Yield: 1.89 g (90%). ^{1}H-NMR (DMSO-d6): δ 1.37 (s, tBu, 9H), 3.06 (m, CH_2NH, 2H), 3.51 (s, PEG, 180H), 4.52 (t, HO-PEG, 1H) ppm.

[33] T. M. Allen, E. Brandeis, C. B. Hansen, G. Y. Kao, and S. Zalipsky, *Biochim. Biophys. Acta* **1237,** 99 (1995).
[34] S. Furukawa, N. Katayama, T. Iizuka, I. Urabe, and H. Okada, *FEBS Lett.* **121,** 239 (1980).
[35] S. Zalipsky, E. Brandeis, M. S. Newman, and M. C. Woodle, *FEBS Lett.* **353,** 71 (1994).

Preparation of Boc-hydrazide(Hz)-PEG-OH.[36] The ω-hydroxy acid derivative of PEG[37] (3, 5 g, 2.38 mmol) and *tert*-butyl carbazate (0.91 g, 6.9 mmol) are dissolved in CH_2Cl_2-ethyl acetate (1:1, 7 ml). The solution is cooled on ice and treated with (DCC) (0.6 g, 2.9 mmol) predissolved in the same solvent mixture. After 30 min the ice bath is removed and the reaction is allowed to proceed for an additional 3 h, filtered from dicyclohexylurea, and evaporated. The product is recovered and purified by two precipitations from ethyl acetate-ether (1:1) and dried *in vacuo* over P_2O_5. Yield: 5.2 g, 98%. ^{1}H NMR ($CDCl_3$): δ 1.46 (s, tBu, 9 H), 3.64 (s, PEG, ≈180 H), 3.93 (br d, J = 4.5 Hz, HNCH2-CO, 2 H), 4.24 (t, J = 4.5 Hz, CH_2OCONH, 2H) ppm. ^{13}C NMR ($CDCl_3$): δ 28.1 (tBu), 43.4 (CH_2 of Gly), 61.6 (CH_2-OH), 64.3 (CH_2OCONH), 69.3 (CH_2CH_2OCONH), 70.5 (PEG), 72.4 (CH_2CH_2OH), 81.0 (CMe_3), 155.1 (C=O of Boc), 156.4 (C=O of urethane), 168.7 (C=O of Gly hydrazide) ppm.

Preparation of End Group-Protected PEG-Succinimidyl Carbonates, Y-PEG-SC (Representative Procedure for Hydroxyl Group Activation)

Dry Y-PEG-OH (1 mmol) dissolved in acetonitrile (1 ml/mmol of PEG-2000) is treated with a 2-fold molar excess of DSC and pyridine (5 mmol/mequivalent OH) overnight at room temperature. The crude product is precipitated with a 10-fold volume excess of ethyl ether at 4°, recrystallized from isopropanol, filtered, and dried *in vacuo* over P_2O_5. Yields are typically 80–90%. The identity of the product is corroborated by ^{1}H-NMR (δ in DMSO-d6) that shows the characteristic signals of PEG (s, 3.5, ≈180H for PEG-2000) and SC-group (s, 2.8, SC, 4H and t, 4.45, CH_2-SC, 2H), as well as the signals of the Y moiety and absence of the characteristic OH signal of the starting material (t, 4.55, 1H). The following derivatives of PEG-SC were prepared in our laboratory: Y = PDP,[33] MP,[17] Boc-NH,[35] Boc-Hz.[36]

Attachment of Y-PEG-SC to DSPE (Representative Procedure for Preparation of Lipopolymers, Y-PEG-DSPE)

1,2-Distearoyl-3-*sn*-glycerophosphoethanolamine (DSPE, 200 mg, 0.27 mmol) is added to a solution of Y-PEG-SC (700 mg, 0.3 mmol) in chloroform (4 ml), followed by triethylamine (TEA, 0.1 ml, 0.72 mmol). As poorly soluble DSPE is being consumed by the reaction, the suspension gradually

[36] S. Zalipsky, *Bioconj. Chem.* **4,** 296 (1993).

[37] S. Zalipsky and G. Barany, *Polym. Prepr. Am. Chem. Soc. Div. Polym. Chem.* **27**(1), 1 (1986); S. Zalipsky and G. Barany, *J. Bioact. Compatible Polym.* **5,** 227 (1990); S. Zalipsky, J. L. Chang, F. Albericio, and G. Barany, *Reactive Polym.* **22,** 243 (1994).

turns into a clear solution during incubation at 40–45° for 10 min with mixing. The solvent is evaporated and replaced with acetonitrile (5 ml). The cloudy solution is kept at 4° overnight, filtered to remove trace of insoluble DSPE, if any is still present, and then evaporated to dryness. The lipopolymer product is often purified further by either silica gel chromatography (eluted with 5–20% methanol in chloroform)[17] or dialysis via MWCO 300,000 tubing.[36] Pooled and evaporated product containing chromatography fractions or the contents of the dialysis bag are lyophilized and dried *in vacuo* over P_2O_5. Yields are in the 65–100% range.

Acidolytic Removal of Boc-NH, tert-BuO, and Benzyl Ether (BnO) Groups Exposing H_2N- and HO-End Groups of PEG-DSPE

Boc removal off protected amino- or hydrazido-PEG-DSPE is carried out in 4 *M* HCl in dioxane for 1–2 h. The product (**1** or **3**) is recovered as a white solid after removal of the volatiles and thorough drying *in vacuo* [TLC ($CHCl_3$-CH_3OH-H_2O, 90:18:2), ninhydrin-positive product]. Yield: 98%. Disappearance of the *tert*-butyl peak at 1.45 ppm of ^1H-NMR spectra indicates that the deprotection goes to completion.

In order to remove the *tert*-Bu ether group, *tert*-BuO-PEG-DSPE is dissolved in 4 *N* HCl in dioxane and left overnight. Thin-layer chromatography reveals that some starting material remains unreacted. The product, HO-PEG-DSPE, is purified on a silica gel column (0–12% methanol in $CHCl_3$) in 78% yield. Alternatively, *tert*-Bu ether groups, as well as benzyl ethers (BnO), are removed quantitatively with $TiCl_4$ (1 *M* solution in CH_2Cl_2, 2 h at 25°). ^1H-NMR (DMSO-d6) confirms the disappearance of the *tert*-BuO-signal (s, 1.07) or the BnO-signals (s, 4.48; m, 7.2–7.4) of the starting material and appearance of the characteristic $HOCH_2$-triplet at 4.5 ppm. With the noted exceptions, the rest of NMR spectra of the products are very similar to those of the protected lipopolymer precursors.

Preparation of Various Lipopolymer Derivatives and Conjugates from H_2N-PEG-DSPE (**1**)

Synthesis of Maleimidopropionamide-PEG-DSPE (MP-PEG-DSPE) (**5**).[17]
Amino-PEG-DSPE (**1**, 500 mg, 0.18 mmol) and *N*-succinimidyl-3-(*N*-maleimido)propionate (62.6 mg, 24 mmol) are dissolved in CH_2Cl_2 (3 ml) and DMF (0.75, μl) followed by TEA (76, μl, 0.54 mmol). After confirming by TLC ($CHCl_3$-CH_3OH-H_2O, 90:18:2, visualized with ninhydrin spray) that the reaction is complete (15 min), the product mixture is purified on a silica gel column with a stepwise gradient of methanol (0–14%) in chloroform. The pure product-containing fractions (eluted in

CHCl$_3$-CH$_3$OH, 88:12) are combined and evaporated and then further dried *in vacuo* over P$_2$O$_5$, yielding MP-PEG-DSPE (**5**) as a white solid (203 mg, 45%). ^{1}H-NMR (CD$_3$OD): δ 0.88 (m, 6H), 1.26 (s, CH$_2$, 56H), 1.58 (br m, CH$_2$CH$_2$C=O, 4H), 2.31 (2 t, CH$_2$C=O, 4H), 2.48 (t, MP-CH$_2$CH$_2$C=O), 3.53 (t, CH$_2$N, 2H), 3.63 (s, PEG, $\approx$180 H), 3.88 and 3.98 (q and t, CH$_2$PO$_4$-CH2, 4H), 4.20 (t, CH$_2$O$_2$CN, 2H), 4.17 and 4.39 (2 dd, OC*H*$_2$CHCH$_2$OP, 2H), 5.2 (m, PO$_4$CH$_2$C*H*CH$_2$O, 1H), 6.69 (s, maleimide, 4H). In a similar way, MMC-PEG-DSPE is prepared from *N*-succinimidyl 4-(*N*-maleimidomethyl)-cyclohexane-1-carboxylate with 70% yield.

Synthesis of PDP-PEG-DSPE (4).[33] A solution of H$_2$N-PEG-DSPE (**1**, 198 mg, 0.07 mmol) in chloroform (2 ml) is treated with SPDP (26 mg, 0.085 mmol) and TEA (30 ml, 0.42 mmol) at 25° overnight. The product (200 mg) is purified on a silica gel (11 g) column with a gradient of CH$_3$OH in chloroform (5–20%). Yield: 183 mg (83%). ^{1}H-NMR (CDCl$_3$): δ 0.89 (t, J = 6.8 Hz, CH$_3$, 6H), 1.26 (s, CH$_2$, 56H), 1.58 (br m, CH$_2$CH$_2$C=O, 4H), 2.28 (2 overlapping t, CH$_2$C=O), 2.62 (t, J = 7 Hz, CH$_2$SS, 2H), 3.09 (t, J = 7 Hz, CH$_2$C=O, 2H), 3.36 (br m, OCH$_2$CH$_2$N, 2H), 3.44 (m, CH$_2$N, 2H), 3.64 (s, PEG, $\approx$180 H), 3.94 (br m, CH$_2$CH$_2$OP, 2H), 4.17 (dd, J = 7.0, 12 Hz, glycerol CH$_2$OP, 2H), 4.21 (m, CH$_2$-O$_2$CN, 2H), 4.39 (dd, J = 3.2, 12 Hz, glycerol CH$_2$OC=O, 2H), 5.20 (m, CH, 1H), 6.73 (br, NH, 1H), 7.10, (m, pyr, 1H), 7.66 (m, pyr, 2H), 8.48 (m, pyr, 1H) ppm. ^{13}C-NMR (CDCl$_3$): δ 14.0 (CH$_3$), 22.7 (CH$_2$CH$_3$), 24.9 (CH$_2$CH$_2$C=O), 29.7 (CH$_2$CH$_2$CH$_2$), 31.9 (CH$_2$CH$_2$CH$_3$), 34.1 and 34.3 (CH$_2$C=O), 34.7 (SCH$_2$CH$_2$C=O), 35.6 (SCH$_2$CH$_2$=O), 39.4 (CH$_2$NHC=O), 42.4 (CH$_2$NHCO$_2$), 62.8 (CH$_2$OC=O), 63.4 (CH$_2$OPO$_3$), 64.2 (CH$_2$OC=ON), 69.9 (CHOC=O), 70.6 (PEG), 120.0 (C2 pyr), 137.0 (C3 pyr), 120.8 (C4 pyr), 149.7 (C5 pyr), 156.6 (C=O of urethane), 159.8 (C1 pyr), 170.7 (C=O of amide), 173.0 and 173.3 (C=O of esters) ppm.

Synthesis of Chloroacetamido-PEG-DSPE (ClCH$_2$CONH-PEG-DSPE) (6). Chloroacetic acid (57 mg, 0.6 mmol) and dicychlohexylcarbodimide (62 mg, 0.3 mmol) in chloroform (2 ml) are stirred at 25° for 2 h. The solution is filtered and added to H$_2$N-PEG-DSPE (**1**, 250 mg, 0.18 mmol) solution in chloroform (2 ml) followed by diisopropyl ethylamine (2 equivalents). After confirming by TLC (CHCl$_3$-CH$_3$OH-H$_2$O, 90:18:2, ninhydrin) that the reaction is complete ($\approx$2 h) the product mixture is purified on a silica gel column (1–13% methanol in CHCl$_3$). The product-containing fractions are combined, evaporated, and then dried *in vacuo* over P$_2$O$_5$ to yield a white solid (137 mg, 55%). ^{1}H-NMR (CD$_3$OD): δ 0.88 (m, 6H.), 1.26 (s, CH$_2$, 56H), 1.58 (br m, C*H*$_2$CH$_2$C=O, 4H) 2.31 (2× t, CH$_2$C=O, 4H), 3.63 (s, PEG 180H), 4.17 (dd, CH$_2$PO$_4$C*H*$_2$, 2H), 5.2 (m, PO$_4$CH$_2$C*H*, 1H), 4.6 (s, ClC*H*$_2$CONH, 1H).

Preparation of Folate-PEG-DSPE.[30] To a solution of folic acid (FA, 100 mg, 0.244 mmol) in DMSO (4 ml) is added H_2N-PEG-DSPE (**1**, 400 mg, 0.14 mmol) and pyridine (2 ml) followed by dicyclohexylcarbodiimide (130 mg, 0.63 mmol). After completion of the reaction ($\approx$4 h) [monitored by TLC on silica gel GF (chloroform-methanol-water 75:36:6) and ninhydrin spray] pyridine is removed by rotary evaporation. The remaining solution is diluted with water (50 ml), centrifuged to remove trace insolubles, and then dialyzed in Spectra/Por CE (Spectrum, Houston, TX) tubing (MWCO 300,000) against saline (50 mM, 2 × 2000 ml) and water (3 × 2000 ml). The dialyzate, containing only the product (single spot by TLC), is lyophilized, and the residue is dried *in vacuo* over P_2O_5. Yield: 400 mg, 90%. Folate content values determined by quantitative UV spectrophotometry of the conjugates in methanol (0.05 mg/ml) using the FA extinction coefficient $\varepsilon_{285} = 27,500\ M^{-1}\ cm^{-1}$ were within 94–104% of the theoretical value. ^{1}H-NMR (DMSO-d6 / CF_3CO_2D, ~10/1, v/v): δ 0.84 (t, CH_3, 6H); 1.22 (s, CH_2, 56H); 1.49 (m, CH_2CH_2CO, 4H); 2.1–2.3 (overlapping 2× t, CH_2CH_2CO & m, CH_2 of Glu, 8H); 3.2 (m, CH_2CH_2N, 4H); 3.50 (s, PEG, $\approx$180 H and $\approx$300 H for derivatives of PEG2000, and −3350 respectively); 4.02 (t, CH_2OCONH, 2H); 4.1 (dd, trans-PO_4CH_2CH, 1H); 4.3 (dd, cis-PO_4CH_2CH, 1H); 4.37 (m, α-CH, 1H); 4.60 (d, 9-CH_2-N, 2H); 5.15 (m, PO_4CH_2CH, 1H); 6.65 (d, 3′, 5′-H, 2H); 7.65 (d, 2′,6′-H, 2H); 8.77 (s, C7-H, 1H) ppm. MALDI spectra yield bell-shaped distributions of 44-Da-spaced lines centered at 3284 for FA-PEG2000-DSPE (calculated molecular mass 3200 Da); 4501 for FA-PEG3350-DSPE (calculated molecular mass 4540 Da).

Preparation of Biotin-PEG-DSPE.[29] A solution of H_2N-PEG-DSPE (**1**, 300 mg, 0.11 mmol) and TEA (46 ml, 0.33 mmol) in chloroform (1.7 ml) is treated with the succinimidyl ester of biotin (41.31 mg, 0.121 mmol) predissolved in dimethylformamide (300 ml). After completion of the reaction ($\approx$20 min) [determined by TLC (silica gel G, $CHCl_3$-CH_3OH-H_2O, 90:18:2) R_f values 0.47 and 0.51 for H_2N-PEG-DSPE and biotin-PEG-DSPE, respectively], the reaction mixture is filtered and loaded onto the silica gel column. The product is eluted with a stepwise gradient of methanol (0 to 14%, v/v) in chloroform (2% increments every 50 ml). The product-containing fractions are pooled and evaporated to dryness. The residue is dissolved in *tert*-butanol and lyophilized, yielding a white powder, 196 mg (60%). ^{1}H-NMR (CD_3OD): δ 0.88 (t, CH_3, 6H), 1.26 (s, CH_2, 56H), 1.45 (m, CH_2 biot., 2H), 1.58 (br m, $CH_2CH_2C=O$, 4H of each lipid and biot.), 3.20 (m, SCH biotin and CH_2NH of PEG, 3H), 3.53 (t, CH_2NH of PE, 2H), 3.64 (s, PEG, $\approx$180 H), 3.88 and 3.98 (q and t, $CH_2PO_4CH_2$, 4H), 4.20 (t, CH_2O_2CN, 2H), 4.17 and 4.39 (2× dd, OCH_2CHCH_2OP, 2H), 4.3 and 4.5 (2× dd, CHNH-CONHCH, 2H), 5.2

(m, $PO_4CH_2CHCH_2OP$, 1H) ppm. MALDI spectra yield bell–shaped distribution of ions spaced at equal 44-Da intervals and centered at 3080 Da (theory 2986 Da).

Preparation of para-Nitrophenyl Carbonate-PEG-DSPE (NPC-PEG-DSPE, 8) and Conjugates Thereof

Synthesis of NPC-PEG-DSPE (8). HO-PEG-DSPE (1.2 g, 0.43 mmol) dissolved in dry $CHCl_3$ (3.5 ml) is reacted with p-nitrophenyl chloroformate (130.6 mg, 0.648 mmol) and TEA (422 μl, 3.024 mmol) at 25° for 4 h. After confirming by TLC the completion of the reaction, the solvent is rotary evaporated and the residue is treated with ethyl acetate ($\approx$ 5 ml). After filtering off the insoluble TEA hydrochloride, an equal volume of ethyl ether is added to precipitate the product. The white precipitate is filtered, washed with cold ethyl acetate, and dried *in vacuo* over P_2O_5. Yield: 1.1 g (87%) ^{1}H NMR (DMSO-d6): δ 0.88 (t, CH_3, 6H); 1.26 (s, CH_2, 58H); 1.58 (m, CH_2CH_2CO, 4H); 2.28 (2× t, CH_2CO, 4H); 3.6 (s, PEG $\approx$180H); 4.02 (t, CH_2OCONH, 2H); 4.1 (dd, trans PO_4CH_2CH, 1H); 4.27 (dd, cis PO_4CH_2CH, 1H); 5.2 (m, PO_4CH_2CH, 1H); 7.51 (d, arom, 2H); 8.3 (d, arom, 2H).

Preparation of Urethane-Linked Peptide Conjugate, Example of DSPE-PEG-GYIGSR-NH$_2$. NPC-PEG-DSPE (**8**, 40 mg, 0.0137 mmol) and GYIGSR-NH$_2$ peptide (16 mg, 0.0206 mmol) are dissolved in (DMF) (200 ml) and reacted at 25° in the presence of TEA (2 ml, 0.0137 mmol) for approximately 2 h. Formation of the peptide–lipopolymer conjugate is confirmed by TLC ($CHCl_3$-CH_3OH-H_2O, 90:18:2, R_f = 0.15). The reaction mixture is acidified with acetic acid, diluted with water (1:10), dialyzed at 4° through a Spectropor CE membrane (MWCO 12,000) against saline (50 mM, 3 × 2000 ml, 4–16 h per period) and against deionized water, and then lyophilized. The residue is dried *in vacuo* over P_2O_5, yielding a white solid product (20 mg, 42%). ^{1}H NMR (CD$_3$OD) shows the characteristic peaks of the peptide, lipid, and PEG in a 1:1:1 ratio: δ 0.89 (overlapping signals, lipid-CH_3 and Ile, 12H), 1.26 (s, lipid-CH_2, 56H), 1.56 (m, CH_2CH_2C=0, 4H), 2.33 (2× t, CH_2CH_2C=O, 4H), 3.65 (s, PEG, $\approx$180H), 4.25 (m, CH_2OCONH, 4H), 6.71 (d, arom-Tyr, 2H each), 7.05 (d, arom-Tyr, 2H).

Preparation of Aldehyde-PEG-DSPE (7) and Its Thiazolidine-Linked[38] *Conjugate with DPYIGSR-NH$_2$.* An NPC-PEG-DSPE (**8**, 200 mg, 0.07 mmol) solution in DMF (0.7 ml) is reacted with 3-amino-1,2-propanediol (62 mg, 0.68 mmol) at 25° for 4 h. After confirming by TLC

---

[38] L. T. Zhang and J. P. Tam, *Anal. Biochem.* **233,** 87 (1996).

the completion of the reaction product, 1,2-diol-PEG-DSPE is precipitated with isopropanol (3 ml), dissolved in water, and dialyzed to remove the last traces of aminopropanediol. A white solid (155 mg, 75% yield) is obtained after lyophilization of the dialyzate. ^{1}H-NMR (DMSO-d6): d 0.84 (t, CH3, 6H), 1.22 (s, CH2, 56H), 1.50 (m, CH_2CH_2C=O, 4H), 2.24 (t, CH_2CH_2C=O, 4H), 2.9 (m, 1H), 3.1 (m, 3H), 3.5 (s, PEG, 180H), 4.0 (m, 5H), 4.24 (m, 1H), 4.4 (m, sec-OH, 1H), 4.55 (t, HO-CH2, 1H), 5.04 (m, 1H), 6.9 (m, NH, 1H), 7.5 (m, NH, 1H).

The aldehyde derivative of PEG-DSPE (**7**) is generated by sodium periodate (39 mg, 0.18 mmol) oxidation of the 1,2-diol-PEG-DSPE (100 mg, 0.035 mmol) in sodium acetate buffer (50 mM, pH 5.0) for 10 min. Water (5 ml) is added to the reaction mixture and it is dialyzed (6000 MWCO membrane) against the same buffer (2 × 21, 4 h each). The peptide, CDPGYIGSR-NH$_2$ (36 mg, 0.03 mmol), is then mixed with the dialysate in a test tube and incubated at 25° for 5 h. After confirming the reaction completion by TLC [(chloroform:methanol:water 75:36:6) R_f = 0.76 (aldehyde-PEG-DSPE, **7**) and R_f = 0.59 (peptide conjugate)], the reaction mixture is lyophilized, the remaining solid is dissolved in chloroform (1 ml), and the conjugate is purified on a silica gel column (methanol 5–45% in CHCl$_3$, followed by CHCl$_3$-MeOH-H$_2$O, 80:18:2 and 65:30:5). The conjugate-containing fractions are pooled and evaporated, and the residue is lyophilized from *tert*-butanol to yield a white solid (50 mg, 42%). ^{1}H NMR (CD$_3$OD) shows the characteristic peaks of the peptide, lipid, and PEG in a 1:1:1 ratio: δ 0.89 (overlapping signals, lipid-CH$_3$ and Ile, 12H), 1.26 (s, lipid-CH$_2$, 56H), 1.6 (m, CH_2CH_2C=O, 4H), 2.33 (2× t, CH_2CH_2C=O, 4H), 3.63 (s, PEG, ≈180H), 4.35 (m, CH$_2$OCONH, 4H), 4.9 (m, 1H), 5.2 (m, lipid-CH, 1H), 6.71 (d, arom-Tyr, 2H each), 7.08 (d, arom-Tyr, 2H). MALDI produced a distribution of signals centered at 3750 Da (calculated 3742 Da).

Coupling of Ligands to Hz–PEG–Liposomes

Preparation of YIGSR-Bearing Liposomes.[26] The TYIGSR-NH$_2$ solution (0.2 ml, in HEPES buffer 25 mM, 0.9% saline, pH 7.2) is treated with a freshly prepared stock solution of sodium periodate (20 μl) for 5 min in the dark (100 mM NaIO$_4$ stock solution is used for oxidation of the 5 mM peptide solution and 200 mM periodate for 10 mM peptide). The excess periodate is consumed by the addition of N^α-acetyl-methionine solution (NAM, 20 μl, 1 M). The stock solution of unilamellar liposomes [100 nm, HSPC-cholesterol-Hz-PEG-DSPE (**3**), 56:39:5 mol%, 2.2 ml, ≈50 μmol of phospholipid (PL)/ml] in acetate buffer (0.1 M, pH 4.8) is added to the oxidized peptide. The resulting solution is incubated overnight at $\approx 6°$ and is then dialyzed extensively against HEPES buffer (25 mM, 0.9% NaCl, pH 7.2) using a Biodesign dialysis membrane MWCO 8000. Aliquots are sent for amino acid analysis and for phosphate analysis for determination of the peptide/PL ratio. The preparation obtained from 5 mM TYIGSR-NH$_2$ results in a peptide/PL mole ratio of 2.6×10^{-3}. Assuming 75,000 molecules of PL per 1000 Å liposome, this corresponds to ~200 peptides per vesicle. Similarly, the preparation derived from 10 mM TYIGSR-NH$_2$ results in 7.3×10^{-3} mol of YIGSR/mol of PL, corresponding to ~500 peptide residues per vesicle.

Preparation of Immunoliposomes.[39] A solution of C225 antibody (5.5 ml, at 3 mg/ml) is incubated with sodium periodate (2 mM, this concentration requires optimization depending on the IgG) in acetate buffer (0.1 M, pH 5.5) at 6°, for 20 min. NAM (0.5 ml of 500 mM) is added to quench the excess periodate. Hz–PEG–liposomes [HSPC-cholesterol, 55:40, containing 5 mol% of both lipopolymers: Hz–PEG–DSPE (3) and mPEG-DSPE, 5.6 ml at 47 μmol phopholipid/ml] in the same buffer are added to the solution, and the mixture is incubated overnight at 6°. Separation of the immunoliposomes from the free antibody is achieved on a 2.5×80-cm Sepharose CL-4B size exclusion column (Pharmacia, Uppsala, Sweden). The immunoliposome fractions are pooled, and the average number of 18 IgG residues/liposome is determined from amino acid analysis and phosphate analysis results.

Preparation of Ligand–PEG–DSPE by Coupling of Thiol-Containing and Haloacetylated Substrates[23]

Preparation of HS-PEG-DSPE.[22,23] Tributylphosphine (80 μl, 0.32 mmol) is added under N$_2$ to a clear solution of PDP-PEG-DSPE (4, 200 mg, 0.064 mmol) in 2-propanol:water (1:4, 6 ml) containing EDTA (10 mM). After completion of the reduction (≈ 2 h), as confirmed by TLC [CHCl$_3$-CH$_3$OH-H$_2$O, 90:18:2], the reaction mixture is lyophilized. The yellowish solid residue is triturated with ether (3×3 ml) to remove 2-thiopyridone. The residual white solid (196 mg, 99% yield) is dried under water aspirator pressure and then *in vacuo* over P$_2$O$_5$. It is used directly in the next step.

Preparation of Sialyl-Lewis^x (SLX)-PEG-DSPE. The bromacetylated derivative of SLX[22] (98 mg, 0.08 mmol) is added to a DMF (2.8 ml) solution of HS-PEG-DSPE (197 mg, 0.07 mmol) followed by potassium iodide (14.3 mg, 0.086 mmol). After approximately 5 min, a NaHCO$_3$ (0.4 M)/

[39] J. A. Harding, C. M. Engbers, M. S. Newman, N. I. Goldstein, and S. Zalipsky, *Biochim. Biophys. Acta* **1327,** 181 (1997).

Na$_2$EDTA (10 mM) solution (pH 8.0, 328 μl, 0.13 mmol bicarbonate) is added. The reaction mixture is stirred at room temperature overnight. After 21 h, TLC (CHCl$_3$-CH$_3$OH-H$_2$O, 75:36:6) shows the SLX-PEG-DSPE product ($R_f = 0.37$) and weak spots of the starting materials: SLX derivative ($R_f = 0.0$) and HS-PEG-DSPE ($R_f = 0.85$). Water (12 ml) is added to the reaction mixture and it is loaded onto a C8 silica column (8 g, 1 $\times$ 40 cm, LC-8 Supelclean, Supelco). The column is eluted with a stepwise gradient of methanol (0–70%, v/v) in water (10% increments every 20 ml) and then H$_2$O-CH$_3$OH-CHCl$_3$, 15:80:5 (60 ml). The product-containing fractions are pooled and evaporated. The thick liquid residue is then dissolved in *tert*-butanol (5 ml) and lyophilized to yield a white fluffy solid (203 mg, 74%). ^{1}H-NMR (CD$_3$OD): δ 0.88 (t, CH$_3$-lipid, 6H), 1.15 (d, H-6 Fuc, 3H), 1.19 (t, OCH$_2$CH$_3$, 3H), 1.29 (bs, (CH$_2$,)$_n$, 56H), 1.61 (bm, CH$_2$CH$_2$C=O, 4H), 1.73 (t, H-3ax-Sial, 1H), 2.01 (s, NHAc, 3H), 2.32 (2$\times$ t, CH$_2$CO, 4H), 2.58 (t, -SCH$_2$CH$_2$CONH, 2H), 2.87 (dd, H-3eq-Sial, 1H), 2.93 (t, -SCH$_2$CH$_2$CONH, 2H), 3.35 (m, NHCH$_2$, 4H), 3.64 (s, PEG, 180H), 3.7–4 (m, overlapping sugar peaks & CH$_2$PO$_4$CH$_2$, 8H), 4–4.31 (m, overlapping, 5H), 4.43 (dd, CHCH$_2$CO, 1H), 4.54 (d, H-1 Gal, 1H), 4.83 (m, H-5 Fuc, 1H), 4.90 (d, H-1 GlcN, 1H), 5.06 (d, H-1 Fuc, 1H), 5.2 (m, POCH$_2$CHCH$_2$, 1H), 7.70 & 7.84 (2$\times$ d, phenyl, 4H). MALDI-TOFMS yields a bell-shaped distribution of ions spaced at equal 44-Da intervals and centered at 4056 Da (theory 4012 Da).

Preparation of DSPE-PEG-YIGSR-NH$_2$. Coupling of N$^\alpha$-bromoacetyl-YIGSR-NH$_2$ and HS-PEG-DSPE under the same conditions results in the formation of the title conjugate with a 69% recovered yield. ^{1}H-NMR (CD$_3$OD): δ 0.9 (3$\times$ t, CH$_3$ of Ile & lipid, 12H), 1.2 (m, 1H), 1.26 (s, (CH$_2$), 56H), 1.58 (bm, CH$_2$CH$_2$C=O, 4H), 1.7 (m, 1H), 2.31 (2$\times$ t, CH$_2$C=O, 4H), 2.45 (t, SCH$_2$CH$_2$CONH), 2.7 (t, SCH$_2$CH$_2$CONH), 2.9 & 3.1 (dd, peptide, 2H), 3.2 (m, 2H), 3.35 (m, NHCH$_2$, 4H), 3.45 (t, 2H), 3.64 (s, PEG, $\approx$180H), 3.7 (m, 4H), 3.9 (m, CH$_2$PO$_4$, 4H), 4.0 (m, 3H), 4.15–4.2 (overlapping peaks, 4H), 4.3–4.45 (overlapping peaks, 3H), 4.6 (m, 2H), 5.2 (m, PO$_4$CH$_2$CH, 1H), 6.7 & 7.1 (2$\times$ d, phenyl of Tyr, 4H). MALDI-TOFMS yields a bell-shaped distribution of ions spaced at equal 44-Da intervals and centered at 3540 Da (theory: 3496 Da).

Coupling of Cysteine-Containing Peptide to Haloacetylated-PEG-DSPE (*6*). The same procedure just described is used to couple N$^\alpha$-Ac-YIGSRC-NH$_2$ to ClCH$_2$CONH-PEG-DSPE at a 65% isolated yield). ^{1}H NMR (CD$_3$OD): δ 0.88 (m, Ile & lipid CH$_3$, 8H), 1.26 (s, CH$_2$, 56H), 1.56 (br m, CH$_2$CH$_2$C=O), 1.92 (s, Ac, 1H), 2.31 (2$\times$ t, CH$_2$C=O, 4H), 3.64 (s, PEG, $\sim$180H), 4.17 (dd, CH$_2$PO$_4$CH$_2$, 2H), 5.2 (m, PO$_4$CH$_2$CH, 1H), 6.7 (d, Tyr, 2H), 7.08 (d, Tyr, 2H). MALDI confirmed the average molecular moss of the conjugate 3670 Da (calculated: 3776 Da).

Insertion of Ligand–PEG–DSPE into Preformed Liposomes[23]

Unilamellar liposomes (100 nm, partially hydrogenated phosphatidyl-choline (PHPC):cholesterol:mPEG-DSPE, 55:40:3) are obtained by thin film hydration followed by extrusion. They are incubated at ambient temperature or at 37° for up to 48 h with amounts corresponding to 1.2 mol% of SLX-PEG-DSPE or DSPE-PEG-YIGSR-NH$_2$. To monitor the insertion of the ligand–lipopolymer conjugates at various time points, the conjugates (micelles) are separated from inserted ligands (liposomes) by size exclusion chromatography. For SLX-liposomes, a Biogel A50M (BioRad, Hercules, CA) column equilibrated with 10 mM sodium phosphate, 140 mM sodium chloride, 0.02% NaN$_3$, pH 6.5, is used. For YIGSR-containing liposomes, a Sepharose 4B column is used with 10% sucrose, 10 mM HEPES, pH 7.0, as the eluent. Fractions (1 ml) are collected, diluted 1:10 in methanol, and analyzed for ligand content by HPLC. In addition to HPLC, the content of the peptide YIGSR in the liposomal product is determined by amino acid analysis. Quantification of the ligand attached to the outer leaflet of the liposomal membrane is achieved by enzymatic cleavage (sialidase for SLX and trypsin for YIGSR) followed by HPLC. These assays show that both ligand conjugates are incorporated into the liposomes by insertion within 4 h and are positioned on the outer monolayer of the vesicles.

Discussion

Various approaches to ligand-bearing PEG-liposomes are depicted schematically in Fig. 1. Several attempts to link ligands directly to polar head groups of lipids incorporated into mPEG–liposomes have been described in the literature.[12,14,40] It was found that the resulting arrangement of ligands and mPEG moieties is unfavorable for binding to the target receptors (Fig. 1A). It was even suggested that the ligand positioning depicted in Fig. 1A is advisable for hiding the ligand moieties on the PEG-grafted liposomal surface.[19,31,41] The highly mobile PEG chains also inhibit the conjugation reactions between reactive groups on the surface of a liposome and a ligand moiecule. In light of these observations, it became conceptually more appealing to link potential ligands to the far end of PEG chains. For this purpose, a number of end group-functionalized derivatives of general formula X–PEG–DSPE were developed,[42,43] where X represents

[40] C. B. Hansen, G. Y. Kao, E. H. Moase, S. Zalipsky, and T. M. Allen, *Biochim. Biophys. Acta* **1239,** 133 (1995).
[41] F. J. Martin, S. Zalipsky, and S. K. Huang, *U.S. Patent 6,043,094* (2000).
[42] S. Zalipsky, *Adv. Drug Delivery Rev.* **16,** 157 (1995).
[43] S. Zalipsky, *Bioconj. Chem.* **6,** 150 (1995).

a reactive functional group-containing moiety (Table I). The nature of the hydrophobic anchor is very important for retention of the macromolecular lipid–conjugates in the liposomal bilayer.[44] DSPE seems to offer optimal bilayer retention properties as well as a primary amino group, which is convenient for attachment of PEG. Several approaches to end group-functionalized PEG–lipids are described in the literature, where homobifunctional PEG derivatives are used as starting materials.[13,18,45–47] In our experience, these methods produce product mixtures containing lipid–PEG–lipid adducts in addition to the sought X–PEG–lipid. Although the latter can be purified usually from these mixtures by chromatography, this was rarely done and there was often ambiguity regarding the purity of the X–PEG–lipids. As illustrated in Scheme 1, most of the end group-functionalized PEG–lipids were synthesized from a few heterobifunctional PEG derivatives, where one of the end groups was hydroxyl and the other end group contained a suitably protected form of a second functionality.[17,36,37,43,48] According to this scheme, the hydroxyl end group of PEG is converted into the reactive form of succinimidyl carbonate, which is then utilized to form a urethane attachment with the lipid anchor, DSPE. Both steps are clean and facile reactions. Activation of the hydroxyl group is achieved easily by treatment with an excess of disuccinimidyl carbonate in acetonitrile/pyridine, while the DSPE coupling to SC-PEG is accomplished readily in warm chloroform in the presence of triethylamine. Once the lipopolymer skeleton is assembled, the second end group of PEG can be deprotected and utilized for conjugation or for further functionalization reactions.[33,35,36,43,48] Among the various functionalities that were carried through these transformations were *tert*-butoxycarbonyl (Boc)-protected amino and hydrazido groups, as well as *tert*-Bu-ether and benzyl ether groups. Note that all the strategies illustrated in Figs. 1B–1D are dependent on the availability of such end group-functionalized PEG–lipids.

Three general strategies have been employed to assemble ligand-bearing STEALTH liposomes, in which the ligand moiety is linked to the far end of the PEG chain. In some instances, end group-functionalized PEG–lipids (1–8) were incorporated into liposomes and then conjugated

[44] J. R. Silvius and M. J. Zuckermann, *Biochemistry* **32**, 3153 (1993).

[45] G. Blume and G. Cevc, *Biochim. Biophys. Acta* **1146**, 157 (1993).

[46] G. Bendas, A. Krause, U. Bakowsky, J. Vogel, and U. Rothe, *Int. J. Pharm.* **181**, 79 (1999).

[47] V. P. Torchilin, T. S. Levchenco, A. N. Lukyanov, B. A. Khaw, A. L. Klibanov, R. Rammonohan, G. P. Samokhin, and K. R. Whiteman, *Biochim. Biophys. Acta* **1511**, 397 (2001).

[48] S. Zalipsky, *in* "Stealth Liposomes" (D. Lasic and F. Martin, eds.), p. 93. CRC Press, Boca Raton, FL, 1995.

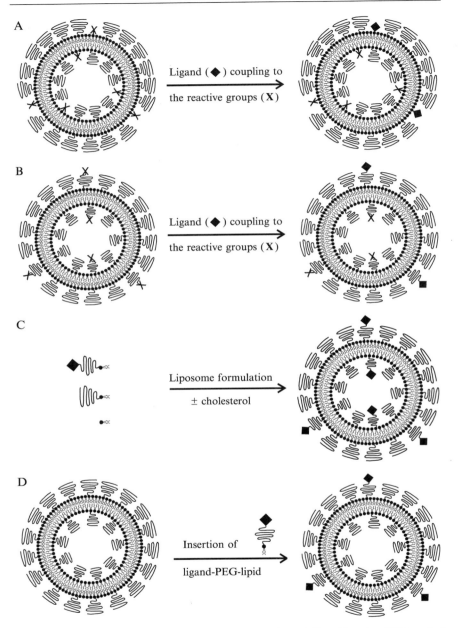

Fig. 1. Four different approaches to the preparation of ligand-bearing PEG-grafted unilamellar liposomes. See the text for a discussion of the scope and limitations of each method. Drawings courtesy of Yuanpeng Zhang, ALZA Corp. (A) Ligands are coupled to reactive polar head groups of lipids, **X**, on the surface of PEG–liposome. (B) Ligands are

TABLE I

SUMMARY OF VARIOUS USEFUL END GROUP-FUNCTIONALIZED
POLYMER–LIPIDS OF GENERAL STRUCTURE X–PEG–DSPE[a]

No.	X	Comments
1	$H_2N—$	Used for preparation of end group functionalized- and ligand–PEG–DSPE via amino group modification[29,30] Amino–PEG–DSPE forms long circulating liposomes, behaving as positively charged particles[35]
2	$HO_2C—$	Useful for further modification and conjugation reactions.[13,20,45] The carboxyl group of the lipopolymer has only a minimal effect on the longevity of circulation of its liposomes
3		Used for conjugation of antibodies, oxidized on their carbohydrate residues, to PEG chains of liposomes. Conjugation of N-terminal Ser or Thr peptides[15,26,36]
4		Used for binding thiol-containing ligands through disulfide linkage. Precursor for HS–PEG–DSPE, for attachment of maleimide containing ligands to liposomes[23,33,49]
5		Used for attachment of Fab′ fragments and other free thiol, Cys-containing peptides[17,18,50]
6		Haloacetylated PEG–lipids were used for attachment of Fab′ and other thiol-containing ligands (see Experimental Procedures)[48]
7		Useful reactive PEG–lipid for reductive alkylation of primary amine-containing ligands and for conjugation via thiozolidine linkage (see Experimental Procedures)
8		Reactive carbonates of p-nitrophenol are useful for further functionalization at the end of PEG chains (see Experimental Procedures) and for linking amino-containing ligands (e.g., peptides, proteins) forming urethane attachments[47]

[a] The PEG residue of average molecular mass 2000 or 3400 Da was used in most cases.

linked to reactive groups (**X**) positioned at the distal ends of liposomal surface grafted PEG chains. (C) Preparation of unilamellar liposomes from a mixture of lipids ± cholesterol and lipopolymers, including the appropriate percentage of a liganded lipopolymer. (D) Preformed liposome containing some surface-grafted PEG is incubated with pure ligand–PEG–lipid to facilitate its insertion into the outer monolayer of the lipid vesicle, thus positioning the ligands on the outer surface at the tips of the PEG chains.

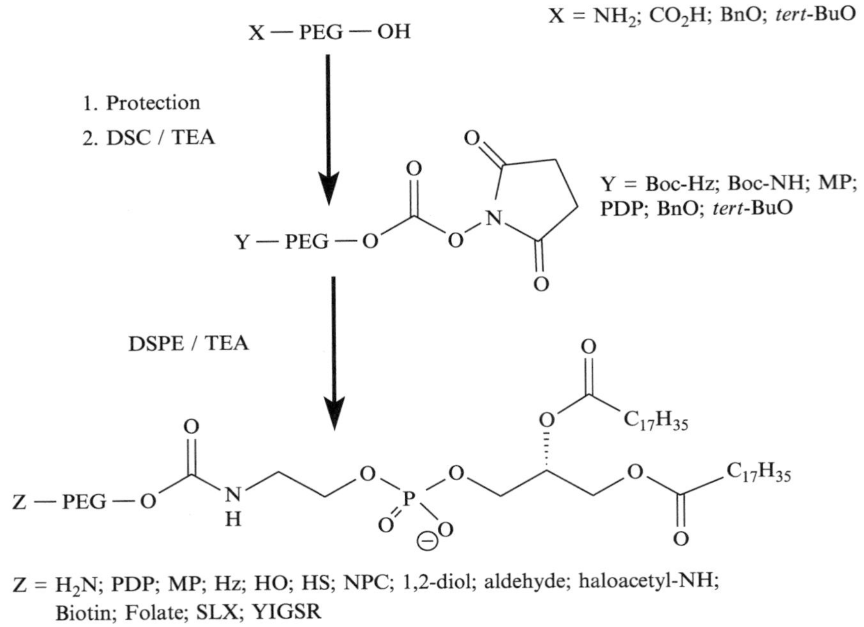

X — PEG — OH   X = NH_2; CO_2H; BnO; *tert*-BuO

1. Protection
2. DSC / TEA

Y — PEG — O   Y = Boc-Hz; Boc-NH; MP;
                                     PDP; BnO; *tert*-BuO

DSPE / TEA

Z — PEG — O ... N/H ... O—P—O ... $C_{17}H_{35}$ ... $C_{17}H_{35}$

Z = H_2N; PDP; MP; Hz; HO; HS; NPC; 1,2-diol; aldehyde; haloacetyl-NH;
    Biotin; Folate; SLX; YIGSR

SCHEME 1. General method used for the preparation of end group-functionalized PEG–DSPE derivatives and their conjugates from heterobifunctional derivatives X-PEG-OH. After assuring the appropriate protection (X converted to Y or X = Y) of one end group, the hydroxyl is converted into a reactive succinimidyl carbonate (SC), which is then coupled to the amino group of DSPE. Usually this is followed by further manipulation of the end groups (transformation of Y to Z) and/or conjugation of ligands. Boc, butyloxycarbonyl; DSC, disuccinimidyl carbonate; NPC, *p*-nitrophenyl carbonate; Hz, hydrazide; BnO, benzyl ether; *tert*-BuO, tertiary butyl ether; MP, maleimidopropinate; PDP, pyridyldithiopropionate; TEA, triethylamine; SLX, sialyl Lewis[X]; YIGSR, Tyr-Ile-Gly-Ser-Arg.

to a specific ligand (Fig. 1B). This approach seems more suitable for macromolecular ligands, e.g., immunoglobulins.[15,16,20,21,39,47] However, due to the imperfections of most conjugation protocols, this approach leads to a ligand-bearing liposome containing some unreacted end groups. In other cases, ligand–PEG–lipids were synthesized first and then formulated into ligand-bearing vesicles with other commonly used lipids, e.g., phosphotidylcholine (PC), and cholesterol. (Fig. 1C). This latter approach is

[49] M. Mercadal, J. C. Domingo, J. Petriz, J. Garcia, and M. A. de Madariaga, *Biochim. Biophys. Acta* **1418,** 232 (1999).
[50] S. Shahinian and J. R. Silvius, *Biochim. Biophys. Acta* **1239,** 157 (1995).

more attractive for low molecular weight ligands such as peptides,[23] oligosaccharides,[22,23] and vitamins,[29,30] whose conjugates are amenable to purification and characterization by common organic chemistry techniques. In comparing the two approaches represented in Fig. 1B and C, it is important to recognize that both methods have their advantages and disadvantages.

The former approach involves the conjugation of preformed liposomes and thus does not require preparation of pure ligand–PEG–lipids. Synthesis of such conjugates is still a formidable challenge, and purification of the products often necessitates employment of tedious and labor-intensive protocols. This approach (Fig. 1B) involves conjugation after the formation of liposomes. Therefore, for a protein ligand, this creates a possibility of cross-linking through the multiple attachments to a single ligand molecule. Additionally, the reactive groups at the termini of the PEG chains on the inner surface of liposomes remain unutilized (Fig. 1B). The potential of the unused reactive functionalities on either the inner or the outer monolayers to undergo side reactions with water, drug molecules, if present, or other lipid components cannot be neglected. In some instances, quenching of the unreacted end groups is advisable. For example, after conjugation of Fab-thiols to maleimido–PEG–liposomes, an excess of β-mercaptoethanol has been used to consume the remaining maleimide residues.[17]

All of the issues mentioned in the preceding paragraph are circumvented by the approach depicted in Fig. 1C in which pure ligand–PEG–lipid (liganded lipopolymer) is mixed with other liposomal matrix-forming components, e.g., lecithin and cholesterol, and then formulated into unilamellar vesicles. Various three-component conjugates containing ligands such as vitamins,[29,30] peptides,[23] and oligosaccharides[22,23] were synthesized by reacting the appropriate X–PEG–DSPE with suitably functionalized ligands. The conjugates were purified, chromatographically and characterized by NMR, HPLC, and matrix-assisted laser desorption / ionization time-of-flight mass spectrometry (MALDI-TOFMS). MALDI, in particular, is a powerful technique for characterization of this type of conjugate.[23,25,30] A straightforward formulation process is among the main attractive features of the "liganded lipopolymer and lipid mixing followed by extrusion" approach to ligand-bearing PEG–liposomes. However, incorporation of the preformed three-component conjugates into liposomes by this method results in slightly less than half of all the ligands facing the inner aqueous compartment. The remaining 55–60% are positioned on the external surface (for a 100-nm liposome),[23] where the ligands should exert their biological activity. Although the partition of the liganded-lipopolymer between the inner and the outer monolayers of liposomal membrane depends, among other variables, on the vesicle size, charge, and bulkiness of the ligand residue, a substantial portion of the total ligand–PEG–lipid that is inner leaflet bound, is wasted in this approach.

It was demonstrated that mPEG-DSPE can be inserted into preformed liposomes achieving external surface densities and *in vivo* performance similar to those of PEGylated liposomes prepared by mixing the lipopolymer and other lipids followed by extrusion.[51] We applied a similar insertion strategy to several ligand–PEG–DSPE conjugates. We found that incubation of ligand–PEG–DSPE (1.2–1.5 mol%) with preformed liposomes containing mPEG-DSPE (2–3 mol%) at 37° resulted in complete insertion of the three-component conjugate.[23] This aggregation-free process positioned the PEG-tethered ligands, including oligosaccharide and peptide conjugates, exclusively on the outer leaflet of the liposomal bilayer. This was demonstrated by a combination of a specific enzymatic cleavage of outer monolayer-bound ligand and HPLC-mediated quantification of both starting and enzymatically "clipped" ligand–PEG–lipids. The insertion methodology allows for achievement of the same external surface densities of both PEG-tethered ligands and mPEG chains as the lipid mixing/extrusion process.[23] Therefore, this approach, depicted schematically in Fig. 1D, overcomes the drawbacks of both of the methods shown in Fig. 1B and C.

Concluding Remarks

In the Experimental Procedures section, we have compiled a variety of protocols for the preparation of end group-functionalized PEG–DSPE derivatives and their conjugation with various ligands. The conjugation reactions can be performed either before or after incorporation of the X–PEG–DSPE into liposomes, according to Fig. 1B. The synthesis of ligand–PEG–lipids is particularly attractive when the ligand is a relatively small molecule such as a vitamin, peptide, or saccharide. Such three-component conjugates can be conveniently purified, characterized, and incorporated into liposomes by mixing with various lipids, with or without additional cholesterol and mPEG–DSPE, and then extruded to form unilamellar vesicles (Fig. 1C). Among the various methods of preparation of ligand-bearing liposomes, the insertion of liganded PEG–lipids into preformed lipid vesicles (Fig. 1D) appears to be the method of choice. This method overcomes all of the drawbacks of the previous approaches

---

[51] P. S. Uster, T. M. Allen, B. E. Daniels, C. J. Mendez, M. S. Newman, and G. Z. Zhu, *FEBS Lett.* **386,** 243 (1996).

(Fig. 1A–C). The insertion method is compatible with the incorporation of more than one ligand into the same liposome or with the generation of target cell-sensitized liposomes by separately incubating the appropriate liganded lipopolymer with an apportioned solution of drug-loaded liposomes.[52] The method leads to vesicles where all of the ligands are positioned on the outer monolayer. In addition, none of the liganded lipopolymer, which is often the most precious of the formulation components, is buried on the interior of the lipid vesicles. It is pertinent to emphasize that, provided that the ligand–PEG–lipid is purified to homogeneity, the insertion method leads to liganded liposomes without any extraneous functional groups, such as in the approach depicted in Fig. 1B. Considering that the PEG-grafted surface is ideal for minimizing nonspecific interactions *in vivo*, and exposure of only the minimal required number of ligand moieties is needed for specific target binding and/or internalization, any additional extraneous residues could only lead to undesirable events, such as faster clearance and increased immunogenicity. The issue of immunogenicity deserves a particularly careful consideration in the case of immunoliposomes, which comprise the most popular class of liganded PEG–liposomes.[11,12] Humoral immune responses leading to the formation of ligand-specific neutralizing antibodies[39,53] are likely to be obstacles limiting the development of PEG–immunoliposomes, particularly if the product requires repeated administrations.[39,54] Judging from the initial reports,[39,53] the immunogenicity of ligand-bearing liposomes is not alleviated by the presence of PEG on their surface and, according to some observations, is actually enhanced.[53,55]

[52] T. Allen, P. Uster, F. Martin, and S. Zalipsky, *U.S. Patent 6,316,024* (2001).

[53] N. C. Phillips, L. Gagne, C. Tsoukas, and J. Dahman, *J. Immunol.* **152,** 3168 (1994).

[54] G. Bendas, U. Rothe, G. L. Scherphof, and J. A. A. M. Kamps, *Biochim. Biophys. Acta* **1609,** 63 (2003).

[55] W. M. Li, M. B. Bally, and M. P. Schutze-Redelmeier, *Vaccine* **20,** 148 (2001).

Section II

Environment-Sensitive Liposomes

[5] Temperature-Sensitive Liposomes

By KENJI KONO *and* TORU TAKAGISHI

Introduction

Temperature-sensitive liposomes are those that release contents in response to environmental temperature. Liposomes that exhibit the response a few degrees above physiological temperature are considered to be especially useful in achieving site-specific delivery of drugs in the body because such liposomes can release drugs specifically at a target area where heat is applied. Since Yatvin *et al.*[1] and Weinstein *et al.*[2] first presented the idea of thermal control of drug delivery using temperature-sensitivity liposomes, these liposomes have attracted much attention for their potential to achieve tumor targeting in conjunction with hyperthermia. Indeed, many studies have revealed their usefulness and significance in cancer chemotherapy.[3,4]

Temperature-sensitive liposomes have been prepared using dipalmitoylphosphatidylcholine (DPPC) as the primary lipid. This design is based on the fact that phospholipid membranes become highly leaky to small water-soluble molecules at a gel-to-liquid crystalline phase transition (T_m).[5,6] Because DPPC membranes undergo the phase transition at 41°, liposomes made of DPPC exhibit enhancement of contents release at this clinically attainable temperature.

Another approach for the production of temperature-sensitive liposomes has been employed using thermosensitive polymers.[7-10] Various synthetic and natural polymers are known to exhibit a lower critical solution temperature (LCST).[11,12] These polymers are highly hydrophilic and hence soluble in water below LCST. However, when the temperature increases

[1] M. B. Yatvin, J. N. Weinstein, W. H. Dennis, and R. Blumenthal, *Science* **202,** 1290 (1978).
[2] J. N. Weinstein, R. L. Magin, M. B. Yatvin, and D. S. Zaharko, *Science* **2041,** 88 (1979).
[3] G. Kong and M. W. Dewhirst, *Int. J. Hyperther.* **15,** 345 (1999).
[4] D. Needham and M. W. Dewhirst, *Adv. Drug Deliv. Rev.* **53,** 285 (2001).
[5] D. Papahadjopoulos, K. Jacobsen, S. Nir, and T. Isac, *Biochim. Biophys. Acta* **311,** 330 (1973).
[6] M. C. Blok, L. L. Van Deenen, and J. De Gier, *Biochim. Biophys. Acta* **433,** 1 (1976).
[7] X. S. Wu, A. S. Hoffman, and P. Yager, *Polymer* **33,** 4659 (1992).
[8] K. Kono, H. Hayashi, and T. Takagishi, *J. Control. Release* **30,** 69 (1994).
[9] J.-C. Kim, S. K. Bae, and J.-D. Kim, *J. Biochem. (Tokyo)* **121,** 15 (1997).
[10] K. Kono, *Adv. Drug Deliv. Rev.* **53,** 307 (2001).
[11] H. G. Schild, *Prog. Polym. Sci.* **17,** 163 (1992).
[12] D. W. Urry, *Angew. Chem. Int. Ed. Engl.* **32,** 819 (1993).

through the LCST, these polymers become hydrophobic and water-insoluble. At the molecular level, chains of these polymers undergo a coil-to-globule transition at LCST. Therefore, if thermosensitive polymers are fixed onto a liposomal membrane, temperature-dependent alteration of characteristics of the polymer chains can provide the liposome with temperature-controlled functionalities. For example, highly hydrated polymer chains fixed to the liposome stabilize the liposome below the LCST, but above the LCST, the dehydrated polymer chains induce destabilization of the liposome, resulting in contents release.[13–15] In addition, because surface properties of the liposome change due to the temperature-dependent alteration of hydrophobicity and conformation of the polymer chains, the liposome exhibits temperature-controlled fusion behavior[16] or affinity to cells.[17] Therefore, in addition to the concept of site-specific delivery by temperature-controlled drug release, thermosensitive polymer-modified liposomes may offer another strategy for targeted drug delivery, namely targeting by the temperature-induced control of interaction with cells.

This article describes experimental procedures for the preparation and characterization of temperature-sensitive liposomes.

Lipid Phase Transition-Based Design of Temperature-Sensitive Liposomes

Since Yatvin et al.[1] first prepared temperature-sensitive liposomes using DPPC as the primary liposomal lipid, many studies have been done using this type of liposome from basic and practical aspects. While temperature-sensitive liposomes composed of DPPC enhance the contents release near T_m (41°), it has been shown that many factors affect their temperature sensitivity. Thus, improvement of their temperature sensitivity has been attempted. Ueno et al.[18] reported that small unilamellar vesicles of DPPC show a low sensitivity to temperature, but liposomes of the same lipid with a larger size exhibit a sharper enhancement of contents release around T_m, consistent with earlier observations.[18a] Optimization of lipid composition has been attempted to improve the temperature sensitivity of DPPC

[13] H. Hayashi, K. Kono, and T. Takagishi, *Biochim. Biophys. Acta* **1280,** 127 (1996).

[14] H. Hayashi, K. Kono, and T. Takagishi, *Bioconj. Chem.* **10,** 412 (1999).

[15] K. Kono, R. Nakai, K. Morimoto, and T. Takagishi, *Biochim. Biophys. Acta* **1416,** 239 (1999).

[16] H. Hayashi, K. Kono, and T. Takagishi, *Bioconj. Chem.* **9,** 382 (1998).

[17] K. Kono, R. Nakai, K. Morimoto, and T. Takagishi, *FEBS Lett.* **456,** 306 (1999).

[18] M. Ueno, S. Yoshida, and I. Horikoshi, *Bull. Chem. Soc. Jpn.* **64,** 1588 (1991).

[18a] N. Düzgüneş, J. Wilschut, K. Hong, R. Fraley, C. Perry, D. S. Friend, T. L. James, and D. Papahadjopoulos, *Biochim. Biophys. Acta* **732,** 289 (1983).

liposomes by the inclusion of various colipids, such as distearoylphosphatidylcholine (DSPC) and cholesterol.[2,19–22] In addition, from the standpoint of extension of the blood circulation time, the influence of inclusion of poly-(ethylene glycol) (PEG)-bearing lipids on the sensitivity of temperature-sensitive liposomes has been investigated.[23] Needham and co-workers[24] reported that inclusion of a lysophosphatidylcholine in the membrane of PEG-bearing, DPPC-based liposomes improves their temperature-sensitivity greatly.

A typical procedure of preparation of calcein-loaded DPPC liposomes is described. Fluorescent dyes, such as calcein and 5(6)-carboxyfluorescein, are used often to evaluate the temperature sensitivity of the liposomes because of their self-quenching property. A dry thin film of DPPC (10 mg) is formed on the wall of a glass vessel by evaporation of the lipid in a chloroform solution and subsequent drying under vacuum for more than 3 h. An aqueous calcein solution (63 mM, pH 7.4, 1 ml) is added to the vessel and the membrane is suspended by vortex mixing above T_m (50°) for 5–10 min. At this point, multilamellar vesicles (MLV) containing calcein are formed. When this MLV suspension is sonicated using a probe-type sonicator for 10–20 min at 50° under argon, small unilamellar vesicles (SUV) are obtained. The suspension is centrifuged at low speed to remove titanium particles. Also, if the MLV suspension is extruded through polycarbonate membranes with various pore sizes above the T_m, SUV, large unilamellar vesicles (LUV), or MLV with a desired size are obtained. Free calcein is removed by gel permeation chromatography on a Sepharose 4B column eluting with 140 mM NaCl, 10 mM Tris–HCl, pH 7.4. Details of the fluorescence measurements may be found in Düzgüneş.[24a]

Liposomes Modified with Thermosensitive Polymers

While a number of polymers are known to exhibit LCST, poly(N-isopropylacrylamide) [poly(NIPAM)] is the most extensively studied.[11] Poly(NIPAM) shows a sharp change in hydrophilicity/hydrophobicity in a

[19] J. L. Merlin, *Eur. J. Cancer* **27**, 1026 (1991).

[20] K. Iga, Y. Ogawa, and H. Toguchi, *Pharm. Res.* **9**, 658 (1992).

[21] T. Tomita, M. Watanabe, T. Takahashi, K. Kumai, T. Tadakuma, and T. Yasuda, *Biochim. Biophys. Acta* **978**, 185 (1989).

[22] G. R. Anyarambhatla and D. Needham, *J. Liposome Res.* **9**, 491 (1999).

[23] M. H. Gaber, K. Hong, S. K. Huang, and D. Papahadjopoulos, *Pharm. Res.* **12**, 1407 (1995).

[24] D. Needham, G. Anyarambhatla, G. Kong, and M. W. Dewhirst, *Cancer Res.* **60**, 1197 (2000).

[24a] N. Düzgüneş, *Methods Enzymol.* **387**, 134 (2004).

very narrow temperature region around 32°.[25] In addition, the LCST of this polymer can be adjusted to a desired temperature by copolymerization with commonomers with varying hydrophilicity or hydrophobicity.[26,27] Thus, modification of liposomes with NIPAM copolymers has been attempted to obtain liposomes with temperature sensitivity.[15,17,28] We have used N-acryloylpyrrolidine (APr) and acrylamide (AAM) as commonomers to prepare thermosensitive polymers with the LCST around physiological temperature.[14,15]

We have examined two approaches for the design of temperature-sensitive liposomes using thermosensitive polymers. One is to use them to destabilize stable liposomes through hydrophobic interaction above their LCST. We have examined the temperature sensitivity of poly(NIPAM)-fixed liposomes of phosphatidylcholines previously.[8] However, we found that contents release from the liposomes was enhanced only to a limited extent even above the LCST. This result suggests that the interaction between poly(NIPAM) and the phosphatidylcholine membrane is not strong enough to destroy the liposomes.

The other approach is to use thermosensitive polymers to stabilize unstable liposomes. It is well known that dioleoylphosphatidylethanolamine (DOPE) forms liposomes under alkaline conditions but takes on a hexagonal II phase under neutral and acidic conditions. Therefore, liposomes of this lipid are unstable under physiological conditions. However, when poly(NIPAM) are fixed onto liposomes of DOPE, the liposomes become stable even at neutral pH below its LCST. It is likely that the hydrated polymer chains covering the liposome surface prevent the occurrence of the lamellar-to-hexagonal II transition and stabilize the lamellar phase of DOPE membrane.[13] However, when the ambient temperature rises above the LCST, a prompt disintegration of the liposomes takes place. Thus, a complete and very fast contents release is achieved in this case.

Synthesis of Thermosensitive Polymers Having Anchors

We have shown that copolymerization of APr and NIPAM is useful in obtaining a thermosensitive polymer with LCST around physiological temperature.[15] Copolymerization of these monomers is carried out using an appropriate free radical initiator, such as azobis(isobutyronitrile) (AIBN). As shown in Fig. 1, LCST of the copolymers decreases with increasing NIPAM content and the copolymer with the APr/NIPAM unit molar ratio

[25] M. Heskins and J. E. Guillet, *J. Macromol. Sci. Chem.* **A2,** 1441 (1968).
[26] L. D. Taylor and L. D. Cerankowski, *J. Polym. Sci. Polym. Chem. Ed.* **13,** 2551 (1975).
[27] H. Feil, Y. H. Bae, J. Feijen, and S. W. Kim, *Macromolecules* **26,** 2496 (1993).
[28] K. Kono, K. Yoshino, and T. Takagishi, *J. Control. Release* **80,** 321 (2002).

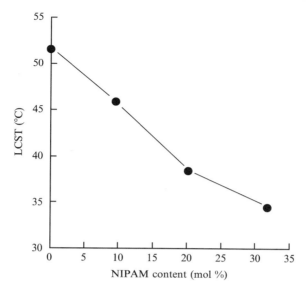

FIG. 1. Lower critical solution temperature (LCST) of copoly(APr-NIPAM) as a function of NIPAM content. LCSTs of the copolymers were detected by cloud point. Transmittance of aqueous copolymer solutions (5 mg/ml) at 500 nm was monitored when the temperature of the solution was raised at 0.3°/min. From Kono et al.[15]

of 4/1 exhibits the LCST at about 40°, which is a clinically achievable temperature.

It has been shown that hydrophobic groups, such as long alkyl groups, connected to hydrophilic polymers can act as anchors that fix the polymer chain onto a liposomal membrane.[29] We have synthesized two types of anchor-bearing thermosensitive polymers, which exhibit the hydrophilicity/hydrophobicity change at about 40° on a lipid membrane, as shown in Fig. 2. A copolymer of APr, NIPAM, and N,N-didodecylacrylamide (NDDAM) are examples of thermosensitive polymers having anchors at random positions of the polymer backbone. The copoly(APr-NIPAM)-$2C_{12}$ is another type of thermosensitive polymer that has an anchor at the chain end.

Copoly(APr-NIPAM-NDDAM) is synthesized as follows: APr (34.7 mmol), NIPAM (8.7 mmol), NDDAM (0.7 mmol), and AIBN (0.22 mmol) are dissolved in freshly distilled dioxane (88 ml) and then heated at 60° for 18 h in N_2 atmosphere. The copolymer is recovered by

[29] H. Ringsdorf, B. Schlarb, and J. Venzmer, *Angew. Chem. Int. Ed. Engl.* **27,** 113 (1988).

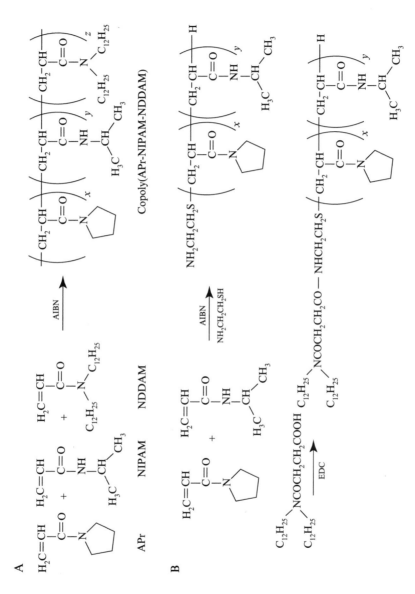

FIG. 2. Synthetic routes for thermosensitive polymers with anchors. (A) Copoly(APr-NIPAM-NDDAM). (B) Copoly(APr-NIPAM)-2C$_{12}$. From Kono et al.[15]

precipitation with diethylether and dried under vacuum. In our experiment, copolymers with the weight average molecular weight of 10,000–20,000 and the number average molecular weight of 5000–8000 were obtained by the aforementioned procedure, as estimated by gel filtration on a Showdex KD-803 column (Showa Denko) using poly(ethylene glycol) as the standard.

Copoly(APr-NIPAM)-2C$_{12}$ is obtained by the synthesis of copoly(APr-NIPAM) having an amino group at the chain terminus and attachment of the anchor moiety to the terminal amino group (Fig. 2B). APr (27.2 mmol), NIPAM (6.8 mmol), 2-aminoethanethiol (1.9 mmol), and AIBN (0.36 mmol) are dissolved in dimethylformamide (15 ml) and heated at 75° for 15 h in N$_2$ atmosphere. The amino group-terminated copolymer is recovered by precipitation with diethyl ether and is then purified using an LH-20 column, eluting with methanol. We obtained the copolymer with the weight average and the number average molecular weights of 9200 and 4100, respectively, by the aforementioned procedure. The copolymer (1.0 g), N,N-didodecylsuccinamic acid (6 × 10^{-4} mol),[30] and 1-ethyl-3-(3-dimethylaminopropyl)carbodiimide (EDC) (6 × 10^{-4} mol) are dissolved in dimethylformamide (10 ml) and stirred at 4° for 2 days. The copolymer is recovered by evaporation and purified by an LH-20 column eluting with methanol.

Preparation of Copolymer-Modified Liposomes Encapsulating Calcein

Liposomes modified with the copolymers are prepared by hydration of a mixture of the copolymer and lipids and subsequent extrusion through a polycarbonate membrane with a desired pore size. A typical procedure is as follows: 10 mg of DOPE, egg yolk phosphatidylcholine (EYPC), or their mixture is dissolved in chloroform (10 ml) in a vessel, and the copolymer in chloroform solution (10 mg/ml, 0.5–1.0 ml) is added to the solution. The mixed solution is evaporated to remove chloroform and dried under vacuum for more than 3 h. The dry thin film formed on the wall of the vessel is suspended in an aqueous calcein solution (63 mM) of pH 7.4 (for EYPC-based liposomes) or 9.0 (for DOPE-based liposomes) at 0° using a bath-type sonicator. The liposome suspension is extruded through a polycarbonate membrane with a pore diameter of 100 nm. Free calcein and free copolymer are removed by gel permeation chromatography on a Sepharose 4B column eluting with a 10 mM Tris–HCl-buffered solution containing 140 mM NaCl (pH 7.4) at 4°.

[30] Y. Okahata and T. Seki, *J. Am. Chem. Soc.* **106,** 8065 (1984).

Contents Release Behavior of Thermosensitive
Polymer-Modified Liposomes

Calcein and carboxyfluorescein have been used widely as fluorescent markers to examine the contents release behavior of liposomes because of their self quenching at high concentrations. When these molecules are encapsulated by a liposome at a high concentration, their fluorescence is negligible. Thus, only when released to the outside of the liposome do these marker molecules exhibit strong fluorescence. Therefore, the percentage release of these markers is usually defined as the following equation:

$$\% \text{ Release} = (F^t - F^i)/(F^f - F^i) \times 100$$

where F^i and F^t are the initial and intermediary fluorescence intensities of the liposome suspension, respectively. F^f is the fluorescence intensity of the liposome suspension after the addition of Triton X-100 [final concentration 0.15% (w/v) in our experiments] or other detergents that disrupt the liposomes completely and cause the release of all marker molecules into the medium.

Figure 3 shows temperature-sensitive contents release behaviors of DOPE liposomes modified with the two types of anchor-bearing

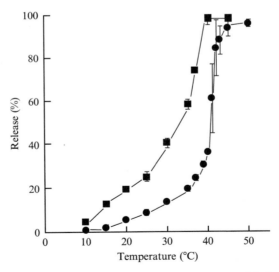

Fig. 3. Percentage release of calcein from DOPE liposomes modified with copoly(APr-NIPAM-NDDAM) (■) and copoly(APr-NIPAM)-2C$_{12}$ (●). The APr/NIPAM/NDDAM (mol/mol/mol) and APr/NIPAM (mol/mol) ratios of these polymers are 79.4:18.5:2.1 and 81.6:18.6, respectively. Percentage release after 15 min is shown. From Kono et al.[15]

thermosensitive polymers, namely copoly(APr-NIPAM-NDDAM) and copoly(APr-NIPAM)-2C$_{12}$. Both of the liposomes hardly release their contents under low temperature conditions, but the contents release is enhanced with temperature. When the copolymer chains are fixed by the anchors connected to the polymer backbone at random positions, namely in the case of the copoly(APr-NIPAM-NDDAM)-modified DOPE liposome, the liposome exhibits gradual enhancement of contents release with temperature. In contrast, when the copolymer chains are fixed by the anchor connected to the chain end, namely in the case of copoly(APr-NIPAM)-2C$_{12}$, the liposome shows a sharper increase in contents release around LCST of the copolymer (39.6°). Differences in the conformational freedom of these copolymer chains probably affect the range of temperature where the enhancement of the contents release takes place.[15]

Similar temperature-sensitive contents release behaviors are obtained by the modification of liposomes with copolymers of NIPAM, AAM, and

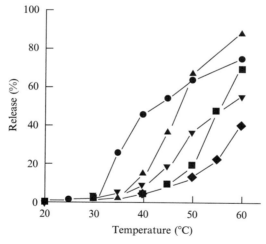

FIG. 4. Percentage release of calcein from DOPE/EYPC (6:4, w/w) liposomes modified with copoly(NIPAM-AAM-NDDAM)s with varying NIPAM/AAM/NDDAM (mol/mol/mol) ratios: ●, 98.9/0/1.1; ▲, 86.7/12.3/1.0; ▼, 73.5/25.5/1.0; and ◆, 63.0/35.9/1.1. Percentage release from plain DOPE/EYPC (6:4, w/w) liposome (■) is also shown. Percentage release after a 1-min incubation is shown. The copolymer-modified liposomes were prepared by incubating calcein-loaded liposomes with an aqueous copolymer solution (10 mg/ml) containing 63 mM calcein at 5° for 12 h and subsequent gel permeation chromatography using a Sephacryl-400 column. Note that the copolymers are fixed on the outer surface of the liposome membrane in this case. LCSTs of copolymers of NIPAM and AAM with the NIPAM/AAM (mol/mol) ratios of 100/0, 12/88, 26/74, and 36/64 are estimated to be 32, 40, 49, and 55°, respectively. From Hayashi et al.[14]

NDDAM.[14] The LCST of the copolymers can be adjusted by controlling the NIPAM/AAM ratio of the copolymers. These copolymers can be prepared by free radical copolymerization of NIPAM, AAM, and NDDAM according to the aforementioned procedure. Also, liposomes modified with these copolymers are prepared in the same way. Figure 4 shows calcein release behaviors of NIPAM-AAM-NDDAM copolymer-modified DOPE/EYPC (6:4, w/w) liposomes. The contents release is accelerated around the LCST of copoly(NIPAM-AAM) with the same NIPAM/AAM ratio of the copolymer fixed onto the liposome.

We found that the incorporation of short poly(ethylene glycol) chains into this type of thermosensitive polymer-modified liposomes suppresses effectively the leakage of the contents below the LCST and, as a result, their temperature-induced regulation of contents release is improved greatly.[28]

[6] The Materials Engineering of Temperature-Sensitive Liposomes

By Jeffrey K. Mills and David Needham

Introduction

The motivation for designing and creating triggered-release micro- and nanocarrier systems has come mainly from the desire to deliver drugs, especially anticancer drugs to tumors. Many of the liposome development programs of the last three decades have focused on anticancer applications. This activity has resulted in methods for making, loading with drug, and characterizing liposome performance, both *in vitro* and *in vivo*. This article describes methods that have been used historically and that have been adapted in our characterization and testing of a new liposome formulation designed to rapidly release its drug contents at temperatures just above body temperature. In addition to describing the methods used to form, load, and characterize drug release from the liposomes, this article shows briefly how a rational materials design follows a composition–structure–property approach in achieving a desired performance in service and outline procedures necessary to create and characterize this particular liposome formulation.

With problems of drug loading and achieving long circulation half-lives essentially solved, the single biggest challenge facing drug delivery for liposomes (and other drug carriers) was to initiate and produce release of the encapsulated drug only at the diseased site and at controllable rates. For this to happen, the carrier/drug relationship must be triggered to change from the stable requirement of the injection/delivery phase to one of gross instability at the site. In the new temperature-sensitive liposome, heat (hyperthermia) can be used exquisitely to control release from the lipid vesicles. When coupled with hyperthermia (the use of microwave or radio frequency energy or phased array ultrasound, energy focused on the tumor resulting in an increase in local temperature)[1] it represents a new and potentially important treatment strategy that can deposit energy noninvasively at a specific site within the body, such as solid tumors, and trigger the release of drug from circulating and accumulated liposomes. Hyperthermia has been noted to increase the effectiveness of radiation treatments and some chemotherapeutic regimens[1] and to increase the local accumulation of liposomes in tumors.[2] By developing lipid-based carriers that release contents at slightly elevated, clinically attainable temperatures (39–42°[3,4]), targeting, as well as specific release, can be achieved.

Materials Engineering Approach and Design Methodology

The ultimate performance of a liposome (or any) system in the body depends on engineering properly the composition, structure, and properties of the carrier to optimize the performance *in vivo*.[5] The properties of the material(s) employed originate from the structures of the material, which range in scale from the atomic, molecular, through microstructural (e.g., dislocations, defects, grain boundaries), up to the size of the product itself. These structures depend, to a large extent, on the atomic and molecular composition of the material, including the organizing role that water plays in these hydrated self-assembled lipid systems. Thus, it is crucial to understand the interrelationships between the composition of the liposome, its structure, and how these characteristics impart certain properties to the carrier and ultimately dictate how it will perform in the body.

[1] M. W. Dewhirst and T. V. Samulski, "Hyperthermia in the Treatment of Cancer: Current Concepts." The Upjohn Company, Kalamazoo, MI, 1988.

[2] M. Gaber, N. Z. Wu, H. Keeling, A. K. Huang, M. W. Dewhirst, and D. Papahadjopoulos, *Int. J. Radiat. Oncol. Biol. Phys.* **36,** 1177 (1996).

[3] M. S. Anscher, T. V. Samulski, G. Rosner, R. Dodge, L. Prosnitz, and M. W. Dewhirst, *Int. J. Radiat. Oncol. Biol. Phys.* **37,** 1059 (1997).

[4] K. A. Leopold, J. R. Oleson, D. Clarke-Pearson, J. Soper, A. Berchuck, T. V. Samulski, R. L. Page, J. Blivin, J. K. Thomberlin, and M. W. Dewhirst, *Int. J. Radiat. Oncol. Biol. Phys.* **27,** 1245 (1993).

[5] D. Needham, *in* "Materials Science of the Cell" (D. Wirtz and E. Evans, eds.). Materials Research Society, 1999.

Our laboratory has developed a temperature-sensitive liposome [lysolipid-thermosensitive liposome (LTSL)] that releases drug contents in the range clinically attainable by mild hyperthermia (39–42°).[6,7] By incorporating a small amount (10 mol%) of the water-soluble lysolipids, monopalmitoylphosphatidylcholine (MPPC) or monostearoylphosphatidylcholine (MSPC), into dipalmitoylphosphatidylcholine (DPPC) liposomes, the phase transition temperature of the liposomes is decreased slightly from 41.9 to ∼40.5° (MPPC formulation) and ∼41.3° (MSPC formulation). More importantly, the onset of release of entrapped contents 6-carboxyfluorescein [(6-CF) or doxorubicin] occurs more rapidly than for the pure DPPC lipid at temperatures just above 39°. These LTSLs release over 80% of drug in only tens of seconds at 40–41°.[7] The initial hypothesis and mechanism, which is now appearing to be true, was that incorporation of an acyl chain-compatible lysolipid (MPPC or MSPC) into the solid gel state of the host DPPC membrane would form an ideal solid solution. The phase transition temperature would be reduced slightly due to the lipid chain defect of the lysolipid. As the bilayer is heated up through the phase transition region, the grain boundaries would melt first and the lysolipid would enhance the permeable defect structures, either by desorbing rapidly from the boundary regions or by stabilizing porous defects. It appears from recent experiments that the latter is the case.[8] Heating the liposomes that were circulating in the body at tumor sites then would release drug just in the tumor. Studies of the formulation in mice with a human tumor xenograft, in combination with local hyperthermia (42°), showed unprecedented local control (11 out of 11 complete regressions).[6]

This article only discusses the methods used to produce these carriers and characterize them *in vitro*. Information regarding *in vivo* performance can be found in a review by Kong and Dewhirst.[9] Techniques concerning liposome formation, characterization and drug loading, and release assays are described in the context of the thermal-sensitive liposome. These methods include measuring thermal profiles via differential scanning calorimetry and permeability studies using both contents release of 6-CF and a simple binding reaction assay, where dithionite crosses only permeabilized membranes and so allows direct ion permeability to be measured as a function of temperature. Assays drug loading and methods that allow determination, extent, and rate of drug release from the liposomes as a function of temperature are also presented.

[6] D. Needham, G. Anyarambhatla, G. Kong, and M. W. Dewhirst, *Cancer Res.* **60**, 1197 (2000).
[7] G. R. Anyarambhatla and D. Needham, *J. Liposome Res.* **9**, 491 (1999).
[8] J. K. Mills, Dissertation thesis. Duke University, 2002.
[9] G. Kong and M. W. Dewhirst, *Intl. J. Hyperthermia* **15**, 345 (1999).

Composition

Initial material choice for temperature-sensitive liposomes focused on lipids that have a transition temperature above body temperature ($37°$). It was noted in the early 1970s that lipid bilayer ion permeability rates show an anomalous peak as the membrane goes through its gel-to-liquid crystalline phase transition at its transition temperature (T_m).[10] When a gel phase membrane is heated, melting initiates at previously formed grain boundaries,[11] forming interfacial regions between still solid domains and melted liquid domains. It is the "soft" and "leaky" nature of these interfaces that leads to the enhanced permeability to ions, drugs, and other small molecules at this temperature.[12] Because the transition temperature (T_m) of the host lipid (DPPC) dictates where the maximum drug release rates will be achieved, the first step in the design of the temperature-sensitive liposome system is to determine the transition temperature of the liposome formulation, and to assess how, and to what extent, the addition of other components like the monoacyl lysolipids might change this temperature.

Measuring the Transition Temperature: Differential Scanning Calorimetry (DSC)

Differential scanning calorimetry is one of the most widely used methods used to examine the thermal behavior of bilayer lipid systems.[13–16,16a] Differential thermograms measuring excess heat flow, normally in milliwatts versus temperature, are used to determine the transition temperature of lipid dispersions. Thermistors in the calorimeter monitor the temperature of a sample pan relative to a reference pan that contains deionized water (or buffer) as the samples are heated or cooled. Throughout the calorimetric scans, the temperatures of the samples are required to remain the same. At a melting phase transition, extra heat is needed to keep the temperature of the sample pan containing the melting lipid equal to the temperature of the reference pan. It is this excess heat that accounts for the endothermic (heat added to the sample) peak at the transition

---

[10] D. Paphadjopoulos, K. Jacobsen, S. Nir, and T. Isac, *Biochim. Biophys. Acta* **311,** 330 (1973).

[11] O. G. Mouritsen and M. J. Zuckermann, *Phys. Rev. Lett.* **58,** 389 (1987).

[12] L. Cruzeiro-Hansson and O. G. Mouritsen, *Biochim. Biophys. Acta* **944,** 63 (1988).

[13] B. D. Ladbrooke and D. Chapman, *Chem. Phys. Lipids* **3,** 304 (1969).

[14] A. Blume, *Thermochim. Acta* **193,** 299 (1991).

[15] R. N. McElhaney, *Chem. Phys. Lipids* **30,** 229 (1982).

[16] G. Lee, *Biochim. Biophys. Acta* **472,** 285 (1977).

[16a] P. Kinnunen, J.-M. Alakoskela, and P. Laggner, *Methods Enzymol.* **367,** 129 (2003).

temperature.[15,16a,17] First, a lipid solution in chloroform is prepared at a concentration of approximately 20 mg of total lipid per milliliter of chloroform. To examine a mixed lipid system, all of the liposome components must be first codissolved in chloroform at the appropriate ratios. For example, our LTSL formulation is composed of DPPC:MPPC:distearoyl-phosphatidylcholine-poly(ethylene glycol–2000 molecular weight) [DSPE-PEG(2000)] or DPPC:MSPC:DSPE-PEG(2000) at 86:10:4 mol% (all lipids are from Avanti Polar Lipids, Alabaster, AL). A small amount of methanol may be needed to completely solubilize certain lipids, for example, those with positively charged head groups. This mixture (5 mg total lipid in 250 μl) is then dried on the bottom of a glass vial, test tube, or round-bottom flask under a stream of nitrogen and is stored under vacuum overnight. The lipid solution can also be lyophilized to remove excess solvent and stored overnight. The next day, the lipid film is scraped off the bottom of the vial using a clean, metal spatula. One milligram of the film scrapings is placed directly into the base of an aluminum Perkin-Elmer DSC pan. Ten microliters of deionized water is then added to the lipid sample, and the sample pan is sealed using the aluminum top and a Perkin-Elmer sample pan crimper press. The entire sample pan is then heated to 10–20° above the estimated transition temperature of the liposome to fully hydrate the sample. Transition temperatures for lipid mixtures can be approximated from reported data on the main lipid component (see Marsh[18]). Without this preheating, the initial thermal scan may not reveal the true thermal characteristics of the fully hydrated formulation. The transition temperature of a partially hydrated sample (<20% water) can be as many as 5–10° above a fully hydrated liposome suspension.[19]

Figure 1 shows the results of thermal analyses of various DPPC:MPPC formulations as endothermic heat flow versus temperature. The addition of MPPC (0–20 mol%) to pure DPPC bilayers begins to shift the transition temperature of the lipid down from approximately 42° (pure) to near 40.5–41°. Note that the transition does not broaden with the addition of MPPC up to 20 mol%, indicating that the lipids do mix ideally (further supported by calorimetric and nuclear magnetic resonance data from Van Echteld et al.[20] of mixtures of DPPC and MPPC). Figure 2 shows the thermal profiles for DPPC and both DPPC:MPPC (10%) and DPPC:MSPC

[17] R. R. C. New, in "Liposomes: A Practical Approach" (R. R. C. New, ed.). Oxford University Press, New York, 1990.

[18] D. Marsh, "CRC Handbook of Lipid Bilayers." CRC Press, Boca Raton, FL, 1990.

[19] M. J. Janiak, D. M. Small, and G. G. Shipley, J. Biol. Chem. **254,** 6068 (1979).

[20] C. J. A. Van Echteld, B. De Kruijff, J. G. Mandersloot, and J. De Gier, Biochim. Biophys. Acta **649,** 211 (1981).

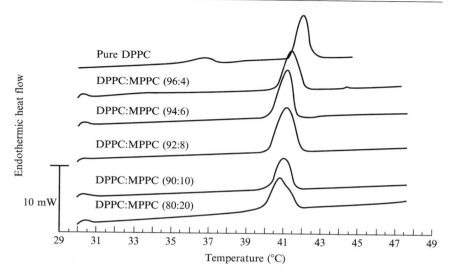

FIG. 1. Differential scanning calorimetry scans ($2°$/min heating rate) showing endothermic heat flow (bar indicates 10 mW) versus temperature for various mixtures of DPPC and MPPC (0–20 mol% MPPC). The peak in the heat flow indicates the transition temperature of the mixture. Reprinted from Anyarambhatla and Needham[7] with permission.

(10%) formulations. The addition of 10 mol% MPPC shifts the T_m down from ∼$42°$ of DPPC to near $40.5°$, and the incorporation of 10 mol% MSPC shifts the transition down to ∼$41.3°$. In both cases, the transition temperature of these formulations falls within the range attained easily in clinical hyperthermia—an important feature of the formulation for its intended clinical use.

Size and Structure

The Encapsulating Membrane

The addition of an aqueous buffer to a dried lipid film and simple shaking produces self-assembled, multilamellar vesicles with a broad size distribution and diameters between ∼100 and 800 nm (our own size measurements). Although these large, multimembrane vesicles have found applications, including sustained release depot formulations, topical delivery, and as imaging agents,[21] for most intravenous drug delivery applications,

[21] W. R. Perkins, *in* "Liposomes: Rational Design" (A. S. Janoff, ed.). Dekker, New York, 1999.

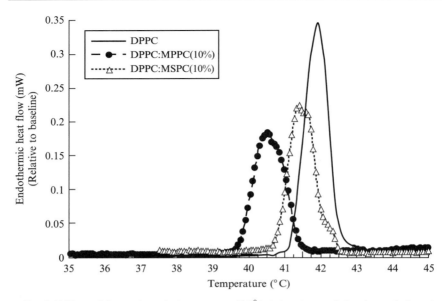

FIG. 2. Differential scanning calorimetry scans (0.2°/min heating rate) showing endothermic heat flow relative to baseline (mW) versus temperature for DPPC:DSPE-PEG(2000)(4%) and our lysolipid temperature-sensitive liposomes, DPPC:MPPC(10%): DSPE-PEG(2000)(4%) and DPPC:MSPC(10%):DSPE-PEG(2000)(4%).

it is necessary to produce single-walled liposomes with diameters of 100–200 nm. Vesicles of this size can pass freely through capillary beds,[17] and several studies have shown enhanced extravasation and accumulation of 100 nm liposomes in tumor tissue, especially in combination with local hyperthermia.[22,23] For vesicles much smaller than ∼50 nm, low encapsulation efficiency begins to limit the amount of drug that can be sequestered within the vesicles.[17] Several methods have been developed to produce single-walled vesicles in the required size range of ∼50–20 nm. Hydration followed by centrifugation,[24] probe or bath sonication,[25] and reverse-phase evaporation[26] have all been used to produce liposomes for use in various

[22] G. Kong, R. D. Braun, and M. W. Dewhirst, *Cancer Res.* **60,** 4440 (2000).

[23] M. W. Dewhirst and D. Needham, *in* "Stealth Liposomes" (D. Lasic and F. Martin, eds.). CRC Press, Boca Raton, FL, 1995.

[24] J. P. Reeves and R. M. Dowben, *J. Cell Physiol.* **73,** 49 (1969).

[25] Y. Barenholz, D. Gibbes, B. J. Litman, J. Goll, T. E. Thompson, and F. D. Carlson, *Biochemistry* **16,** 2806 (1977).

[26] F. Szoka and D. Papahadjopoulos, *Proc. Natl. Acad. Sci. USA* **75,** 4194 (1978).

applications. This article describes the method of thin film hydration followed by extrusion to produce unilamellar vesicles.[27,28] We (as well as others)[29] have found this to be the most reproducible method to prepare 100 nm liposomes suitable for dye and drug loading.

Hydration of Temperature-Sensitive Liposomes

The first step in the preparation of temperature-sensitive liposomes by the hydration/extrusion method is to dry down a lipid film from its organic solution. Lipid components are codissolved in chloroform at the appropriate mole percentages [for LTSL formulations, DPPC:MPPC:DSPE-PEG(2000) or DPPC:MSPC:DSPE-PEG(2000) at 86:10:4 mol%]. Depending on the experiment and desired lipid concentration, up to a few hundred milligrams of total lipid can be dried easily in a 100-ml round-bottom flask. The appropriate amount of lipid–chloroform solution is placed in a clean, round-bottom flask, and the solvent is removed using either a stream of nitrogen or a rotavapor setup. The temperature of the rotavapor water bath is set to 38–40°, ice is placed in the condenser, and the flask is rotated at 40 rpm. rpm. A vacuum line is attached to the evaporator, and with the flask rotating slowly, the vacuum tap is opened gradually to allow the chloroform in the flask to evaporate. Care should be taken to not boil the chloroform (the bath is kept between 35 and 40°, and the vacuum is increased slowly) because this can produce clumps of dried lipid high up in the flask that are difficult to hydrate. The resulting film should be thin and uniform on the inside of the flask. This film is then dried fully in a desiccator under vacuum overnight.

The lipid is then hydrated with the appropriate aqueous buffer (see section on Properties for a description of buffers, e.g., dye solution or low pH solution for drug loading). For lysolipid formulations, the hydrating media must always contain 1 μM of the appropriate lysolipid. These single chain lipids are water soluble [with a critical micelle concentration (CMC) = 4–7 μM[30,31]]. In addition, uptake and desorption of MOPC lysolipid from the outer monolayer of liquid state bilayers have been observed in

[27] L. D. Mayer, M. J. Hope, and P. R. Cullis, *Biochim. Biophys. Acta* **858**, 161 (1986).

[28] M. J. Hope, R. Nayar, L. D. Mayer, and P. R. Cullis, *in* "Liposome Technology: Liposome Preparation and Related Techniques" (G. Gregoriadis, ed.). CRC Press, Boca Raton, FL, 1993.

[29] L. D. Mayer, L. C. L. Tai, D. S. C. Ko, D. Masin, R. S. Ginsberg, P. R. Cullis, and M. B. Bally, *Cancer Res.* **49**, 5922 (1989).

[30] M. E. Haberland and J. A. Reynolds, *J. Biol. Chem.* **250**, 6639 (1975).

[31] W. Kramp, G. Pieron, R. N. Pincard, and D. J. Hanahan, *Chem. Phys. Lipids* **35**, 49 (1984).

micropipette bathing experiments.[32,33] Hydration with a lysolipid solution at a concentration around its critical micelle concentration provides an equilibrating amount of lysolipid during self-assembly that retains the lyso- lipid in the liposome membranes. During the entire hydration process, it is crucial that the solution temperature be maintained at 5–10° above the transition temperature of the main lipid component (DPPC in this case) to ensure that the liposomes are always in the liquid-crystalline phase, and the lipids are well mixed. The easiest way to do this is to suspend the round-bottom flask in a circulating water bath during hydration. The flask is then swirled/shaken gently for at least 15–20 min at this elevated temper- ature until all of the lipid has been removed from the sides of the flask. A small Teflon stir bar is added to help with the removal of lipid from the sides of the flask. Vigorous shaking should be avoided, as it produces foam and can limit the retrieved volume. Several groups have reported the benefits of freeze-thaw cycling the lipid solution several times to aid hydra- tion and increase encapsulation efficiency.[34] After removing the lipid from the flask walls, the sample is placed in a cryogenic vial. The sample is placed in alternating baths of liquid nitrogen and warm water (approxi- mately 40°) for 10 min at each temperature to allow for complete thermal equilibration. Five to seven freeze-thaw cycles are sufficient.

Extrusion of Temperature-Sensitive Liposomes

Once the lipid has been hydrated completely it can be extruded imme- diately or stored for several days at 4°. For extrusion, a Lipex Biomem- branes thermobarrel extruder (Lipex Biomembranes, Vancouver, BC) is assembled according to the instructions, using a polyester drain disc and two of 100 nm pore diameter polycarbonate membrane filters (Poretics, Livermore, CA). The liposomes must be in a liquid crystalline state during extrusion, so the temperature of the extruder barrel is maintained at approximately 5–10° above the liposome transition temperature using a cir- culating water bath. The polycarbonate membranes are washed with a small amount of the hydrating buffer. The lipid solution is then added to the extruder barrel and allowed to equilibrate for approximately 10 min at the elevated temperature. The solution is then extruded through the fil- tration system 5–10 times using a maximum of 400 psi of pressure from a compressed nitrogen tank. If high lipid concentrations are used (greater than approximately 50 mg/ml), it may be necessary to stage the liposome

[32] D. Needham, N. Stoicheva, and D. V. Zhelev, *Biophys. J.* **73,** 2615 (1997).
[33] D. Needham and D. V. Zhelev, *Ann. Biomed. Eng.* **23,** 287 (1995).
[34] L. D. Mayer, M. J. Hope, P. R. Cullis, and A. S. Janoff, *Biochim. Biophys. Acta* **817,** 193 (1986).

formation by first extruding the sample through two filters of 200 nm pore diameter and then a subsequent extrusion through two 100-nm filters.

Following extrusion, the size of the vesicles is measured by photon correlation spectroscopy (Coulter N4 Plus submicrometer particle sizer, Miami, FL) following the manufacturer's recommendations for sample concentration and runparameters. Analysis yields a size distribution of 120.7 ± 27.7 nm, as shown in Fig. 3. The top portion of Fig. 3 depicts the distribution of four size-measuring scans of the same sample. Each scan measures particle size at a scattering angle of 90° by measuring the rate of diffusion of particles. The chart at the bottom of Fig. 3 shows the mean and standard deviation of each run, as well as the mean of the four repetitions. The polydispersity index is a measure of the "breadth" of the sample, and an index below 0.1 indicates a narrow size distribution. The extruded sample can be stored at room temperature for several hours or at 4° for several days. During prolonged storage, the appearance of larger particles may be noted, indicating liposome aggregation or possibly liposome–liposome fusion. Size measurements should be made periodically to make sure that liposomes are still within the desired size range.

Size-Exclusion Column Filtration

Following extrusion, it may be necessary to filter the liposomes through a size-exclusion column to remove unencapsulated contents or exchange the external buffer (see section on Properties for a more detailed description of where this filtration system is used.). Size-exclusion minicolumns are prepared by spinning the excess solvent out of a slurry of gel beads. When hydrated, the gel beads form a porous matrix that retains solutes up to a certain molecular weight, but lets larger molecules, including 100-nm liposomes, pass through the column for collection.[35]

To prepare minicolumns, a small amount of glass wool or filter paper is placed in the bottom of a 5-ml syringe suspended in a 15-ml centrifuge tube. The syringe is filled with a slurry of Sephadex G-50 medium beads (Pharmacia Biotech Products) in deionized water prepared according to the manufacturer's directions (approximately 5 g beads/200 ml deionized water). These beads have a defined pore size that will retain dextrans between 500 and 10,000 Da and globular proteins between 1500 and 30,000 Da. The column is then spun down at 2000 rpm for 10 min. After this spin, beads form a tightly packed column that has pulled away from the sides of the syringe. A volume of buffer, equivalent to the volume of liposome solution to be filtered, is added to the column. Care should be taken

[35] C. S. Wu, "Handbook of Size Exclusion Chromatography." Dekker, New York, 1995.

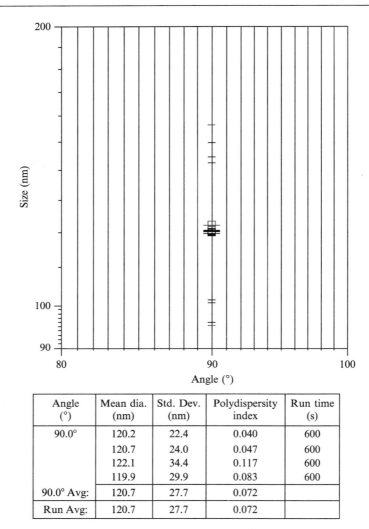

Angle (°)	Mean dia. (nm)	Std. Dev. (nm)	Polydispersity index	Run time (s)
90.0°	120.2	22.4	0.040	600
	120.7	24.0	0.047	600
	122.1	34.4	0.117	600
	119.9	29.9	0.083	600
90.0° Avg:	120.7	27.7	0.072	
Run Avg:	120.7	27.7	0.072	

FIG. 3. Size analysis of 100-nm extruded vesicles from photon correlation spectroscopy.

to add all solutions dropwise, directly to the center of the column and not down the sides of the syringe. The equilibration buffer is then spun down for 10 min at 2000–2200 rpm. Between 200 μl and 1.5 ml of temperature-sensitive liposome solution can be added to the column for filtration. This solution and the centrifuge are kept between 4 and 25° so that no leakage of dye or drug from the temperature-sensitive liposomes (i.e., near T_m) occurs during the filtration process. The liposome solution is spun down at 2200–2400 rpm for 10 min. If the columns are prepared and preequilibrated properly, the entire volume of the original solution will be returned. It may be necessary to preequilibrate the column with a solution of non-dye/drug-containing liposomes if loss of solution volume is noted after spinning.

Properties

The next step in characterization of the temperature-sensitive liposome formulation is to measure its encapsulation stability and temperature-triggered release properties. An ideal temperature-sensitive liposome will maintain stability at 37° during transport through the body, but then initiate a rapid, triggered release of contents when it reaches the tissue site heated to the elevated temperature. Many of the traditional temperature-sensitive formulations, especially those that contain cholesterol, demonstrate sufficient stability at body temperature.[36–39] However, not only is the release temperature often too high and difficult to attain clinically (42–45°), but the rate of contents release is often too slow (on the order of 60% over 30 min[36]) to be effective as a chemotherapeutic delivery system for drugs such as doxorubicin that, optimally, require a rapid release.

The lysolipid temperature-sensitive liposome formulation that we developed not only exhibits the required stability at body temperature, but also the desired rapid release of drug (>80% released within tens of seconds).[6,7,40] The following sections outline the methods used to make measurements of the properties of this formulation, including contents stability, ion permeability, and temperature-triggered dye and drug release, as well as the kinds of results that were obtained.

Contents Stability: 6-Carboxyfluorescein Leakage

Encapsulation and leakage of the fluorescent dye 6-CF is one of the most straightforward and widely used methods to investigate liposome stability/release.[17,40a] At high concentrations (greater than 30 mM), the dye is

[36] M. H. Gaber, K. Hong, S. K. Huang, and D. Papahadjopoulos, *Pharm. Res.* **12,** 1407 (1995).
[37] K. Maruyama, S. Unezaki, N. Takahashi, and M. Iwatsuru, *Biochim. Biophys. Acta* **1149,** 209 (1993).
[38] J. N. Weinstein, R. L. Magin, R. L. Cysyk, and D. S. Zaharko, *Cancer Res.* **40,** 1388 (1980).
[39] J. B. Bassett, R. U. Anderson, and J. R. Tacker, *J. Urol.* **135,** 612 (1986).
[40] D. Needham and M. Dewhirst, *Adv. Drug Deliv. Rev.* **53,** 285 (2001).
[40a] J. N. Weinstein, S. Yoshikami, P. Henkart, R. Blumenthal, and W. A. Hagins, *Science* **195,** 489 (1977).

self-quenched and emits a very low fluorescent signal. When diluted 10,000–1000× to 3–30 μM, the fluorescence signal increases on the order of 5- to 20-fold and is linear over this concentration range.[17] Dye can be encapsulated in liposomes at >30 mM and release can be monitored by measuring the increase in the fluorescence signal with time due to dilution into a large, external volume.[17,41] The addition of detergent and subsequent lysis of the entire vesicle population releases all the contents, and results can be reported as percentage release.[17]

This stability assay is performed as follows. 6-CF Fluka (Buchs, Switzerland) is prepared in 20 mM phosphate-buffered saline (PBS) [16.8 mM sodium phosphate dibasic +3.2 mM sodium phosphate monobasic +0.8% NaCl (8 g/1 liter deionized H$_2$O)] at a dye concentration of 50 mM. It is necessary to first raise the pH of the PBS buffer from pH 7.4 with NaOH (one tablet) to approximately 11.0 in order to solubilize the dye completely and then bring it back down to approximately pH 7.4 with HCl (several drops). A lipid film is prepared as described earlier (20 mg of total lipid) and hydrated with 1 ml of this dye solution. After extrusion to 100-nm, the vesicles are chromatographed on a Sephadex G-50 column preequilibrated with PBS to remove all unencapsulated 6-CF.

Liposomes are diluted at a ratio of 5 μl of liposomes to 1 ml of 20 mM PBS (pH 7.4) so that after lysis, the total concentration of released dye will be 20–30 μM. A 1-ml aliquot of this solution is placed in a quartz cuvette, and a fluorescence reading is made (F_0) using an excitation wavelength of 470 nm and an emission wavelength of 519 nm. Fluorescence measurements are made using a Shimadzu RF-1501 spectrofluorophotometer. Stability can now be monitored at several temperatures simply by incubating the sample at, for example, 4, 25, 37° or other desired temperatures. Fluorescence measurements can be made at any desired time interval (F_t), for example every 10 min up to 3 h, and then every hour for several days. The most ideal way to perform these experiments is to place the sample cuvette into a fluorimeter with a temperature-controlled cuvette holder. If this is not possible, the samples can be incubated by suspending them in a water bath. At the end of the experiment, the sample(s) is lysed with 5 μl of 10% Triton X-100. This micellizes the vesicles and releases all of the encapsulated dye, and a final, 100% release measurement is made (F_{TX}). The percentage release of encapsulated dye is the calculated as

$$\text{Percentage release} = [(F_t - F_0)/(F_{TX} - F_0)] \times 100 \qquad (1)$$

Figure 4 shows the percentage release of 6-CF versus time at 37° from the DPPC:MPPC(10%):DSPE-PEG(2000)(4%) LTSL formulation.

---

[41] J. K. Mills, G. Eichenbaum, and D. Needham, *J. Liposome Res.* **9**, 275 (1999).

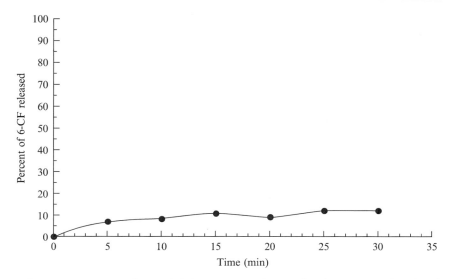

FIG. 4. Percentage release of 6-CF from DPPC:MPPC(10%):DSPE-PEG(2000)(4%) vesicles at 37° demonstrating formulation stability.

Approximately 10% of the dye is released within the first 10–15 min, with minimal additional release up to the 30-min end point, demonstrating fairly good contents stability at body temperature over this time course. For comparison, Doxorubicin leakage from the temperature-sensitive formulation prepared by Gaber et al.,[36] which incorporated cholesterol to enhance liposome stability, was also approximately 10% at 37° over the same time period.

Dithionite Ion Permeability

Several groups have measured the permeability of anions and cations through lipid vesicle membranes. Most of the classical methods utilize ion-sensing electrodes or radiolabeled ions that require very specialized equipment and laboratory environments.[10,42,43] This section describes a method to measure the permeability of the dithionite ion ($S_2O_2^{2-}$) through liposome membranes using simple absorbance measurements.

In aqueous solution, the $S_2O_2^{2-}$ is in equilibrium with the SO_2^- radical. SO_2^- is a reaction intermediate in the nitro reduction of

[42] R. T. Hamilton and E. W. Kaler, J. Phys. Chem. **94**, 2560 (1990).
[43] D. Papahadjopoulos and A. Bangham, Biochim. Biophys. Acta **126**, 185 (1966).

N-(7-nitro-2,1,3-benzoxadiazol) (NBD) to its corresponding amine.[44] This reduction results in the irreversible quenching of NBD fluorescence, as well as the elimination of the absorbance peak of NBD at 465 nm.[44] This fluorescence-quenching property has been used to measure dithionite ion permeability through multilamellar[45,46] and unilamellar[47] vesicle membranes, but there are several concerns with this method that can complicate permeability measurements. Continuous excitation of the fluorophore causes it to photobleach, which can be mistaken for the NBD-dithionite quenching reaction. Second, it is unclear whether self-quenching occurs between the labeled lipids if and when they come in close proximity to each other at the bilayer surface (unpublished observations). The following method, based on absorbance decay, measures the temperature-dependent, dithionite ion permeability through lipid vesicles, completely eliminating any photobleaching or self-quenching concerns.

Liposome Preparation for Dithionite Permeability Measurements. Two sets of lipid samples are prepared. They are identical in composition, except that one "blank" sample does not contain labeled lipid, whereas the other sample, "NBD labeled," contains 1 mol% of N-(7-nitrobenz-2-oxa-1,3-diazol-4-yl)-1,2-dihexadecanoyl-*sn*-glycero-3 phosphoethanolamine, triethylammonium salt (NBD-PE, from Molecular Probes, Eugene, OR). Incorporation of 1 mol% NBD-PE into large, unilamellar liposomes results in a slightly asymmetrically labeled bilayer.[48] The curvature of a 120-nm-diameter vesicle makes the surface area of the outer monolayer larger than the inner monolayer by a ratio of 54:46. The NBD lipids partition evenly, but because there are overall less lipids in the inner monolayer than the outer, the total number of NBD molecules inside relative to outside reflects the absolute numbers of lipids, and therefore partition ≈54%:46% outer:inner.

For each set, 30 mg of total lipid is dried into a thin film using the technique described earlier. Each film is hydrated with 1.5 ml of a solution of 100 mM NaCl with 709 mM sucrose (20 mg/ml lipid). The salt provides ion screening, while the sucrose is added to increase the solution osmolarity to ~900 mOsm. This elevated osmolarity ensures that when dithionite ions (30 mM ions) are added, the increase in external solution osmolarity relative to inside the liposomes is less than 10%. In the case of our lysolipid

[44] C. R. Wasmuth, C. Edwards, and R. Hutcherson, *J. Phys. Chem.* **68,** 423 (1964).
[45] M. Langner and S. W. Hui, *Chem. Phys. Lipids* **65,** 23 (1993).
[46] J. Risbo, K. Jorgensen, M. M. Sperotto, and O. G. Mouritsen, *Biochim. Biophys. Acta* **1329,** 85 (1997).
[47] M. Langner and S. W. Hui, *Biochim. Biophys. Acta* **1463,** 439 (2000).
[48] C. Balch, R. Morris, E. Brooks, and R. G. Sleight, *Chem. Phys. Lipids* **70,** 205 (1994).

temperature-sensitive formulation, this NaCl/sucrose buffer also contains 1 μM of the appropriate lysolipid (MPPC or MSPC). The liposomes are extruded through two, 100-nm pore-diameter filters to yield a homogeneous sample of unilamellar lipid vesicles.

The liposomes are then cycled five times through their transition temperature, allowing the sample to equilibrate for 5 min above (50°) and below (10°) T_m. This cycling ensures that vesicles that may initially rupture due to the approximately 20% reduction in bilayer area during the first freezing cycle[49] can reform into intact liposomes with minimal tension in their membranes on reheating. Finally, after cooling the liposomes to room temperature, they are passed through a Sephadex G-50 column, preequilibrated with the NaCl/Sucrose buffer, to remove free lysolipid.

Absorbance Measurements. After column filtration, both "blank" and "NBD-labeled" liposomes are diluted in NaCl/sucrose buffer to a final concentration of 3 mM lipid. Absorbance measurements are made using a Shimadzu UV-1601 spectrophotometer. This instrument uses a dual cuvette system, where the "sample" cuvette (NBD-labeled sample) is always measured relative to a "blank" cuvette. Initially, *both* cuvettes are filled with 1 ml of "blank" liposomes that do not contain NBD-labeled lipid. Both "sample" (front) and "blank" (rear) cuvettes are equilibrated to the proper recording temperature with the use of a circulating water bath connected to the water circulation system of the spectrophotometer. The "sample" cuvette is stirred using an Instech stirring system that fits on the top of the cuvette and uses a short stirring rod to agitate the solution. The temperature is monitored with a small thermocouple wire inserted into the cuvette. After 10 min of equilibration, the spectrophotometer is "zeroed" at 465 nm, i.e., any absorbance at this wavelength is subtracted from the "sample" reading. The "sample" cuvette (front cuvette) is washed out and then filled with 1 ml of "NBD-labeled" liposome solution. The "blank" cuvette and solution are left in the instrument as a reference.

The "sample" is then equilibrated to the appropriate temperature for 10 min with constant stirring. Absorbance readings at 465 nm are then started, with the instrument programmed to automatically record measurements every 10 s. These readings are always within 1% of each other, yielding a stable absorbance trace. After 2 min, 30 μl of 1 M dithionite ions is added to both cuvettes. The 1 M dithionite solution is prepared from sodium hydrosulfite ($Na_2S_2O_4$) in 1 M Trizma buffer (both from Sigma St. Louis, MO). The addition of ions to both cuvettes ensures that any decrease in absorbance is due to the dithionite-NBD reaction and not simply to dilution of the "sample" volume. The stirring mechanism is replaced, and the

[49] D. Needham and E. Evans, *Biochemistry* **27**, 8261 (1988).

absorbance of the "sample" is recorded for a total of 30 min, with measurements recorded automatically every 10 s. After 30 min, 20 μl of 10% Triton X-100 is added to both cuvettes. After 5 min to allow for complete lysis, a final absorbance reading is taken. This reading corresponds to 100% reacted NBD molecules (additional dithionite ions do not reduce this final value).

The addition of dithionite to the outside of the liposomes results in an immediate reaction with the NBD molecules on the exterior monolayer (complete reaction within about 10–20 s). Permeation of the $S_2O_4^{-2}$ ion or SO_2^- radical through the bilayer results in a subsequent reaction with the inner monolayer.[50] This occurs at a rate of approximately 0.25% reacted/min (for dioleoyl-PC vesicles at 2°),[50] and therefore bilayer permeability is the rate-limiting factor in the reaction. By monitoring the optical absorbance at 465 nm, we are able to obtain temperature-dependent measurements of the rate at which dithionite ions cross bilayer membranes. Although rates of permeability are temperature and composition dependent, all membranes reach 100% reacted NBD over some timescale (hours to days).

Results of Permeability Measurements: Rate Constants

We have used this absorbance decay method to measure the permeability rate enhancement that our lysolipid temperature-sensitive formulations provide in comparison to DPPC (C16) and palmitoyloleoylphosphatidylcholine [POPC (C16:C18:1), $T_m = -2°$[18]] liposomes.

To measure ion permeability rates and make comparisons between the various compositions, readings from the absorbance spectrophotometer must be converted to relative absorbance values. The initial 2 min of steady-state absorbance readings are averaged (12 total readings), resulting in a value that corresponds to "100% *unreacted* NBD" (Abs_{Init}). The addition of dithionite causes the absorbance to drop quickly (within 20 s) as all of the NBD molecules on the outside monolayer react. Any subsequent decay of the absorbance signal (Abs_{Dith}) is due to permeation of the dithionite ion through the lipid membrane and reaction with NBD molecules on the inner monolayer. After 30 min, the addition of TritonX-100 destroys all of the vesicles, allowing dithionite to react with all NBD molecules, and a final reading is made (Abs_{TX}). This value is never zero due to some residual absorbance signal, but is only a small percentage of the initial absorbance. Data collected after the addition of dithionite are corrected by subtracting this residual "100% reacted" value. Dividing all of

---

[50] J. C. McIntyre and R. G. Sleight, *Biochemistry* **30**, 11819 (1991).

the data by the 100% *unreacted* NBD (ABS_{init}) value gives relative absorbance readings.

$$\text{Relative absorbance} = \frac{Abs_{Dith} - Abs_{TX}}{Abs_{Init}} \qquad (2)$$

Dithionite permeability through DPPC:DSPE-PEG(2000)(4%) membranes is represented in Fig. 5 as a plot of relative absorbance at 465 nm versus time in minutes for five temperatures in the examined range (30–48°). Data points and error bars represent the mean and standard deviation of three separate experiments. Only every fourth data point is shown to help with symbol identification. For the first 2 min, the relative absorbance remains steady at "1" for all traces. Following the addition of dithionite at 2 min, the relative absorbance drops quickly to ∼0.46 (the 130-s time point is dropped for all traces, as it often takes slightly longer than 10 s to add the dithionite ion to both cuvettes). This value of 0.46 corresponds to the 46% remaining NBD molecules that have not reacted on the inner monolayer. Any subsequent loss of the absorbance signal is due to dithionite ions crossing the bilayer and reacting with the NBD lipids on the inner monolayer.

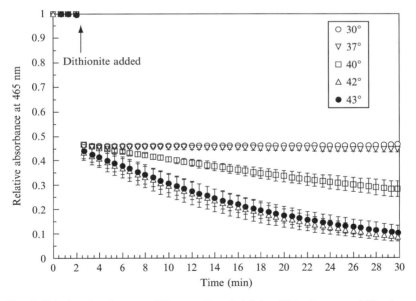

FIG. 5. Relative absorbance at 465 nm vs time (min) for dithionite permeability through DPPC:DSPE-PEG(2000)(4%) membranes shown for five temperatures in the examined range. Data points and error bars represent the mean and standard deviation, respectively, of three separate experiments.

In Fig. 5 at 30°, there is no dithionite ion permeability through the gel phase DPPC membranes, as the absorbance readings remain stable at 0.46 over the 30-min time course. Increasing the temperature to 37°, a gradual decay in the absorbance trace with time is noted, indicating that ions are permeating slowly through the membrane. The absorbance curve at 40° shows even faster absorbance decay and permeability, whereas at 42° (the transition temperature of DPPC as determined from DSC measurements) the absorbance drops fairly quickly with time. The absorbance trace at 43° shows a slightly slower rate of decay, indicating a peak in the permeability around the actual gel-to-liquid crystalline transition temperature. In all cases, if the experiment is followed for longer times (hours to days), the curves will effectively reach 0 (less than 0.009) relative absorbance.

Figure 6 shows a similar plot of relative absorbance at 465 nm versus time for the DPPC:MPPC (10%) lysolipid temperature-sensitive formulation. At 30° (gel phase) the absorbance remains steady at 0.42. Although a minor effect, it appears that the addition of 10 mol% MPPC creates a preferential distribution of NBD lipid on the outer monolayer from the

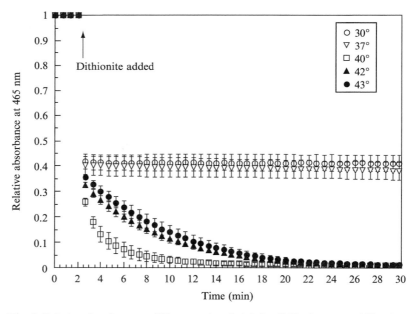

FIG. 6. Relative absorbance at 465 nm vs time (min) for dithionite permeability through DPPC:MPPC(10%):DSPE-PEG(2000)(4%) membranes shown for five temperatures in the examined range. Data points and error bars represent the mean and standard deviation, respectively, of three separate experiments.

expected 54 to 58%, and concomitantly reduces the amount of lipid in the inner monolayer from 46 to 42%. As with DPPC, the permeability of the DPPC:MPPC(10%) membranes increases when the temperature is raised to 37°. As the temperature is raised to 40°, very near to the transition temperature for DPPC:MPPC(10%) (~40.5°), the permeability increases dramatically, dropping to only 5% remaining NBD absorbance within 6 min following the addition of dithionite, and falling to approximately 1% remaining by 15 min. Increasing the temperature for this LTSL, traces at 42 and 43° show slower absorbance decay than at T_m, but faster than DPPC at the same temperatures, again illustrating the permeability maximum at T_m.

To obtain a more quantitative measure of the ion permeability, Fig. 7 shows the method used to determine temperature-dependent ion permeability rates. Data and fits are shown for three DPPC traces as an example (rates for all other membranes were determined in a similar fashion). Only data collected after the addition of dithionite (>2 min) are fit to determine the permeability rate constants. The following equation is fit to data using the curve fit feature of KaleidaGraph (Synergy Software, Reading, PA).

$$\text{Relative absorbance} = m1 * \exp[-m2 * (t)] \qquad (3)$$

where m1 and m2 are constants and t is time in minutes.

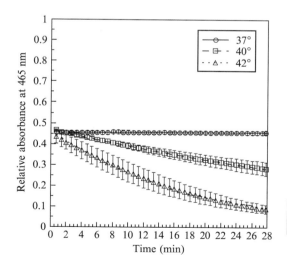

$y = m1* \exp[-m2*(t)]$ 37°		
	Value	Error
m1	0.45837	0.00018189
m2	0.0012057	2.433e-05
Chisq	1.3158e-05	NA
R	0.99196	NA

$y = m1* \exp[-m2*(t)]$ 40°		
	Value	Error
m1	0.47047	0.00028192
m2	0.018586	4.1897e-05
Chisq	2.4776e-05	NA
R	0.9999	NA

$y = m1* \exp[-m2*(t)]$ 42°		
	Value	Error
m1	0.46394	0.0022081
m2	0.057725	0.00046888
Chisq	0.00094171	NA
R	0.99896	NA

FIG. 7. DPPC:DSPE-PEG(2000)(4%) relative absorbance data vs time collected after the addition of dithionite (i.e., 2 min) is fit using an exponential: $y = m1 * \exp[-m2 * (t)]$. Here m1 represents the intercept and m2 (1/min) the temperature-dependent rate constant. In all cases where absorbance decay was observed, correlation was greater than 0.95.

The constant, m1, in Eq. (3) is the intercept of the best fit line with the time axis, and therefore represents the amount of relative absorbance remaining after the outside monolayer has reacted, i.e., approximately 0.46. The constant, m2, represents the rate at which the absorbance is decaying in units of 1/min. Both constants are fit by the KaleidaGraph curve-fitting program, and the correlation coefficient (R) in all cases is greater than 0.95 where exponential decay is noted (i.e., at I > 30°). Figure 8 shows a compilation of the ion permeability rate constants (m2, in units of 1/min) for DPPC:DSPE-PEG(2000)(4%), POPC:DSPE-PEG(2000)(4%) and both lysolipid temperature-sensitive liposomes:DPPC:MPPC(10%):DSPE-PEG(2000)(4%) and DPPC:MSPC(10%):DSPE-PEG(2000)(4%) through the temperature range of 30–48°. The lines joining data in Fig. 8 are shown simply to aid the eye.

The permeability rates for the DPPC membrane below its main transition (i.e., in the bilayer gel state) are very slow, less than 0.0013/min, from 30 to 37°, and then begin to increase near 39° and reach a local maximum of 0.058/min at approximately 42° (T_m). The rates then drop below this maximum when the temperature is increased to 42.5°, remain lower

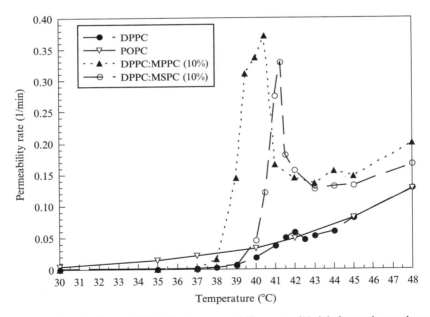

FIG. 8. Compilation of dithionite ion permeability rates (1/min) for each membrane [DPPC:DSPE-PEG(2000)(4%), DPPC:MPPC(10%):DSPE-PEG(2000)(4%), DPPC: MSPC(10%):DSPE-PEG(2000)(4%), and POPC:DSPE-PEG(2000)(4%)] through the temperature range studied. Lines connecting data are presented simply to aid the eye.

through 44°, and then increase above the peak at T_m if the temperature is raised above 45°. Permeability rates for POPC shown in Fig. 8 represent dithionite ion permeation through a bilayer membrane that is in its liquid crystalline phase (i.e., completely melted) over the entire temperature range. A steady increase in dithionite permeability occurs with increasing temperature, but no local maximum is noted because there is no phase transition over the temperature range. It is important to note that the enhancement in ion permeability at T_m of DPPC (42°) is only 1.5 times above the liquid state POPC permeability at the same temperature.

The results in Fig. 8 for the lysolipid temperature-sensitive formulations demonstrate the dramatic ion permeability enhancement that this formulation has achieved. The permeability rates through the DPPC:MPPC(10%) formulation (closed triangles) shown in Fig. 8 indicate minimal gel phase permeability below 37°, virtually identical to DPPC over the same temperatures. For this LTSL, however, permeability rates begin to increase at 38–39° and reach a peak (0.372/min) at its phase transition temperature, 40.5°. Increasing the temperature to 41°, the rates drop below this peak and then continue to increase slightly as the temperature is raised. Unlike DPPC, however, the permeability rates never approach the height of the peak attained at T_m. Rates for DPPC:MSPC(10%) (open circles) in Fig. 8 also show minimal gel phase permeability through 39°, virtually identical to DPPC. As the temperature is raised, a substantial rate enhancement is noted, reaching a peak (0.3283/min) at 41.3° (T_m), slightly lower then the T_m peak for the DPPC:MPPC formulation. Again, increasing the temperature above T_m, the rates drop below the peak and follow the trend noted with the MPPC composition. In both cases, a tremendous enhancement in the ion permeability at T_m is noted. The MPPC formulation shows nearly a 20-fold enhancement in ion permeability above pure DPPC, whereas the MSPC formulation shows a 10-fold enhancement.

Dye and Drug Release Measurements

While ion permeability measurements provide valuable information regarding the transport of single ions, measurements of the temperature-dependent triggered release of contents must be made to fully characterize a temperature-sensitive liposome formulation. We will first describe methods to measure the release of the fluorescent dye 6-CF. We will then describe how to load liposomes with the cancer chemotherapeutic agent doxorubicin (DOX) and how to obtain measurements of drug release rates.

6-Carboxyfluorescein Encapsulation and Release. The methods used to encapsulate and measure the release of 6-CF are virtually identical to the stability measurements described earlier. 6-CF is prepared in 20 mM PBS at a dye concentration of 50 mM. A lipid film is dried down (20 mg of total lipid) and hydrated with 1 ml of this dye solution. After extrusion, the vesicles are chromatographed on a Sephadex G-50 column preequilibrated with PBS to remove all unencapsulated 6-CF. The liposomes are then diluted at a ratio of 5 μl of liposomes to 1 ml of 20 mM PBS (pH 7.4). A 1-ml aliquot of this solution is placed in a quartz cuvette, and a fluorescence reading is made (F_0) using an excitation wavelength of 470 nm and an emission wavelength of 519 nm. Release is monitored by incubating the samples at various temperatures over the desired range, typically 30–50°. The most ideal way to perform these experiments is to place the sample cuvette into a temperature-controlled cuvette holder in a fluorimeter. If this is not possible, the samples can be placed in an external temperature-controlled water bath. Fluorescence measurements can be made at any desired time interval (F_t), e.g., every 5 min up to 30 min. If more frequent sampling is required (i.e., less than every 2–3 min), individual samples must be prepared and recorded separately to avoid photobleaching. (See doxorubicin release assay for a description of this procedure.) At the end of the experiment, the sample(s) is lysed with Triton X-100, and a 100% release measurement is made (F_{TX}). The percentage release of encapsulated dye is the calculated as in the stability section.

Figure 9 shows results from one of the lysolipid formulations [DPPC:MPPC(10%)] as percentage release of 6-CF versus time over 30 min at several temperatures. The formulation remains fairly stable up to 38° (15% released) and begins to release dye fairly quickly above 39°, with very rapid release at 40°. Release can also be measured in the presence of serum to more closely mimic *in vivo* conditions. The procedure given earlier is followed as outlined, but instead of diluting the 6-CF-loaded liposomes in PBS, a solution of 50% PBS and 50% bovine serum is used. Figure 10 shows the percentage of 6-CF released after a 5-min incubation at each temperature in both PBS and 50% serum. There is no effect of serum on the temperature at which release initiates (both near 39°), and the total amount of dye released is similar under both conditions.

Doxorubicin Loading and Release

This procedure is based on the pH gradient loading method described initially by Mayer et al.[51] and later by Madden et al.[52] with slight modifications. An ammonium sulfate gradient has also been used to successfully

[51] L. D. Mayer, M. B. Bally, and P. R. Cullis, *Biochim. Biophys. Acta* **857,** 123 (1986).

[52] T. D. Madden, P. R. Harrigan, L. C. L. Tai, M. B. Bally, L. D. Mayer, T. E. Redelmeier, H. C. Loughrey, C. P. S. Tilcock, L. W. Reinish, and P. R. Cullis, *Chem. Phys. Lipids* **53,** 37 (1990).

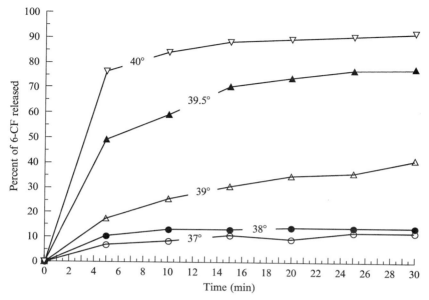

Fig. 9. Percentage of 6-CF released versus time in minutes for the DPPC:MPPC(10%): DSPE-PEG(2000)(4%) formulation over the temperature range 37–40°. Measurements were made every 5 min over the entire 30-min time course.

load DOX.[53] Thirty milligrams of lipid is dried onto a round-bottom flask using the evaporation method described earlier. This lipid film is hydrated with 3 ml (10 mg/ml lipid) of 300 mM, pH 4.0, citric acid buffer (Sigma). It is necessary to increase the pH of the citrate buffer from ~2.5 to 4.0 with NaOH before hydration. The liposomes are extruded through two 100-nm filters. This lipid suspension is cooled to room temperature and maintained below its transition temperature until drug loading. To establish the pH gradient as described in Madden et al.,[52] one of several methods can be used. The pH of the external solution can be increased with the addition of several drops of 1 M NaOH[37] or by diluting or dialyzing with a high pH buffer.[54] The most reproducible method we have found is to pass the liposomes through a Sephadex G-50 mini size exclusion column.[52] In this case, instead of using deionized water to hydrate the gel beads, 20 mM HEPES (Sigma) buffer at pH 7.4 is used (it is necessary to increase

[53] G. Haran, R. Cohen, L. K. Bar, and Y. Barenholz, *Biochim. Biophys. Acta* **1151,** 201 (1993).
[54] D. D. Lasic, B. Ceh, M. C. A. Stuart, L. Guo, P. M. Frederik, and Y. Barenholz, *Biochim. Biophys. Acta* **1239,** 145 (1995).

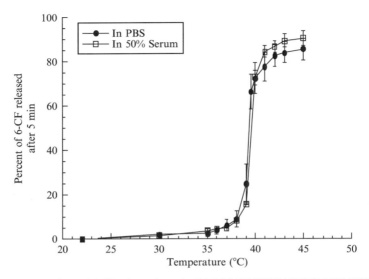

FIG. 10. Comparison of 6-CF release from the DPPC:MPPC(10%):DSPE-PEG(2000)(4%) formulation in PBS and in the presence of 50% bovine serum. Reprinted from Anyarambhatla and Needham[7] with permission.

the pH of the HEPES to 7.4 with a few drops of NaOH). When the lipo-somes are passed through two consecutive, different columns, the external pH 4.0 citrate buffer is exchanged for pH 7.4 HEPES buffer, while main-taining the pH 4.0 interior. The liposomes are now ready for drug loading. A DOX solution is prepared from doxorubicin hydrochloride (Sigma) at 5 mg/ml in 0.8% saline (0.8 g NaCl/100 ml in deionized water). The solu-tion is added to the liposome suspension at a 10:0.5, lipid:drug, weight ratio, i.e., for 20 mg of lipid (2 ml of 10 mg/ml extruded stock), 1 mg of drug (200 μl of a 5-mg/ml stock) is added. The exact amount of lipid in a sample can be determined by using the Bartlett assay, described in detail in New[17] and in Düzgüneş[55] For some formulations, DOX can be loaded up to a weight ratio of 10:2.7 (lipid:drug) and still maintain loading efficiencies of >95%,[56] but a lower ratio was chosen here (10:0.5) to ensure reproducible and maximum loading for the LTSL systems to ~62 mM.

The drug loads into the vesicles in response to the pH gradient that has been established across the liposome membrane. At pH 7.4, assuming a

[55] N. Düzgüneş, *Methods Enzymol.* **367**, 23 (2003).

[56] L. D. Mayer, M. B. Bally, P. R. Cullis, S. L. Wilson, and J. T. Emerman, *Cancer Lett.* **53**, 183 (1990).

Henderson–Hasselbach relationship, DOX (pK_a ~8.6[57]) is present in two species: ~90% membrane insoluble (protonated: DOX^+) and ~10% membrane-soluble (uncharged:DOX). The membrane-soluble form of DOX is able to cross the liposome membrane; once inside, it becomes protonated in the low pH environment and is trapped as DOX^+. It has also been shown that a solid, fibrous bundle crystal forms between DOX^+ and citric a acid molecules, which helps trap the drug inside the liposomes.[58] This loading produces a "sink" for the membrane-soluble DOX species that drives the conversion of DOX^+ to DOX from the equilibrium established outside the liposomes at pH 7.4. Following the drug incubation step (see later), the vesicles are cooled to room temperature and then passed through a Sephadex G-50 (prepared in HEPES) minicolumn to remove any unencapsulated drug.

The efficiency at which DOX can cross the liposome membrane is a function of the solubility of drug in the bilayer. By raising the temperature of the liposome suspension above T_m, the solubility of the drug in the membrane increases. In the case of pure DPPC liposomes (and other liquid crystalline lipid formulations), >95% drug loading can be achieved reproducibly by incubating at 60° for 10 min. In addition to drug solubility, it is also essential to maintain the established pH gradient to achieve maximum drug loading. Collapse of this gradient can occur at high drug-to-lipid ratios (>1.7:1, mol:mol, approximately 1.4:1 weight ratio) where the accumulated drug depletes the interior buffering capacity[59] or if hydrogen ions start to leak out of the liposomes. If this gradient drops to less than two pH units, the efficiency of encapsulation falls dramatically (<50%).[59]

In the case of our temperature-sensitive formulations, maintaining this pH gradient has been a primary concern, especially as a function of temperature. From our ion permeability measurements, we know that as soon as the liposome suspension reaches a temperature within a degree or so of the transition temperature, ions begin to cross the membrane rapidly. Therefore, once a pH gradient has been established, if a liposome suspension is heated above T_m to maximize drug solubility, the pH gradient is lost within a few seconds at T_m. We therefore utilize a method of loading the drug below its transition temperature. The pH gradient is established in the same manner as described previously, but incubation with the drug is carried out

[57] P. R. Harrigan, K. F. Wong, T. E. Redelmeier, J. J. Wheeler, and P. R. Cullis, *Biochim. Biophys. Acta* **1149,** 329 (1993).
[58] X. Li, D. J. Hirsh, D. Cabral-Lilly, A. Zirkel, S. M. Gruner, A. S. Janoff, and W. R. Perkins, *Biochim. Biophys. Acta* **1415,** 23 (1998).
[59] L. D. Mayer, L. C. L. Tai, M. B. Bally, G. N. Mitilenes, R. S. Ginsberg, and P. Cullis, *Biochim. Biophys. Acta* **1025,** 143 (1990).

at room temperature (23–25°) overnight (20–24 h) for the DPPC:MPPC (10%) formulation or at 37° for 20 min for the DPPC:MSPC (10%) formulation. Although the membranes are in a gel phase state, the presence of the lysolipid appears to increase the solubility of DOX in the membrane so that the drug is able to cross into the interior of the liposome. By maintaining the temperature well below the threshold for substantial ion permeability, the pH gradient is maintained and >95% drug loading is achieved.

To determine the amount of drug loaded, a simple fluorescence assay is used. When DOX loads into liposomes, it self-quenches much the same as 6-CF. Release of the drug into a larger, diluting volume results in an increase in the fluorescence signal.[36] This characteristic can be used to measure the release of the drug, as well as the encapsulation efficiency. To measure the percentage loading, 5 μl of the liposome suspension (pre-Sephadex column) is added to 1 ml HEPES buffer, and 5 μl of Triton X-100 is added to lyse the vesicles and release all of the drug. A fluorescence reading is made (F_{TX}) using a Shimadzu RF-1501 spectrofluorophotometer with an excitation wavelength of 470 nm and an emission wavelength of 555 nm. A second 5 μl sample (postcolumn) is then measured in the same manner. Percentage encapsulation is determined as

$$\% \text{ Encapsulation} = F_{TX}(\text{postcolumn})/F_{TX}(\text{precolumn}) \times 100 \quad (4)$$

In all cases using this procedure, the lysolipid temperature-sensitive formulations load greater than 95% of drug to a concentration of ~62 mM.

Doxorubicin Release Measurements

The ideal method to measure drug release from a liposome sample is to monitor a single sample heated to the temperature of interest over a set time interval. Unfortunately, in the case of DOX release from liposomes, this is not possible. DOX is an excellent choice for initial drug encapsulation and release measurements due to its well-established pH gradient-loading method and the fluorescence dequenching assay described earlier. Constant excitation of the drug, however, causes substantial photobleaching, and therefore it is not possible to expose the same sample to several fluorescence measurements in order to obtain time-dependent drug release data. To alleviate this concern, we use the following method. This method should also be used to monitor 6-CF release if time intervals of less than 2–3 min are used.

Drug release is monitored for a total of 30 min, taking measurements every 20 s for the first 5 min, then every 60 s for the next 13 min (18 total minutes), and finally every 120 s for the final 12 min, with each time point being a separate sample. Small (12 × 75-mm) glass test tubes (35 total tubes) are filled with 1 ml of HEPES buffer, pH 7.4, at room temperature (same buffer used during the drug-loading procedure). To each test tube, 5 μl of DOX-loaded liposomes is added and mixed gently by swirling. All of the test tubes are placed into a test tube rack, and the entire rack is submerged into a circulating water bath equilibrated to the temperature of interest. A timer is started, and a single test tube is removed (time 0). For the first 5 min, one test tube is removed every 20 s. Immediately upon removal, the test tubes are placed into an ice water bath to cool them quickly below T_m and to stop any subsequent drug release. In this manner, every test tube is a single time point, all the way up to the last 30-min time point. When the last test tube has been removed, each sample is read separately by placing the entire 1 ml into a quartz cuvette at room temperature and recording the fluorescence with ex$\lambda = 470$ nm, em$\lambda = 555$ nm (F_t). Six random samples are then chosen, lysed with 5 μl of Triton X-100, and a fluorescence recording is made. These six readings are then averaged for a final "100% released" value (F_{TX}). The percentage of drug release is then calculated, where, F_{t0} is the fluorescence of the zero time point sample.

$$\text{Percentage release} = [(F_t - F_{t0})/(F_{TX} - F_{t0})] \times 100 \qquad (5)$$

Percentage release from Eq. (5) is converted into moles of drug released as follows. One hundred-nanometer liposomes have an entrapped volume of approximately 1 μl/μmol of lipid.[58] Based on the initial amount of lipid, the total encapsulating volume can be estimated. Because the amount of drug added to the liposomes and the loading efficiency are both known, it is straightforward to calculate the amount of drug, in moles, encapsulated and then released. For the procedure described earlier, using 20 mg of lipid and 1 mg of drug loaded at 98% efficiency:

> Total volume of liposomes loaded = 27.2 μl
> Concentration of drug = 62.4 mM
> Total number of moles in a 5-μl aliquot = 3.86 × 10^{-9} mol

Data from these release measurements are presented in Fig. 11 as moles of DOX released versus time in seconds. The plot shows the release from DPPC liposomes at 37 and 42°, and from our DPPC:MSPC(10%) temperature-sensitive liposomes (LTSL) at 37, 39, 40, 41.3, and 45°. DPPC liposomes show very little release at 37° and release only approximately 15% of the total encapsulated amount at 42° (T_m) by the end of the experiment (30 min). In comparison, the LTSL formulation demonstrates good stability at 37° and at 39°C, where release is similar to DPPC at 42°. The release of DOX from the LTSL increases rapidly when the liposomes are heated to 40° and releases >80% of the encapsulated contents within 20–40 s at 41.3°

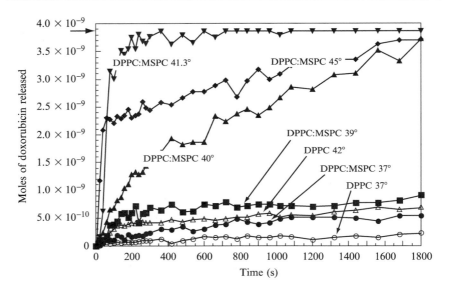

FIG. 11. Drug release measurements represented as moles of doxorubicin released versus time in seconds. Data are shown for DPPC:DSPE0-PEG(2000)(4%) at 37 and 42° and the DPPC:MSPC(10%):DSPE-PEG(2000)(4%) formulation at 37, 39, 40, 41.3, and 45°. The release of 3.86×10^{-9} mol of drug corresponds to 100% release of contents (short arrow).

(T_m). At 45°, a biphasic release is noted, where release is initially very rapid as the liposome passes through the transition region, but then slows when the vesicles melt completely above 42°.

To calculate release rates, the approach represented in Fig. 12 is used. Moles of DOX released (open circles in Fig. 12) are plotted on the left-hand Y axis versus time in seconds on the X axis. Plotted on the right-hand Y axis is the temperature of the sample (corresponds to solid line) showing the rate of heating as the samples are placed from room temperature into the circulating water bath equilibrated to 40° for this trial. Once the sample temperature reaches 40° (in about 40–60 s), a curve fit is used to determine the initial rate of release of the linear portion of data, as demonstrated by the short-dash line. The slope of this line, m1, is the rate of DOX released in moles per second when the sample reached the measuring temperature of 40°.

Figure 13 shows the calculated DOX release rates (moles per second) versus temperature for both DPPC liposomes and the DPPC:MSPC(10%): DSPE-PEG(2000)(4%) temperature-sensitive formulation. For DPPC, a slight peak in the release rate is noted at T_m (42°), which is only 5 times

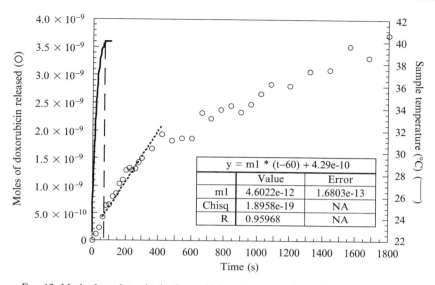

FIG. 12. Method used to obtain doxorubicin release rates in moles per second. Moles of doxorubicin released from DPPC:MSPC(10%):DSPE-PEG(2000)(4%) are plotted (O) versus time in minutes. Along the right-hand Y axis, the sample of the temperature is plotted as it is raised from room temperature to 40°. The long dash line indicates the time at which the sample reaches 40°, approximately 40–60 s. The initial linear portion of the release curve is then fit with a straight line (short dash line). The constant, m1, is the release rate in moles per second.

greater than the release rate at 37°, demonstrating a modest enhancement at the transition temperature above the relatively impermeable solid phase. In contrast, release from the DPPC:MSPC formulation at T_m is 55 times greater than at 37° and 20 times greater than pure DPPC at the same temperature (41.3°). This demonstrates the dramatic enhancement in the drug release rate that is achieved with the LTSL formulation. The same measurements can be repeated in the presence of bovine and human plasma to more closely mimic *in vivo* characteristics, but these methods provide a framework for characterizing any temperature-sensitive liposome.

Summary

This article outlined the methods used to characterize a temperature-sensitive liposome delivery system, including (1) differential scanning calorimetry; (2) liposome hydration and extrusion; (3) contents stability measurements using fluorescent dye; (4) dithionite ion permeability; and

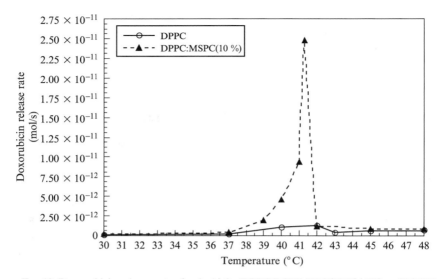

FIG. 13. Doxorubicin release rates (moles/s) for DPPC:DSPE-PEG(2000)(4%) and DPPC: MSPC(10%):DSPE-PEG(2000)(4%) formulation versus temperature. Lines connecting data are presented simply to aid the eye.

(5) fluorescent dye and drug loading and release. Using the materials engineering approach and these methods, we have successfully developed a temperature-sensitive, liposome-based drug delivery system that maintains drug contents stability at body temperature, but initiates a rapid, triggered release of contents at a temperature achieved easily with clinical hyperthermia. In preclinical and clinical tests in animals, 50 to 100 μM drug is deposited in solid tumors heated to $42°$ for just 1 h. With LTSLs circulating in the bloodstream and the tumor heated to $42°$ for 1 h 50% of this drug was bound to DNA and RNA in the tumor.[60] In other experiments, we have observed that tumor blood flow slows down in this first hour, and after 24 h there are no functioning blood vessels in some of the tumors. Although these new data have to be confirmed, the complete regression of tumors in nude mice and blood flow data point to a potentially important new treatment for solid tumors, where rapid drug release from LTSLs, triggered by mild hyperthermia, deposits therapeutically active concentrations of drug in the tumor vasculative and tumor tissue.

[60] G. Kong, G. Anyarambhatla, W. P. Petros, R. D. Braun, O. M. Colvin, D. Needham, and M. W. Dewhirst, *Cancer Res.* **60,** 6950 (2000).

Acknowledgments

We acknowledge support for this project from NIH Grants GM40162 and CA87630 and from NSF grant CDR-8622201 and thank our collaborators and co-workers, Mark Dewhirst, Gopal Anyarambhatla, and Garheng Kong.

[7] Tunable pH-Sensitive Liposomes

By Ismail M. Hafez and Pieter R. Cullis

Overview

Liposomes are membrane-enclosed vesicles composed of a lipid bilayer shell surrounding an aqueous core. Liposomes have historically been studied and developed as biophysical models of cellular membranes and as drug delivery systems—an application for which an appropriate response to a physiological stimulus such as bilayer lysis or lipid chemical breakdown is highly desirable. Smart liposome systems that are able to alter their permeability or change their chemical composition in response to external signals have utility in drug delivery applications[1] and as components of biosensor elements[2] in the emerging fields of nano and biotechnology.

One of the most attractive stimuli to exploit *in vivo* for drug delivery applications is that of local pH change. During the cellular process of endocytosis, early endosomal vesicles become acidified by the vacuolar-ATPase proton pump, producing a large vesicular proton gradient versus cytosolic pH.[3] The pH inside endosomal compartments can become as low as pH 5.5–4.5, sufficiently acidic to be exploited by microbial and viral agents to trigger mechanisms that allow the escape of pathogens and toxins from endosomal membranes and access to the cell interior. Liposome systems that contain pH-sensitive components can be designed to respond appropriately to low pH environments encountered during endocytosis.

pH-sensitive liposome systems are traditionally composed of a mixture of a titratable anionic lipid and a neutral or zwitterionic lipid.[4] The neutral lipids used, such as dioleoylphospatidylethanolamine (DOPE), adopt nonbilayer lipid phases in isolation, but mixtures with titratable anionic lipids, such as cholesteryl hemisuccinate (CHEMS), can be stabilized into

---

[1] O. V. Gerasimov, J. A. Boomer, M. M. Qualls, and D. H. Thompson, *Adv. Drug. Deliv. Rev.* **38,** 317 (1999).

[2] M. B. Esch, A. J. Baeumner, and R. A. Durst, *Anal. Chem.* **73,** 3162 (2001).

[3] B. Tycko and F. R. Maxfield, *Cell* **28,** 643 (1982).

[4] R. M. Straubinger, *Methods Enzymol.* **221,** 361 (1993).

a bilayer phase. The titratable anionic lipid can stabilize the nonbilayer lipid only in the ionized state and thus does so in a pH-dependent manner. The range of pH sensitivity of systems composed of mixtures of anionic and neutral lipids is, in general, dependent on the pKa of the anionic stabilizing lipid[5] and may also be altered slightly based on the molar ratio of the anionic and neutral lipid in the bilayers.[6] An alternative method for the preparation of tunable pH-sensitive liposomes has been described by our group.[7]

This article describes a novel type of pH-sensitive liposome system composed of a mixture of cationic and anionic lipids. In this system the pH of membrane instability can be predictably altered, simply by changing the binary ratio of the cationic and anionic components of the liposome membrane. This article discusses the unique polymorphic and fusogenic properties of tunable pH-sensitive liposome systems that derive from the presence of both cationic and anionic lipids in the same lipid membrane. Details of the design, preparation, and methods for characterization of several formulations of tunable pH-sensitive liposomes are highlighted.

Introduction

Equimolar Mixtures of Cationic and Anionic Lipids Form Nonbilayer Phases

Mixtures of cationic and anionic amphipaths dispersed in aqueous media assemble into aggregates with the charged head groups near the membrane/water interface forming charge neutralized ion pairs.[8] This property has been exploited to prepare bilayer vesicles from mixtures of cationic and anionic detergents such as cetyl trimethylammonium tosylate (CTAT) and sodium dodecyl benzene sulfonate (SDBS), each of which adopt micellar phases in isolation.[9] Bilayer vesicles can be formed from mixtures of single chain cationic and anionic surfactants in a range of charge ratios, making vesicles that are neutral, anionic, or cationic in net charge.[9]

Surprisingly, only until very recently have the phase properties of mixtures of bilayer-forming cationic lipids in mixtures with bilayer-forming anionic lipids been described. Physicochemical studies have revealed that

[5] D. F. Collins, F. Maxfield, and L. Huang, *Biochim. Biophys. Acta* **987,** 47 (1989).
[6] H. Ellens, J. Bentz, and F. C. Szoka, *Biochemistry* **24,** 3099 (1985).
[7] I. M. Hafez, S. Ansell, and P. R. Cullis, *Biophys. J.* **79,** 1438 (2000).
[8] S. Bhattacharya and S. S. Mandal, *Biochemistry* **37,** 7764 (1998).
[9] E. W. Kaler, A. K. Murthy, B. E. Rodriguez, and J. A. Zasadzinski, *Science* **245,** 1371 (1989).

mixtures of bilayer-forming cationic and anionic lipids exhibit unique poly-morphic phase behavior.[7,10–12] Specifically, mixtures of bilayer-forming cationic and anionic lipids adopt nonbilayer phases such as the inverted hexagonal (H_{II}) phase when mixed under conditions of zero surface charge. Figure 1 shows that this phenomenon holds true for mixtures of different cationic and anionic lipids.[12] Alternatively, mixtures of cationic and anionic lipid can exist as bilayer vesicles if stabilized sufficiently by strong bilayer forming lipids such as phosphatidylcholine[12] or in the presence of an excess of either the anionic or the cationic lipid component.[7] In the case of ioniz-able lipid species comprising either vesicular cationic or anionic lipid com-ponents, the net charge of the vesicle systems and the resultant phase properties are directly a function of pH and the cationic-to-anionic lipid molar ratio.

Tunable pH-Sensitive Lipid Phases and Liposomes from Mixtures of Cationic and Anionic Lipids

Ionizable Anionic Lipid Present in Excess over a Permanently Charged Cationic Lipid. Consider the properties of a binary mixture of an ionizable anionic lipid in molar excess over a monovalent cationic lipid containing a quaternary amine. In this case, the pH at which the liposome surface charge will be zero (and thus favors the formation of nonbilayer phases) is determined by the amount of ionized anionic lipid and the molar ratio of the cationic and anionic lipids. The pH at which the lipid mixture is neu-tralized (pH_n) occurs when the molar fraction of the anionic lipid ($X_{anionic}$ with an ionization constant pK_a) that is negatively charged equals the cat-ionic lipid ($X_{cationic}$) fraction of the membrane. Numerically, this is given by the modified Henderson–Hasselbach equation,[7]

$$pH_n = pK_a + \log_{10}[X_{cationic}/(X_{anionic} - X_{cationic})] \tag{1}$$

Equation (1) may be used to predict the pH when a mixture of an ionizable anionic lipid and a permanently charged cationic lipid become neutralized and thus prefer to adopt nonbilayer lipid phases over bilayer phases. Equation (1) shows that pH_n is theoretically in the range of $pK_a \pm 1.99$ for liposome formulations prepared from a mixture of a permanently charged cationic lipid and an ionizable anionic lipid with cationic/anionic lipid molar ratios ranging from 0.01 to 0.99. This suggests that tunable bilayer systems prepared in this manner can be designed to respond to a

[10] R. N. Lewis and R. N. McElhaney, *Biophys. J.* **79,** 1455 (2000).
[11] Y. S. Tarahovsky, A. L. Arsenault, R. C. MacDonald, T. J. McIntosh, and R. M. Epand, *Biophys. J.* **79,** 3193 (2000).
[12] I. M. Hafez, N. Maurer, and P. R. Cullis, *Gene Ther.* **8,** 1188 (2001).

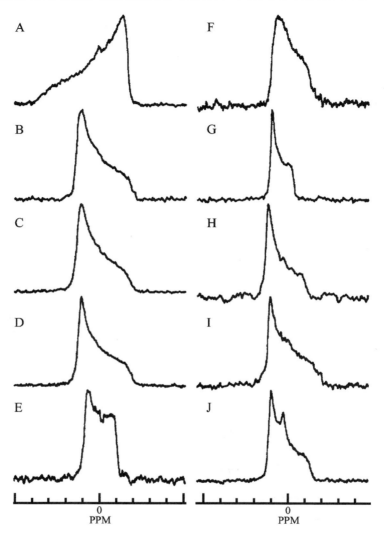

FIG. 1. Equimolar mixtures of cationic lipids and anionic phospholipids adopt the inverted hexagonal (H_{II}) phase. ^{31}P NMR spectra obtained from aqueous dispersions of (A) DOPS, (B) DODAC/DOPS, (C) DOTMA/DOPS, (D) DOTAP/DOPS, (E) DC-Chol/DOPS, (F) DODAC/DOPA, (G) DODAC/LBPA, (H) DODAC/CL (2:1; molar ratio), (I) DODAC/PI, and (J) DODAC/POPG. Cationic lipid abbreviations: N,N-dioleoyl-N,N-dimethylammonium chloride (DODAC), N-[1-(2,3-dioleoyloxy)propyl]-N,N,N-tri-methylammonium chloride (DOTMA), 1,2-dioleoyloxy-3-(trimethylammonio) propane (DOTAP), 3α-[N-(N',N'-dimethyl-aminoethane)-carbamoyl] (DC-Chol). Anionic lipid abbreviations: dioleoylphosphatidylserine (DOPS), dioleoylphosphatidic acid (DOPA), sn-(3-oleoyl-2-hydroxy)-glycerol-1-phospho-sn-

range of proton concentrations that differs up to 10,000×. It will be demonstrated that lipid vesicles composed of the anionic lipid CHEMS and the cationic lipid *N, N*-dioleoyl-*N,N*-dimethylammonium chloride (DODAC) adhere very closely to the behavior predicted by Eq. (1).

Ionizable Cationic Lipid in Excess over an Anionic Lipid. Ionizable cationic lipids such as 3α-[*N*-(*N',N'*-dimethylaminoethane)carbamoyl] cholesterol hydrochloride (DC-Chol) may be included into bilayer phases with mixtures of an anionic lipid such as dioleoylphosphatidic acid (DOPA). In this case, DOPA may be considered as the permanently charged lipid having an ionizable phosphate group with at least one negative charge at pH values above ∼4.0. With the DC-Chol in excess over the anionic lipid, bilayer phases are favored at acidic pH values when the net charge of the system is positively charged. In this system, bilayer stability is maintained at acidic pH and nonbilayer phases are favored under neutral and basic conditions. Interestingly with this system, some mixtures of DC-Chol/DOPA (e.g., 1.6 molar ratio) will adopt bilayer phases at both acidic and basic pH values when the system is either positively or negatively charged, and nonbilayer phases when neutralized.[7] Vesicle and bilayer phases that are prepared from these lipid mixtures may have application not only in the area of smart drug delivery systems, but also in the development of biosensor elements based on supported bilayer membranes.[13] However, this article restricts the content to the preparation of tunable pH-sensitive liposomes that are acid sensitive, as these present the most potential for intracellular drug delivery applications.

Materials

Lipids and Chemicals

DODAC is from Steven Ansell of Inex Pharmaceuticals Corporation (Burnaby, BC). 1,2-Dioeoyl-*sn*-glycero-3-phosphoethanolamine (DOPE), 1,2-dioleoyl-*sn*-glycero-3-phosphoserine (DOPS), 1,2-dioleoyl-3-trimethylammonium propane (DOTAP), 1-palmitoyl-2-oleoyl-*sn*-glycero-3-phosphocholine (POPC), 1,2-dioleoyl-*sn*-glycero-3-phosphoethanolamine-*N*-(7-nitro-2-1,3-benzoxadiazol-4-yl) (NBD-PE), 1,2-dioleoyl-*sn*-glycerol-3-phospho-L-serine-

[13] E. Sackmann, *Science* **271**, 43 (1996).

3'-(1'-oleoyl-2'-hydroxy)-glycerol (lysobisphosphatidic acid/LBPA), tetraoleoylcardiolipin (CL), phosphatidylinositol liver derived (Pl), 1-palmitoyl-2-oleoylphosphatidylglycerol (POPG). Reproduced from Hafez *et al.*[12] with permission.

N-(7-nitro-2-1,3-benzoxadiazol-4-yl) (NBD-PS), and 1,2-dioleoyl-*sn*-glycerol-3-phosphoethanolamine-*N*-lissamine rhodamine b sulfonyl (Rh-PE) are from Avanti Polar Lipids (Alabaster, AL). Cholesterol, cholesteryl hemisuccinate (morpholine salt), 2-[*N*-morpholino]ethanesulfonic acid (MES), Triton X-100, HEPES, and sodium dithionite are from Sigma Chemical Co. (St. Louis, MO). Lipids are checked for purity by thin-layer chromatography and are used if samples produce only one spot by phospholipid detection or sulfuric acid char.

Methods

General Preparation of Large Unilamellar Vesicles

Lipids are obtained as dry powders, dissolved in chloroform, and stored at $-20°$ until use. Large unilamellar vesicles (LUVs) of uniform size distribution are prepared by a freeze-thaw and extrusion technique.[14] Positive displacement pipettes that utilize glass capillary-like tips (Digital Microdispenser, VWR Scientific, West Chester, PA) are used to dispense accurately chloroform solutions of lipids in the 1- to 100-μl range. Lipids are codissolved in chloroform at desired molar ratios and dried to a thin film under a stream of nitrogen gas. Fluorescent lipids used for the preparation of labeled LUVs for the lipid-mixing assays are included at 1 mol% each; combinations are either NBD-PE/Rh-PE or NBD-PS/Rh-PE. Typically, between 5 and 10 μmol of lipid is dried in 13 × 100-mm borosilicate glass test tubes. The resulting thin films are placed in high vacuum for 1 to 2 h to remove residual solvent. Lipid films are hydrated in 1 to 2 ml of buffer with vigorous vortex mixing to form multilamellar vesicles (MLVs). Occasionally, some lipid mixtures do not hydrate well using this method. In these cases, the test tubes containing the dried lipid and buffer are frozen either by careful swirling in liquid nitrogen or placement in a $-80°$ freezer until the buffer and suspended lipid are frozen. Vortex mixing during the thawing of frozen samples allows the lipids to be removed from the walls of the test tubes. This preliminary freeze-thaw and vortex procedure is repeated several times if required. This method is very effective for hydrating lipid films that adhere strongly to the walls of the glass test tubes. Lipid suspensions are extruded through two stacked 0.2-μm pore-size polycarbonate filters using the extruder (Lipex Biomembranes, Vancouver, BC). The size of extruded samples is determined by quasi-elastic light scattering with a Nicomp C270 submicrometer particle sizer operating in particle size mode. LUVs are stored at $4°$ and used within 2 weeks. Standard freeze-fracture

---

[14] L. D. Mayer, M. J. Hope, and P. R. Cullis, *Biochim. Biophys. Acta* **858,** 161 (1986).

transmission electron microscopy methods are also used to characterize the liposome systems.

Preparation of Tunable pH-Sensitive Liposomes

Tunable pH-sensitive liposomes are prepared from binary mixtures of cationic and anionic lipids. The majority of our studies have used the mixture of cholesteryl hemisuccinate (CHEMS) and DODAC as the anionic and cationic amphipaths, respectively, although it is possible to prepare vesicles with similar properties from CHEMS and the commercially available DOTAP. Binary lipid mixtures contain only one species of anionic and cationic lipid with the exception of fluorescently labeled vesicles used in lipid mixing experiments. To prepare labeled LUVs, NBD-PE and LUVs, NBD-PE, and Rh-PE are included at 1 mol% each. Dried lipid films composed of the various formulations of DODAC/CHEMS are, hydrated with an aqueous buffer containing 150 mM NaCl, 10 mM HEPES, pH 8.1 and processed into LUVs using the freeze-thaw and extrusion methods described earlier.

Preparation of "Target" Liposomes

LUVs are prepared from mixtures of zwitterionic, neutral, and anionic lipids for use as target membranes in lipid mixing fusion assays for tunable pH-sensitive LUVs. Target LUVs are used to determine the extent of lipid mixing of DODAC/CHEMS vesicles with stable membranes that share no compositional similarity. Target membranes are composed of POPC, DOPS/POPC (30/70 mol%), and DOPS/cholesterol/POPC (30/30/40 mol%). LUVs are prepared as described earlier by freeze-thaw and extrusion in an aqueous buffer containing 150 mM NaCl, 10 mM HEPES, pH 7.8. For the preparation of labeled vesicles for lipid mixing assays, fluorescent lipids are included at 1 mol% each; the combinations are either NBD-PE/Rh-PE or NBD-PS/Rh-PE. Caution should be observed when using the fluorescent lipids NBD-PE, NBD-PS, and Rh-PE in target vesicles. These fluorescent lipids have negatively charged head groups and thus impart a negative charge to vesicles harboring them.

Preparation of Cationic and Conventional pH-Sensitive Liposomes

For comparison with tunable pH-sensitive liposomes, cationic liposomes composed of DODAC/cholesterol, DODAC/DOPE, and conventional pH-sensitive liposomes composed of DOPE/CHEMS are prepared. DODAC/DOPE and DODAC/cholesterol lipid films are hydrated in 10 mM Na-HEPES, pH 7.5, whereas DODAC/CHEMS and DOPE/CHEMS systems are prepared with in 150 mM NaCl, 10 mM HEPES, pH 8.1. These systems are processed into LUVs by freeze-thaw and extrusion. Cationic formulations of DODAC/DOPE and DODAC/cholesterol are both made at a molar ratio of 85:15. DOPE/CHEMS is formulated at 70/30 mol% to produce pH-sensitive liposomes, which destabilize at ~pH 5.[6]

Lipid Mixing Fusion Assays

A lipid mixing assay based on fluorescence resonance energy transfer (FRET) is used to assess lipid mixing properties of tunable pH-sensitive liposomes at various pH values.[15–17] To assay "self-fusion," lipid-mixing experiments are preformed with labeled and unlabeled DODAC/CHEMS vesicles of identical composition, with the exception that labeled vesicles contain 1 mol% NBD-PE and Rh-PE. The lipid mixing assay is used to measure the differential acid sensitivity of each of the DODAC/CHEMS formulations. To examine the interaction of DODAC/CHEMS vesicles with target membranes of different compositions, labeled DODAC/CHEMS vesicles are mixed with unlabeled target vesicles or vice versa.

Acid-Induced Lipid Mixing of DODAC/CHEMS Vesicles

NBD-PE/Rh-PE-labeled and unlabeled DODAC/CHEMS vesicles are mixed at a 1:5 ratio (labeled to unlabeled) and are injected into 2 ml of aqueous buffer (150 μM final lipid concentration) containing 140 mM NaCl, 10 mM HEPES, 10 mM MES, and 10 mM acetate set to desired pH values. The injected lipid is less than 2% of the total buffer volume. To prepare easily many buffers with different pH values, two solutions are made, a basic (pH ~ 8.0) buffer and an acidic (pH ~ 3.8) buffer. They are mixed appropriately to prepare buffers with desired pH values in a reproducible matter. NBD-PE fluorescence is monitored at 25° in a stirred cuvette on a Perkin-Elmer LS-50B fluoometer using excitation and emission wavelengths of 467 and 540 nm, respectively, with an emission filter of 530 nm. Lipid mixing is monitored for ~400 s. Complete dilution of the fluorescent probes is estimated by the addition of highly purified Triton X-100 (10%, v/v) to a final concentration of 0.1%. The pH of the buffer does not change with the addition of lipid or Triton X-100. The extent of lipid mixing is calculated using the equation: lipid mixing (%) = $(F_t - F_o)/(F_{max} - F_o) \times 100\%$, where F_t is the NBD-PE fluorescence during the time course, F_o is the initial NBD-PE fluorescence of the labeled sample,

[15] D. K. Struck, D. Hoekstra, and R. E. Pagano, *Biochemistry* **20**, 4093 (1981).
[16] D. Hoekstra and N. Düzgüneş, *Methods Enzymol.* **220**, 15 (1993).
[17] N. Düzgüneş, *Methods Enzymol.* **372**, 260 (2003).

and F_{max} is the NBD-PE fluorescence in the presence of 0.1% Triton X-100. Fluorescence of the NBD fluorophore is not sensitive to pH in the range used in our experiments.

Fusion with Target Membranes: Acid-Induced Lipid Mixing of DODAC/CHEMS Vesicles with Target Membranes

Unlabeled DODAC/CHEMS vesicles and NBD-PE/Rh-PE-labeled target vesicles are added to a final concentration of 75 μM each into 2 ml 140 mM NaCl, 10 mM HEPES, 10 mM MES, 10 mM acetate, with the pH set to desired values. Alternatively, NBD-PE/Rh-PE-labeled DODAC/CHEMS vesicles and unlabeled target vesicles (1:1 ratio, 150 μM total lipid) and mixed in 2 ml of 150 mM NaCl, 5 mM HEPES, pH 7.8, and the buffer is "acid shocked" with an aliquot of 1 M acetic acid. The extent of lipid mixing is assayed and calculated as described previously.

Inner Monolayer Membrane Fusion Assay

To assess whether lipid mixing of DODAC/CHEMS vesicles with target membranes encompasses a complete fusion reaction or only lipid mixing with the outer monolayer (hemifusion), a modification of the lipid mixing assay, which should only be sensitive to the dilution of the lipids of the inner monolayer, is used. The assay employs dithionite to selectively quench the fluorescence of the NBD-lipid in the outer monolayer of LUVs.[18] Changes in NBD-lipid fluorescence should only be due to the dilution of probe in the inner monolayer, provided that the NBD-lipid asymmetry is maintained. For this purpose, NBD-PS, a lipid conjugate that does not undergo membrane flip-flop, is utilized.[19] A method similar to this has been used to show inner monolayer lipid mixing in reconstituted liposomes harboring recombinant exocytotic membrane fusion proteins.[20]

Outer monolayer NBD-PS fluorescence is eliminated by chemical reduction using dithionite. Asymmetrically labeled LUVs are prepared by freeze-thaw and extrusion of a hydrated dispersion of DOPS/POPC/NBD-PS/Rh-PE (30/70/1/1 molar ratios) at a 20 mM lipid concentration in 150 mM NaCl, 10 mM HEPES pH 7.8. Sodium dithionite is dissolved freshly to a concentration of 1 M dithionite in 1 M Tris and added to the LUV suspension within 1 min of preparation. The dithionite solution is added to the 20 mM LUV suspension at a 1:10 volume ratio, resulting in

[18] J. C. McIntyre, D. Watson, and R. G. Sleight, *Chem. Phys. Lipids* **66**, 171 (1993).
[19] B. R. Lentz, J. Talbot, J. Lee, and L. X. Zheng, *Biochemistry* **36**, 2076 (1997).
[20] T. Weber, B. V. Zemelman, J. A. McNew, B. Westermann, M. Gmachl, F. Parlati, T. H. Sollner, and J. E. Rothman, *Cell* **92**, 759 (1998).

a dithionite to NBD-PS ratio of ~500:1. Upon addition of dithionite to the fluorescent lipid suspension, the color of the suspension changes instantly from a bright pink to a dull cherry color. The mixture is mixed by vortex mixing and is allowed to stand in the dark at room temperature for 10 min. Elution of the dithionite/liposome reaction mixture through a Sephadex CL-4B column (1 × 20 cm) with 150 mM NaCl, 5 mM HEPES, pH 7.8, exchanges the external buffer and removes the dithionite from the vesicle suspension. Vesicles elute in the void volume of the column, and the lipid concentration is estimated from the final dilution.

Quantitation of NBD of the outer monolayer NBD-PS reduction by dithionite is assessed using Rh-PE as an internal standard. The ratio of Rh-PE fluorescence emission (excitation 560 nm and emission 590 nm) to NBD-PS fluorescence emission (excitation 467 nm and emission 540 nm), before and after dithionite reduction, in the presence of 0.1% Triton X-100 is measured. For one reduction experiment the relative fluorescence ratio of Rh-PE/NBD-PS was 1.59 for untreated LUVs and 3.08 for dithionite-reduced LUVs. The total NBD-PS reduced in this LUV sample was ~51%. NBD-PS reduction by dithionite typically yielded between 50 and 51% of the original NBD-PS fluorescence. This result correlates well with the nearly equal distribution of lipid in the inner and outer monolayer of LUVs and is in agreement with a previous report using this method to assay bilayer membrane asymmetry.[21]

Inner monolayer-labeled target LUVs are studied with the lipid-mixing assay using mixtures of unlabeled DODAC/CHEMS, DOPE/CHEMS, DODAC/DOPE, and DODAC/cholesterol LUVs vesicles in a 1:1 lipid ratio with inner monolayer-labeled NBD-PS/symmetrically labeled Rh-PE target vesicles. DODAC/CHEMS vesicles and inner monolayer-labeled DOPS/POPC/Rh-PE/NBD-PS vesicles are diluted in 150 mM NaCl, 5 mM HEPES pH 7.5, to a final lipid concentration of 150 μM and the buffer is acidified by the addition of an aliquot of 1 M acetic acid. In the case of non-pH-sensitive cationic liposomes, lipid mixing between DODAC/DOPE or DODAC/cholesterol is initiated by injection of the cationic liposomes into a stirred solution of inner monolayer-labeled target vesicles (1:1 lipid ratio) to a final lipid concentration 150 μM. NBD-PS fluorescence is monitored at excitation and emission wavelengths of 467 and 540 nm, respectively. Inner monolayer lipid mixing is calculated as described earlier using F_{max} values obtained from the addition of Triton X-100, which is 0.5 the fraction of the total NBD fluorescence of the symmetrically labeled target vesicles. Lipid mixing (%) in inner monolayer-labeled

---

[21] C. Balch, R. Morris, E. Brooks, and R. G. Sleight, *Chem. Phys. Lipids* **70,** 205 (1994).

vesicles represents the relative extent of inner monolayer NBD-PS dilution. Lipid mixing (%) is calculated as described previously.

Formation of Liposomes from Mixtures of Cationic and Anionic Lipids

Mixtures of DODAC and CHEMS could be formulated into LUVs in mildly alkaline buffer at cationic-to-anionic lipid molar ratios less than one. The structure of DODAC/CHEMS dispersions was examined by freeze-fracture electron microscopy. Figure 2A shows that MLVs are formed upon hydration of DODAC/CHEMS (0.72 molar ratio), which could be extruded through 0.2-μm pore-size filters to give rise to uniformly sized LUV structures (Fig. 2B).

Table I gives the vesicle sizes of various DODAC/CHEMS formulations following extrusion through 0.2-μm pore-size filters. Lipid films composed of an equimolar mixture of DODAC and CHEMS cannot be extruded, yielding precipitates upon hydration. Vesicles composed of DODAC/CHEMS at a molar ratio of 0.85 have a larger mean diameter and larger size distribution than vesicles composed of lower DODAC/CHEMS molar ratios. The vesicle size is stable for at least 1 month when stored at 4°.

Tunable pH-Sensitive Fusion of DODAC/CHEMS Vesicles

The pH-dependent fusion properties of DODAC/CHEMS vesicles were examined using the lipid-mixing assay. The lipid-mixing properties of DODAC/CHEMS vesicles with molar ratios from 0 to 0.85 were measured as a function of pH. As shown in Fig. 3A, the lipid-mixing properties of DODAC/CHEMS vesicles prepared at a molar ratio of 0.11 show an onset of lipid mixing at pH 4.7 with maximum lipid-mixing occurring at approximately pH 3.8. In contrast, the lipid-mixing properties of vesicles with a DODAC/CHEMS molar ratio of 0.85 show an onset of lipid mixing at pH 6.8 with a maximum occurring at approximately pH 6.4 (Fig. 3B). Figure 3C summarizes lipid mixing data for each formulation; in Fig. 3D the data are normalized to the maximum lipid mixing. From normalized lipid-mixing data from Fig. 3D it is possible to calculate the pH at which half-maximum membrane fusion (pH_f) was observed and to compare this to the predicted pH of charged neutralization (pH_n) determined from Eq. (1). As shown in Fig. 4, the predicted pH_n values correlate well with the experimentally obtained values for pH_f when using value for the apparent pK_a of CHEMS of 5.8.[7] This supports the hypothesis that the fusion of DODAC/CHEMS vesicles proceeds upon surface charge neutralization and that the pH of fusion can be tuned by adjusting the molar ratio of the cationic and anionic lipids in the vesicles.

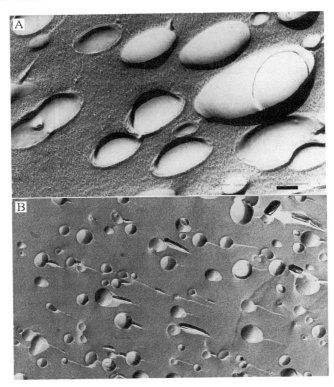

FIG. 2. Structural characteristics of aqueous dispersions of DODAC/CHEMS (0.72 molar ratio). The lipid was hydrated in 50 mM HEPES, 150 mM NaCl (pH 8.1), and freeze-fracture electron micrographs were prepared as in Hafez *et al.*[7] (A) DODAC/CHEMS MLVs formed on hydration of the lipid film. (B) DODAC/CHEMS LUVs formed after extrusion through two stacked filters with 0.2 μm pore size. Scale bar: 200 nm. Figure reproduced from Hafez *et al.*[7] with permission.

Fusion of DODAC/CHEMS Liposomes with "Target" Vesicles

Preparation of vesicles that only exhibit pH-sensitive lipid mixing or fusion with vesicles of identical composition are not particularly useful in terms of designing an efficient delivery vehicle that must destabilize or fuse with biological membranes in order to deliver its cargo to the cell interior. We hypothesized that the presence of the cationic lipid present in mixed DODAC/CHEMS liposomes would enhance the affinity of these vesicles for "target" membranes that have a negative surface charge. The cationic nature of the liposomes would only become apparent once a sufficient amount of anionic lipid was neutralized, thus exposing the positive charge

TABLE I
Size of Various Formulations of DODAC/CHEMS Vesicles
After Extrusion

DODAC/CHEMS (molar ratio)	Size following extrusion (nm)
0.11	142 ± 28^a
0.43	144 ± 30
0.52	152 ± 31
0.61	146 ± 32
0.72	153 ± 35
0.85	274 ± 94
1.0	N/A^b

[a] The size of DODAC/CHEMS LUV formulations following hydration in 10 mM HEPES, 150 mM NaCl, pH 8.1, and extrusion through two-stacked filters with 0.2 μm pore size. The mean diameter of the LUV systems was determined with a NICOMP C270 submicrometer particle sizer operating in the particle mode.
[b] Hydration of lipid films of DODAC/CHEMS mixtures at a molar ratio of 1.0 produced precipitates that could not be extruded.

of the cationic lipid in the membrane. Lipid-mixing assays were therefore performed to evaluate the pH-dependent fusion of DODAC/CHEMS vesicles with pH-stable target vesicles.

Figure 5 illustrates that upon acidification to pH 4.4, fluorescently labeled DODAC/CHEMS/NBD-PE/Rh-PE (DODAC/CHEMS molar ratio of 0.61) liposomes undergo substantial lipid mixing with target vesicles composed mainly of POPC, only if the target vesicle contain the anionic lipid phosphatidylserine (PS). Enhanced lipid mixing is observed with target membranes containing both PS and cholesterol. This result is not unexpected, as cholesterol is known to be able to induce negative membrane curvature and to promote the formation of inverted lipid phases and thus fusion in some lipid mixtures.[22] As a validation of the assay, no fluorescence changes are observed in the absence of target vesicles, indicating that the observed changes in fluorescence are due to dilution of the NBD-PE/Rh-PE probes from the DODAC/CHEMS liposomes into the target vesicle membranes. The pH change to 4.4 for this experiment is lower then that required for the self-fusion of DODAC/CHEMS vesicles (0.61 molar ratio; pH$_f$ 6.11) and was purposely chosen so that a large proportion of the anionic lipid should be neutralized, thus exposing the positive charge of the cationic lipid DODAC. The following set of experiments characterized the

[22] P. R. Cullis and B. de Kruijff, *Biochim. Biophys. Acta* **507,** 207 (1978).

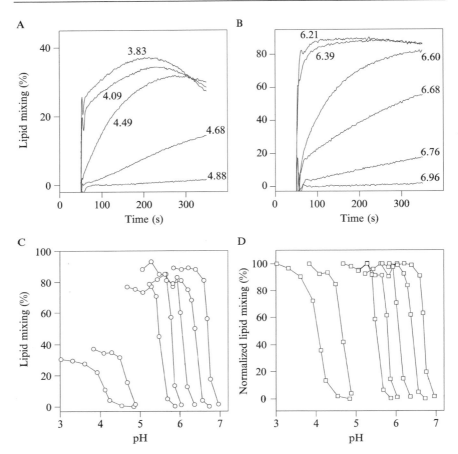

Fig. 3. pH-dependent fusion properties of DODAC/CHEMS LUVs containing increasing amounts of the cationic lipid DODAC determined by lipid mixing. Labeled and unlabeled (5:1 lipid ratio) DODAC/CHEMS LUVs were mixed and introduced into buffer with pH values indicated at 50 s. Lipid mixing traces are presented for DODAC/CHEMS molar ratios of (A) 0.11 and (B) 0.85. Summarized lipid-mixing data presented for DODAC/CHEMS molar ratios of 0, 0.11, 0.43, 0.52, 0.61, 0.72, and 0.85 (from left to right) in (C) actual lipid mixing (%) and (D) normalized lipid mixing (%). Lipid mixing (%) was determined as described in the text. Figure reproduced from Hafez et al.[7] with permission.

lipid-mixing properties of DODAC/CHEMS vesicles with anionic target vesicles as a function of pH and found that the pH required for fusion with a target membrane is lower than that required for self-fusion of DODAC/CHEMS liposomes.

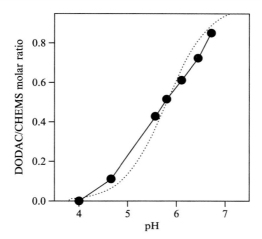

FIG. 4. Correlation between membrane fusion and surface charge neutralization of DODAC/CHEMS LUVs. The pH of half-maximum fusion (pH$_f$) values plotted as a function of the DODAC/CHEMS molar ratio *(solid circles)*. The pH at which the surface charge was predicted to be zero *(dashed line)* was determined from Eq. (1) using pK_{CHEMS} = 5.8. pH$_f$ was determined from Fig. 2 for each DODAC/CHEMS LUV formulation as the pH at which lipid mixing was 50% of the maximum observed lipid mixing value. Figure reproduced from Hafez et al.[7] with permission.

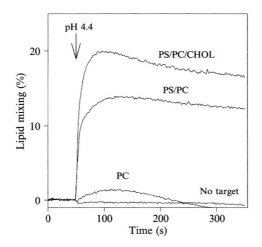

FIG. 5. Fusion properties of DODAC/CHEMS LUVs with target LUVs assayed by lipid mixing. NBD-PE/Rh-PE-labeled DODAC/CHEMS LUVs (0.61 molar ratio) were mixed with unlabeled POPC, DOPS/POPC (30/70 mol%), or DOPS/POPC/cholesterol vesicles (30/30/40 mol%) at a lipid ratio of 1:1 labeled to unlabeled vesicles as indicated on the graph. The pH was adjusted from 7.4 to 4.4 with an aliquot of 1 M acetic acid. Acidification of labeled DODAC/CHEMS LUVs in the absence of unlabeled target was also performed (no target).

Tunable pH-Sensitive Fusion of DODAC/CHEMS Liposomes
  with Anionic Target Membranes

The pH-dependence of lipid mixing of DODAC/CHEMS liposomes
with anionic target vesicles composed of labeled PS/PC (30/70 mol%)
was examined. DODAC/CHEMS (0.43 molar ratio) undergo lipid mixing
with PS/PC target vesicles at ~pH 5 and below (Fig. 6A), whereas
DODAC/CHEMS (0.72 molar ratio) does so below pH 6.2 (Fig. 6B). A
summary of the extent of lipid mixing with anionic target vesicles as a function of pH for each of the DODAC/CHEMS formulations tested is shown
in Fig. 6C. DODAC/CHEMS liposomes undergo lipid mixing with anionic
target vesicles, and the pH threshold for this event depends on the
ratio of cationic to anionic lipid in the membrane. The level of lipid mixing
for DODAC/CHEMS vesicles of molar ratio 0.11 is low and is due to the
small amount of cationic lipid in this formulation. This contention is
supported by the result that purely anionic CHEMS vesicles do not undergo lipid mixing with anionic target vesicles. Above a DODAC/CHEMS
molar ratio of 0.43 the extent of lipid mixing appears to decrease with
an increasing DODAC/CHEMS molar ratio (Fig. 6C). This effect is likely
an artifact of the slow mixing of the two populations of DODAC/CHEMS
and target vesicles. We have found that in the absence of anionic target
vesicles, DODAC/CHEMS vesicles fuse and undergo a phase transition
to the inverted H_{II} phase,[7] thus losing the ability to undergo a lipid
mixing reaction with anionic PS/PC vesicles. DODAC/CHEMS vesicles
likely fuse with membranes of like composition before adequate sample
mixing occurs between the DODAC/CHEMS vesicles and the anionic
target vesicle population. Normalization of lipid-mixing values given in
Fig. 6C allows for the determination of half-maximum lipid-mixing values,
pH_{50}, of the DODAC/CHEMS liposomes with anionic target membranes
(Fig. 6D).

A summary of the half-maximum values from lipid-mixing assays
for self-fusion (pH_f) and fusion with anionic target membranes (pH_{50}) is
given in Table II. Lipid mixing of DODAC/CHEMS with anionic target
membranes occurs with pH_{50} values between 0.4 and 0.53 pH units lower
than that required for self-fusion (pH_f). Fusion of all DODAC/CHEMS
vesicle formulations with anionic target liposomes occurs in a pH range
where DODAC/CHEMS vesicles are predicted to carry a net positive charge.
During the time course of lipid-mixing assays between DODAC/CHEMS
and anionic target vesicles, an increase in solution turbidity signaling
fusion of DODAC/CHEMS vesicles occurs at values near pH_f without lipid
mixing of DODAC/CHEMS vesicles with anionic target vesicles. These

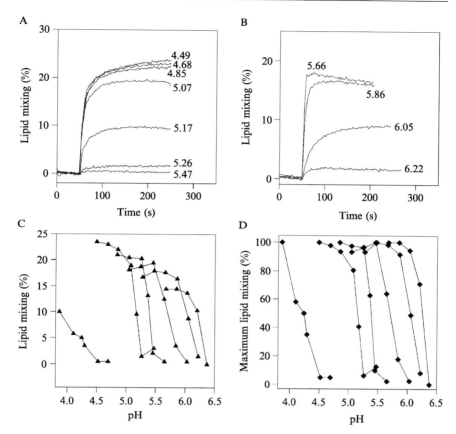

FIG. 6. Fusion properties of DODAC/CHEMS LUVs containing increasing amounts of DODAC with target DOPS/POPC (30/70 mol%) LUVs. Unlabeled DODAC/CHEMS LUVs were mixed with labeled target DOPS/POPC LUVs and introduced into buffer with the pH values indicated at 50 s. Lipid-mixing traces are presented for DODAC/CHEMS molar ratios of (A) 0.43 and (B) 0.72. Summarized lipid-mixing data presented for DODAC/CHEMS molar ratios of 0.11, 0.43, 0.52, 0.61, 0.72, and 0.85 (from left to right) with DOPS/POPC LUVs in (C) actual lipid mixing (%) and (D) normalized lipid mixing (%). Lipid mixing (%) was determined as described in the text.

results suggest strongly that fusion of DODAC/CHEMS vesicles with anionic target vesicles requires the exposure of a cationic lipid, which is masked by the anionic lipid until a sufficiently low pH is reached, allowing the presentation of the positive character of the DODAC/CHEMS vesicles.

TABLE II
Comparison of Predicted and Experimental Fusogenic Properties of
DODAC/CHEMS LUVs

DODAC/CHEMS (molar ratio)	Predicted pH of neutralization (pH_n)	Experimental 50% maximum "self" lipid mixing (pH_f)	Experimental 50% maximum lipid mixing with anionic target[a]
0	3.80	4.00	No lipid mixing
0.11	4.90	4.65	4.22
0.43	5.68	5.57	5.10
0.52	5.83	5.80	5.41
0.61	6.00	6.11	5.71
0.72	6.22	6.45	6.00
0.85	6.56	6.72	6.19

[a] Lipid mixing with an anionic target refers to fusion with anionic LUVs composed of DOPS/POPC (30/70 mol%).

Complete Fusion of DODAC/CHEMS Liposomes with Anionic Target Membranes

It was important to assess whether DODAC/CHEMS vesicles undergo hemifusion (outer monolayer lipid mixing only) with anionic target vesicles or a complete fusion event involving both inner and outer monolayers of target liposomes. To address this issue, a lipid-mixing assay that is sensitive only to probe dilution of the inner monolayer of target vesicles was utilized. The measurement of inner monolayer lipid mixing relies on using target vesicles that are labeled fluorescently with NBD-PS only on the inner monolayer. Inner monolayer lipid-mixing (%) values will reach those of symmetrically labeled vesicles if equal lipid dilution of the inner and outer monolayers occurs. DOPS/POPC/NBD-PS/Rh-PE vesicles are treated with dithionite to eliminate outer monolayer NBD-PS. The fluorescent lipid analogue NBD-PS is not prone to transmembrane redistribution (unlike NBD-PE) and can, therefore, be used to prepare stable asymmetrically labeled vesicles.[19] Changes in the fluorescence dequenching of NBD-PS in anionic vesicles should be due to lipid dilution of the inner monolayer of the target vesicles.

Figure 7 shows that DODAC/CHEMS liposomes are able to cause probe dilution in the inner membrane of target vesicles, but do so to various degrees depending on the DODAC/CHEMS vesicle composition. Vesicles with low cationic lipid content, e.g., DODAC/CHEMS (0.11, 0.42, and 0.52 molar ratios) induce inner monolayer lipid mixing that is not comparable to the total lipid mixing in symmetrically labeled targets, whereas

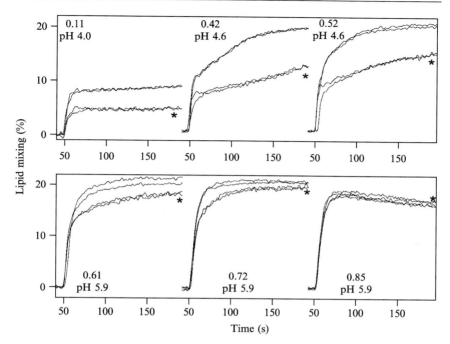

FIG. 7. Inner monolayer lipid mixing fusion assay. Lipid mixing of unlabeled DODAC/CHEMS vesicles was measured with symmetrically and inner monolayer-labeled NBD-PS/symmetrically labeled Rh-PE anionic DOPS/POPC (30/70 mol%) LUVs. The DODAC/CHEMS vesicle formulation is noted on each lipid-mixing trial as the DODAC/CHEMS molar ratio as well as the final pH after acidification. Inner monolayer lipid mixing traces are denoted by an asterisk. Two experimental trials are shown for each DODAC/CHEMS formulation for both inner monolayer labeled and symmetrically labeled target LUVs. Media were acidified at 50 s using an aliquot of 1 M acetic acid to the final pH value as indicated. Lipid mixing (%) was determined as described in the text.

DODAC/CHEMS vesicles with a high cationic lipid content (0.61, 0.72, and 0.85 molar ratios) produce nearly equivalent lipid mixing in both inner monolayer and symmetrically labeled PS/PC target vesicles (Fig. 7). This result indicates that vesicles composed of DODAC/CHEMS with high amounts of cationic lipid undergo a more complete membrane fusion reaction with anionic target vesicles than formulations with less cationic lipid.

. The inner monolayer lipid mixing assay employed was able to detect variances among the fusogenicity of DODAC/CHEMS formulations with target membranes that were not apparent from traditional lipid-mixing assays utilizing the symmetric distribution of fluorescent probes in labeled vesicles. We next turned our efforts toward comparing the fusogenic behavior of tunable pH-sensitive liposomes, which have intrinsic pH-sensitive and cationic characteristics, with commonly employed pH-sensitive liposomes and cationic liposomes.

Comparison of Tunable pH-Sensitive Liposomes with Traditional pH-Sensitive and Cationic Liposomes

We compared the properties of fusion between anionic target vesicles and tunable pH-sensitive DODAC/CHEMS liposomes, cationic liposomes composed of DODAC/cholesterol, DODAC/DOPE, and pH-sensitive liposomes composed of DOPE/CHEMS. The conventional and inner monolayer lipid-mixing assay was employed for this purpose. Data in Fig. 8 show that pH-sensitive DODAC/CHEMS and non-pH-sensitive DODAC/cholesterol and DODAC/DOPE undergo fusion with anionic target vesicles, with dilution of the inner monolayer lipids of target vesicles, indicating complete fusion (Fig. 8A–C). The amount of the cationic lipid present was the same for each of the formulations containing DODAC. As observed previously, DODAC/CHEMS LUVs undergo lipid mixing with anionic target vesicles only after an acidic stimulus is given (Fig. 8A). However, DODAC/cholesterol and DODAC/DOPE vesicles undergo lipid mixing immediately upon addition of the cationic vesicles to the PS/PC vesicle suspension, and this fusion reaction cannot be controlled (Fig. 8B and C). In sharp contrast, lipid mixing does not occur between pH-sensitive DOPE/CHEMS liposomes and anionic target vesicles even upon sample acidification (Fig. 8D). Clearly, DODAC/CHEMS vesicles have the ability to undergo lipid mixing with anionic target vesicles in a pH-sensitive manner, thus encompassing the best of the properties of both cationic liposomes and conventional pH-sensitive liposomes. It can be concluded that tunable pH-sensitive liposomes composed of mixtures of cationic and anionic lipid have properties that are highly desirable in a liposome delivery system. The vesicles exhibit pH-sensitive instability and exposure of a cationic surface charge that can promote fusion with anionic lipid membranes that approximate the composition of cellular membranes.

Conclusions

We have described the design and construction of a tunable pH-sensitive liposome system with the capacity to fuse with stable target anionic membranes. Tunable pH-sensitive liposomes represent a defined, non-phospholipid, synthetic membrane envelope that is able to respond to changes in surrounding pH. Subtle changes to the composition of the

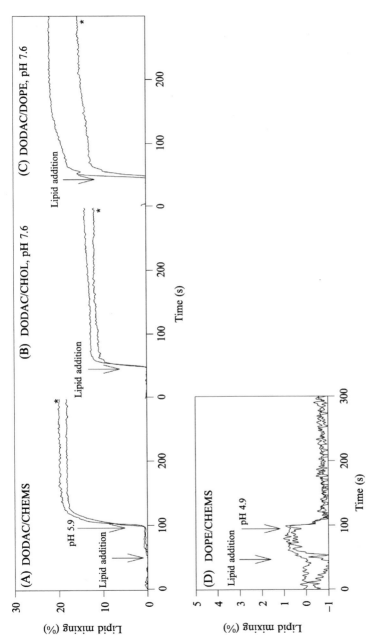

Fig. 8. Fusogenic properties of tunable pH-sensitive, cationic, and conventional pH-sensitive LUVs with anionic target LUVs. Symmetrically or inner monolayer-labeled NBD-PS (both with symmetrically labeled Rh-PE) anionic DOPS/POPC/NBD-PS/Rh-PE LUVs (30/70/1/1 molar ratio) were mixed with unlabeled (A) DODAC/CHEMS (0.85 molar ratio), (B) DODAC/cholesterol (0.85 molar ratio), (C) DODAC/DOPE (0.85 molar ratio), and (D) DOPE/CHEMS (70/30 mol%). Inner monolayer lipid-mixing traces are denoted by an asterisk. Unlabeled fusogenic LUVs were added to labeled target vesicles at 50 s. In the case of pH-sensitive DODAC/CHEMS and DOPE/CHEMS LUVs, the pH was acidified from pH 7.5 as indicated with an aliquot of 1 M acetic acid at 100 s. Lipid mixing (%) was determined as described in the text.

membrane allow for fine-tuning of the pH threshold of membrane instability. Tunable pH-sensitive liposomes are capable of being formulated into vesicles of defined size and are suitable for the entrapment of a variety of water-soluble molecules. Endosomal escape has been defined as a major barrier to successful drug and gene delivery.[23] Tunable pH-sensitive liposomes represent a promising new carrier system with the advantageous property of controllable pH-sensitive fusion with anionic membranes. These systems may be useful as a synthetic molecular device for overcoming the obstacles of endosomal escape and intracellular delivery. Based on a report by Shi *et al.*,[24] pH-sensitive liposomes composed of mixtures of cationic and anionic lipids show promise as efficient intracellular delivery systems.

[23] M. Nishikawa and L. Huang, *Hum. Gene Ther.* **12,** 861 (2001).
[24] G. Shi, W. Guo, S. M. Stephenson, and R. J. Lee, *J. Control. Rel.* **80,** 309 (2002).

[8] Sterically Stabilized pH-Sensitive Liposomes

By Vladimir Slepushkin, Sérgio Simões, Maria C. Pedroso de Lima, and Nejat Düzgüneş

Since their introduction in the 1960s,[1,2] liposomes have been considered as possible vehicles for delivering molecules into cells that would otherwise not be internalized readily. Realization of this potential, however, has been hampered because either (i) the liposomes are removed rapidly from the circulation by the reticuloendothelial system or (ii) they do not deliver large or highly charged molecules into cells.[3–5] The first of these handicaps has been overcome by the inclusion of poly(ethylene glycol)-derivatized phosphatidylethanolamine (PEG-PE) in the liposome membrane, resulting in much longer half-lives *in vivo,* compared to conventional liposomes.[6–8]

[1] A. D. Bangham and R. W. Horne, *J. Mol. Biol.* **8,** 660 (1964).
[2] A. D. Bangham, M. M. Standish, and J. C. Watkins, *J. Mol. Biol.* **13,** 238 (1965).
[3] A. Chonn and P. R. Cullis, *Curr. Opin. Biotechnol.* **6,** 698 (1995).
[4] T. M. Allen, *Trends Pharmacol. Sci.* **15,** 215 (1994).
[5] D. D. Lasic and D. Papahadjopoulos, *Science* **267,** 1275 (1995).
[6] T. M. Allen, *Adv. Drug Delivery Rev.* **13,** 285 (1994).
[7] D. Lasic and F. Martin, eds., "Stealth Liposomes." CRC Press, Boca Raton, FL, 1995.
[8] M. C. Woodle and D. D. Lasic, *Biochim. Biophys. Acta* **1113,** 171 (1992).

The second handicap has been addressed by developing liposomes that destabilize at mildly acidic pH (pH-sensitive liposomes) and deliver highly charged encapsulated materials into cells more efficiently than non-pH-sensitive liposomes.[9–11] After liposomes enter cells via endocytosis, the acidic pH inside the endosomes causes pH-sensitive liposomes to release their aqueous contents into the cytoplasm, most likely by destabilizing the endosome membrane.[9–11] Like other non-sterically stabilized liposomes, however, pH-sensitive liposomes have very short circulation times *in vivo*.[12]

Several groups have reported the development of pH-sensitive, sterically stabilized liposomes that deliver their contents into cells in a pH-dependent manner and have prolonged circulation *in vivo* (reviewed in Drummond *et al.*[13]). These liposomes were formulated by direct inclusion of a stabilizing polymer into a pH-sensitive lipid mixture,[14] by complexation of sterically stabilized liposomes with a pH-sensitive polymer,[15,16] or by mixing the destabilizing lipid, dioleoyl phosphatidylethanolamine (DOPE), with a polymer that degrades at mildly acidic pH.[17]

Preparation of Sterically Stabilized pH-Sensitive Liposomes

Relatively high amounts of aqueous phase materials can be encapsulated into liposomes by reverse-phase evaporation.[18–20] To make liposomes sensitive to low pH, DOPE is usually stabilized with a weak acidic lipid, such as cholesteryl hemisuccinate (CHEMS) or oleic acid (OA). Poly(ethylene glycol)-distearoyl phosphatidylethanolamine (PEG-DSPE) is added to render the liposomes sterically stabilized. Different preparation methods and

[9] N. Düzgüneş, R. M. Straubinger, P. A. Baldwin, and D. Papahadjopoulos, *in* "Membrane Fusion" (J. Wilschut and D. Hoekstra, eds.), p. 713. Dekker, New York, 1991.

[10] V. P. Torchilin, F. Zhou, and L. Huang, *J. Liposome Res.* **3**, 201 (1993).

[11] D. Collins, *in* "Liposomes as Tools in Basic Research and Industry" (J. R. Philippot and F. Schuber, eds.), p. 201. CRC Press, Boca Raton, FL, 1995.

[12] D. Liu, F. Zhou, and L. Huang, *Biochem. Biophys. Res. Commun.* **162**, 326 (1989).

[13] D. C. Drummond, M. Zignani, and J.-C. Leroux, *Prog. Lipid Res.* **39**, 409 (2000).

[14] V. A. Slepushkin, S. Simões, P. Dazin, M. S. Newman, L. S. Guo, M. C. Pedroso de Lima, and N. Düzgüneş, *J. Biol. Chem.* **272**, 2382 (1997).

[15] K. Kono, T. Igawa, and T. Takagishi, *Biochim. Biophys. Acta* **1325**, 143 (1997).

[16] O. Meyer, D. Papahadjopoulos, and J.-C. Leroux, *FEBS Lett.* **421**, 61 (1998).

[17] X. Guo and F. C. Szoka, Jr., *Bioconjug. Chem.* **12**, 291 (2001).

[18] F. Szoka and D. Papahadjopoulos, *Proc. Natl. Acad. Sci. USA* **75**, 4194 (1978).

[19] N. Düzgüneş, *Methods Enzymol.* **367**, 23 (2003).

[20] R. C. MacDonald, R. I. MacDonald, B. P. M. Menco, K. Takeshita, N. K. Subbarao, and L.-R. Hu, *Biochim. Biophys. Acta* **1061**, 297 (1991).

pH-sensitive lipid compositions were reported by others and were referenced earlier. As an example, this chapter describes the preparation of liposomes composed of DOPE, CHEMS, and PEG-DSPE.

Chloroform solutions of lipids [20 μmol total; molar ratio CHEMS: DOPE:PEG-PE (4:6:0.3)] are placed into glass tubes, and the solvent is evaporated under a stream of argon. Residues of solvent are removed in a vacuum oven. Lipids are dissolved in 0.78 ml of diethyl ether, and 0.26 ml of an aqueous solution is added. The osmolality of this solution should be about 300 mOsm [checked with a Wescor (Logan, UT) vapor pressure osmometer] and its pH should be slightly alkaline (7.5–8.0). When making pH-sensitive liposomes, one should take into account that carboxylic groups of the lipids used for liposome preparation can contribute to the pH of the solution. Therefore, higher concentrations of the buffer or a more alkaline initial pH should be used. It is usually convenient to check the pH of the final liposome preparation. Examples of different aqueous solutions that we have encapsulated successfully into liposomes are provided.

Liposomes are prepared by reverse-phase evaporation.[18,19] A water-in-oil emulsion is formed by light sonication in a bath-type sonicator (Laboratory Supply Co., Hicksville, NY) under an argon atmosphere in a tightly capped tube with a Teflon-coated screw cap. The ether is then evaporated under controlled vacuum (see Düzgüneş[19]). When the amount of ether becomes low, a gel phase is formed. The gel is broken by vortexing and the mixture is evaporated further, forming an aqueous suspension of liposomes. To render the liposomes of uniform size and to increase their circulation life *in vivo*, they are filtered through polycarbonate filters of uniform pore diameter. Usually the liposomes are extruded 21 times through two polycarbonate filters of 100 nm pore diameter (Costar, Cambridge, MA) using a LiposoFast device (Avestin, Inc., Ottawa, Canada).[20] Nonencapsulated water-soluble material can be removed if needed by dialysis or by gel filtration.

Characterization of pH-Sensitive Liposomes

In Vitro *Studies*

In vitro assays are generally used to characterize initially liposome stability as a function of the pH or the composition of the milieu in which the liposomes are suspended. Data collected *in vitro* can serve only as a preliminary characterization and must be confirmed by cell culture or *in vivo* experiments. The release of the aqueous phase from the liposomes can be monitored using either fluorescently or radioactively labeled small molecules. Fluorescent assays are more convenient because they can be performed without separation of liposomes from the leaked material. They are based on fluorescence dequenching,[21,22] which occurs when a highly concentrated fluorophore encapsulated inside liposomes leaks out and dilutes in the surrounding buffer. The increase in fluorescence intensity will measure the amount of marker leaked and, hence, the stability of the liposomes under given conditions. However, the radioactive assay is useful when fluorescent dye cannot be used due to its quenching in the media surrounding liposomes, e.g., in high concentrations of serum. This section describes examples of both assays.

Fluorescence Assay. Calcein is encapsulated in liposomes at a concentration of 80 mM in 10 mM TES buffer, pH 8.2, 1 mM EDTA, adjusted to 300 mOsm by NaCl. Nonencapsulated calcein is removed by dialysis at 4° against three changes of 4 liters of TES-buffered saline (TBS: 140 mM NaCl, 10 mM TES, pH 7.4) containing 0.1 mM EDTA. One microliter of calcein-loaded liposomes (final phospholipid concentration between 5 and 6 μM) is added to 2 ml of MES-buffered saline (140 mM NaCl, 10 mM MES) at various pH values in a fluorometer cuvette at 37° under constant stirring. After a 10-min incubation, calcein fluorescence is measured at $\lambda_{ex} = 490$ and $\lambda_{em} = 520$ nm using an LS-5B fluorometer (Perkin-Elmer, Mountain View, CA) operated with a Softways (Morena Valley, CA) computer program. Fluorescence intensities obtained at acidic pH values are corrected for the slight effect of pH on calcein fluorescence. To calibrate the assay, 100% leakage is achieved by the addition of Triton X-100 (final concentration 0.1%). The percentage of calcein leakage is calculated according to the formula: $[(I_{pH} - I_0)/(I_{100} - I_0)] \times 100\%$, where I_0 is the fluorescence at neutral pH, I_{pH} is the corrected intensity at acidic pH before the addition of Triton X-100, and I_{100} is the totally dequenched calcein fluorescence at neutral pH. The release of calcein from liposomes of various composition is shown in Fig. 1. The inclusion of DSPE-PEG at 3 mol% in DOPE:CHEMS liposomes reduces substantially the low pH-triggered release of calcein from liposomes.

Radioactivity Assay. Liposomes are made as described earlier, except that TBS containing 5 mM DTPA is used as the aqueous phase for the subsequent chelation of [111]In. Liposomes are extruded through polycarbonate filters of 100 nm pore size. The mean diameter of liposomes is

[21] J. N. Weinstein, S. Yoshikami, P. Henkart, R. Blumenthal, and W. A. Hagins, *Science* **195**, 489 (1977).

[22] T. M. Allen and L. G. Cleland, *Biochim. Biophys. Acta* **597**, 418 (1980).

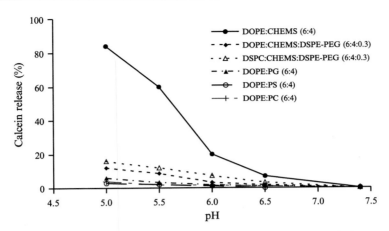

FIG. 1. The pH sensitivity of liposomes of various composition. Calcein-containing liposomes were incubated for 10 min at 37° in MES-buffered saline at different pH values. The fluorescence intensity of calcein was measured before and after the addition of Triton X-100.

measured by light scattering (Coulter Model N450, Coulter Corporation, Miami, FL). Unencapsulated DTPA is removed by dialysis against three changes of 4 liters of TBS overnight. Fifty microliters of ^{111}In-labeled oxine is added to 0.3 ml of liposomes, and the mixture is incubated for 1 h at room temperature for the irreversible chelation of ^{111}In to DTPA inside liposomes. After chelation of excess ^{111}In with 10 mM EDTA, liposomes are purified by gel filtration on a Sephadex G-75 column. Fifty to 95% of the added isotope is encapsulated, depending on the lipid composition.

To test the stability of liposomes in rat plasma, 5 μl of liposomes encapsulating ^{111}In is added to 95 μl of rat plasma and incubated for various times at 37°. The mixture is then applied to a Sephadex G-75 column, and the radioactivity of liposomal and free fractions is measured in a γ counter. Leakage of ^{111}In is calculated as the percentage of the isotope found in the free ^{111}In peak compared to the total radioactivity applied to the column.

Cell Culture Experiments

In vitro testing of liposomes can provide only preliminary guidance for their behavior *in vivo*. When endocytosed by cells, liposomes can interact with cellular proteins that can influence their properties. These interactions may render liposomes pH sensitive inside the cells, despite their low sensitivity to pH *in vitro*.[14]

Several approaches can be used to study the interaction of sterically stabilized pH-sensitive liposomes with cells. Some of these are based on the measurement of the release of a fluorescent marker from liposomes into the cell cytoplasm. The mechanisms of liposome uptake and leakage of their contents into the cytoplasm can be investigated by using inhibitors of endocytosis or of endosome acidification. Some of these inhibitors, including ammonium chloride, are nonspecific,[23] whereas others, such as bafilomycin A, inhibit specifically the proton ATPase in the endosome membrane.[24]

Fluorescence Microscopy. One of the easiest, although qualitative, ways to assess the release of liposomal contents inside the cells is fluorescence microscopy. Liposomes are loaded with an aqueous solution of a fluorescent marker, e.g., calcein. Different cells can be used for this assay. As an example, monocytic human THP-1 cells[25] are cultured in RPMI 1640 medium (Irvine Scientific, Santa Ana, CA) supplemented with 10% fetal bovine serum (FBS) and maintained, as described previously.[26] They are differentiated into macrophage-like cells by incubation with 160 nM phorbol 12-myristate 13-acetate (Sigma, St. Louis, MO) in Lab-Tek chambered cover glasses for tissue culture, obtained from Nunc (Naperville, IL) (10^6 cells per chamber). Five or 6 days after differentiation, the culture medium is replaced with fresh medium. Cells are washed with cold RPMI medium without phenol red, containing 20 mM HEPES buffer, pH 7.4. Liposomes are added to cells at a final phospholipid concentration of 200 μM and incubated for 1 h at 4°. After prebinding, cells are washed with cold medium, and the initial calcein and rhodamine fluorescence images are recorded using a Photon Technology International (PTI) ratio-imaging system. To evaluate the kinetics of calcein dequenching, medium at 37° is added to the chambers and cells are incubated at this temperature for various times. After incubation, cells are washed with cold medium, and the calcein and rhodamine fluorescence images are recorded. Cells are observed in a Nikon Diaphot epifluorescence microscope (Melville, NY) using a 100× objective and a filter for both fluorescein and Texas red. In some experiments, bafilomycin A$_1$ (200 nM) or a mixture of antimycin A

[23] C. J. Chu, J. Dijkstra, M. Z. Lai, K. Hong, and F. C. Szoka, *Pharm. Res.* **7**, 824 (1990).
[24] E. M. Bowman, A. Siebers, and K. Altendorf, *Proc. Natl. Acad. Sci. USA* **85**, 7972 (1988).
[25] S. Tsuchiya, Y. Kobayashi, Y. Goto, H. Okumura, S. Nakae, T. Konno, and K. Tada, *Cancer Res.* **42**, 1530 (1982).
[26] K. Konopka, E. Pretzer, B. Plowman, and N. Düzgüneş, *Virology* **193**, 877 (1993).

(1 μg/ml), NaF (10 mM), and NaN$_3$ (0.1%) is added to the cells 30 min before the liposomes to inhibit the acidification of endosomes[27] or to inhibit endocytosis,[28] respectively.

A fluorescence imaging system (Photon Technology International, Lawrenceville, NJ) is used to quantify the release of the fluorescent marker from the liposomes. This system allows measurement of the ratio between the aqueous florescent marker (calcein) and the lipid phase-embedded dye lissamine rhodamine B phosphatidylethanolamine (Rh-PE) in real time. As described earlier, liposomes encapsulating 80 mM calcein are added to the cells. For the ratio measurements, 1% of Rh-PE is added to the liposomal lipid at the chloroform solution stage. Calcein and rhodamine fluorescence images are acquired at various time intervals and then analyzed by the accompanying software to calculate the calcein-to-rhodamine ratio. Because it is possible to obtain a ratio of images with image analysis software, reflecting the global efficacy of liposome–cell interaction, we developed a novel methodology to quantitate this efficacy (Fig. 2A). Averages of 16 snapshots are taken to reduce the background and ratio images obtained from PTI software. Histograms of the calcein-to-rhodamine ratio are determined using a square of 100×100 pixels, representing the approximate area of one cell at the magnification used. Average histograms for each liposome composition and time point are then calculated using Microsoft Excel spreadsheets. Medians for each average histogram are calculated from cumulative curves and are used to follow the dequenching of calcein (Fig. 2 and Table I). To calculate the growth of the area occupied by dequenched calcein, the highest ratio in the control experiment, i.e., incubation at 4°, is taken as the cutoff. The sum of pixels with ratios higher than the cutoff ratio is taken to estimate the dequenched area at various times.[29] This method can thus be used to estimate the kinetics of leakage of liposome contents inside the cells under different experimental conditions.

Flow Cytometry. Flow cytometry is one of the convenient methods used to quantify the process of fluorescent marker delivery into cells. Again, calcein is encapsulated in liposomes at a high, self-quenching concentration, and incorporation of 1 mol% Rh-PE in the liposome membrane is used as a reference for the total amount of cell-associated liposomal lipid. For

[27] T. Yoshimori, A. Yamamoto, Y. Moriyama, M. Futai, and Y. Tashiro, *J. Biol. Chem.* **266,** 17707 (1991).

[28] K.-D. Lee, S. Nir, and D. Papahadjopoulos, *Biochemistry* **32,** 889 (1993).

[29] S. Simões, V. Slepushkin, N. Düzgüneş, and M. C. Pedroso de Lima, *Biochim. Biophys. Acta* **1515,** 23 (2001).

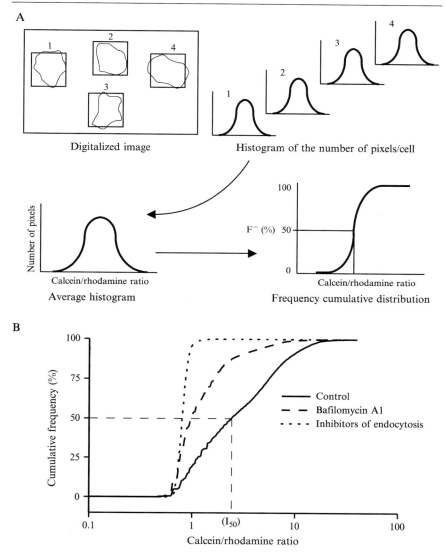

FIG. 2. Image analysis of the intracellular delivery of liposome contents. (A) Scheme of the methodology developed to quantitate intracellular delivery mediated by pH-sensitive liposomes based on the image analysis of fluorescence microscopy of cells. (B) Cumulative distribution curves of calcein/rhodamine ratio fluorescence intensities. DOPE:CHEMS liposomes (final phospholipid concentration of 200 μM) were incubated with THP-1 cells for 30 min at 37° in the absence or presence of bafilomycin A_1 or inhibitors of endocytosis. Curves were based on average histograms of 16 cells obtained from different experiments. The mode of calculation of the median (I_{50}) is illustrated for the case of untreated cells (Control). I_{50} corresponds to a cumulative frequency of 50%, i.e., 50% of the pixels exhibit an intensity equal to or lower than that value. Reproduced with permission from Simões et al.[29]

MEDIANS (I_{50}) OF CUMULATIVE DISTRIBUTION CURVES OF CALCEIN/RHODAMINE
RATIO FLUORESCENCE INTENSITIES

Experimental condition	Median (I_{50})
Control	2.47
Bafilomycin A_1	0.97
Inhibitors of endocytosis	0.8

flow cytometric analysis,[30] the cells are detached from plastic by adding 0.5 ml of dissociation buffer (Gibco BRL, Gaithersburg, MD) and are mixed with 0.5 ml of PBS with divalent cations containing 2% FBS and 1 μg/ml propidium iodide used to assess cell viability. Rhodamine and calcein fluorescence are detected with a Becton Dickinson (San Jose, CA) FACStar Plus flow cytometer, controlled by a Hewlett Packard computer with Lysis II software.[14] Samples are analyzed for lissamine rhodamine using excitation at 528 nm and emission at 575 nm and for calcein using excitation at 488 nm and emission at 520 nm with a 0.1 neutral density filter. Ten thousand events are recorded for each sample. Forward scatter and propidium iodide fluorescence signals are used to gate the cell subset of interest and to eliminate debris, dead cells, and cell aggregates.

Mean rhodamine fluorescence values reflect the binding and uptake of liposomes, whereas the mean calcein fluorescence reflects the intracellular dequenching of the dye. The calculated ratio of calcein to rhodamine fluorescence is taken to measure the amount of aqueous marker released intracellularly per cell-associated liposome. The initial calcein-to-rhodamine fluorescence ratio of liposomes bound to the cells, without any endocytosis, is obtained by pretreating the cells for 30 min with 1 μg/ml antimycin A, 10 mM NaF, and 0.1% NaN$_3$ to inhibit endocytosis.[28] Figure 3 shows the calcein/rhodamine fluorescence intensity ratios obtained with various liposome compositions incubated with macrophage-like differentiated THP-1 cells.[29]

[30] Z. Darzynkiewicz and H. A. Crissman, eds., "Flow Cytometry." Academic Press, San Diego, 1990.

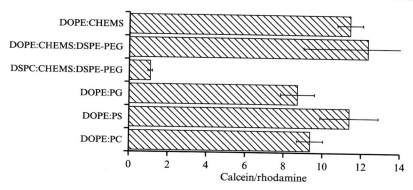

FIG. 3. Calcein/rhodamine fluorescence intensity ratios obtained with various liposome compositions incubated with macrophage-like differentiated THP-1 cells. Liposomes (100 μM final lipid concentration in the medium) encapsulating calcein and containing rhodamine-PE in the membrane were incubated with the cells in RPMI medium with 10% FBS for 30 min at 37°. Cells were washed twice with cell dissociation buffer and transferred to polypropylene tubes for flow cytometry. They were mixed with an equal volume of PBS with 2% FBS and 1 μg/ml propidium iodide (to gate the cell subset of interest and eliminate debris, dead cells and cell aggregates) and were subjected to flow cytometry as described in the text. Data represent the mean ±SD obtained from duplicates of three independent experiments. Reproduced with permission from Simões et al.[29]

In Vivo Studies

Preliminary estimates of liposome stability in vivo can be obtained from the serum stability experiments described earlier. If liposomes are leaking upon incubation with serum, they will most likely have a short half-life in the circulation. However, one cannot guarantee that liposomes stable in plasma in vitro will also have "stealth" properties in vivo.

To study the pharmacokinetics of pH-sensitive liposomes, ^{111}In-loaded liposomes, prepared as described earlier, are injected intravenously into rats.[14] At various times post-injection, animals are anesthetized, and blood samples are obtained. The concentration, $c(t)$, of ^{111}In is calculated as a function of time, t, and is given as the percentage of injected dose, $[c(t)/c(0)] \times 100\%$. A nonlinear, weighted, least-squares curve-fitting program (RSTRIP, Micromath, Salt Lake City, UT) is used to fit mean values of the percentage of injected dose in blood versus time data to calculate values for the area under the curve (AUC) and to estimate the terminal half-lives ($t_{i/2}$).

To evaluate liposome biodistribution, animals are sacrificed 24 h post-injection, and selected tissues are removed, weighed, and counted with a γ

counter to measure the amount of ^{111}In. The total radioactivity of urine collected in 24 h is estimated as well. The total label remaining *in vivo* gives an indication of the label retained in liposomes during circulation or delivered to tissues, as any [^{111}In]-DTPA released from liposomes is removed rapidly from blood by the kidneys. Sterically stabilized ("stealth") pH-sensitive liposomes stay in the circulation for a long time and deliver their contents efficiently into cells *in vivo*.

Use of Sterically Stabilized pH-Sensitive Liposomes for Delivery of Antisense Oligodeoxynucleotides

As one example of the possible utility for sterically stabilized pH-sensitive liposomes, this section describes the delivery of anti-HIV antisense oligodeoxynucleotides into cells. Anti-RRE (REV-responsive element) 15-mer phosphorothioate oligodeoxynucleotides (ODNs) are synthesized and provided by Lynx Therapeutics, Inc. (Hayward, CA). Sterically stabilized pH-sensitive liposomes (CHEMS/DOPE/PEG-PE, 4:6:0.3 molar ratio), regular pH-sensitive liposomes (CHEMS/DOPE, 4:6), and control non-pH-sensitive PG/DOPE liposomes (4:6) are made by the reverse-phase evaporation method.[18,19] Antisense ODNs are dissolved at a concentration of 0.2 mM in 100 mM HEPES buffer, pH 7.5, made isotonic with NaCl. Chloroform solutions of lipids (30 μmol) are mixed in a glass tube and dried under a stream of argon. Residues of the solvent are removed in a vacuum oven at room temperature. The dried lipids are dissolved in 780 μl of diethyl ether and are mixed with 260 μl of the ODN solution. The mixture is sonicated briefly under argon until a stable emulsion is formed, and the ether is removed in a rotary evaporator (Büchi, Flawil, Switzerland). After a gel is formed, an additional 240 μl of ODN solution is added, and the gel is broken by vortexing. Evaporation is continued for 30 min to remove the residues of ether. HEPES buffer without ODNs is used to prepare "empty" liposomes. Liposomes are extruded 21 times through two polycarbonate filters of 100 nm pore diameter using a LiposoFast device (Avestin, Inc., Ottawa, Canada) to obtain a uniform size distribution. Unencapsulated ODNs are removed, and the buffer is exchanged, by dialysis in Spectra/Por (Spectrum, Houston, TX) dialysis bags (MW cutoff: 50,000), against 10 mM HEPES-buffered isotonic saline, pH 7.4 (two changes of 4 liters of buffer for 20 h each at 4°). Liposomes are sterilized by filtration through 0.45-μm syringe filters (MSI, Westboro, MA).

Human macrophages are isolated from HIV-seronegative buffy coats (obtained from a local blood bank) by Ficoll–Hypaque (Histopaque 1077,

Sigma) gradient centrifugation and plastic adherence. Mononuclear cells separated by centrifugation are counted and plated in Dulbecco's modified Eagle's medium–high glucose (DME-HG) (Irvine Scientific, Santa Ana, CA) without serum supplemented with L-glutamine (4 mM), penicillin (100 units/ml), and streptomycin (100 μg/ml). Cells are allowed to adhere overnight, after which they are washed gently, and the medium is replaced with DME-HG with 20% FBS (Sigma) and 10% human AB serum (Advanced Biotechnologies, Inc., Columbia, MD). The cells are left undisturbed in this medium to differentiate for 5–6 days. Subsequent culture and experiments are carried out in DME-HG with 20% FBS and the usual supplements. The medium is replaced three times per week. HIV-1$_{BaL}$ is purchased from Advanced Biotechnologies, Inc., and is then propagated in macrophages as described before.[31] The number of cells remaining at

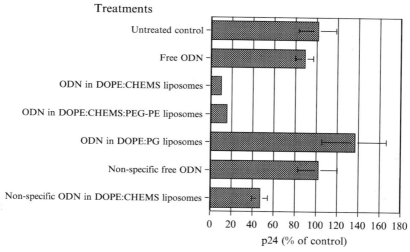

FIG. 4. Inhibition of HIV-1 production in human monocyte-derived macrophages by an anti-RRE 15-mer ODN encapsulated in pH-sensitive liposomes. Macrophages were infected with HIV-1BaL at a multiplicity of infection of 0.1 for 2 h at 37° and were washed with medium. Treatments were added 24 h after the infection and were maintained for 8 days with two medium replacements. Viral p24 levels in the supernatants were determined by ELISA. Reproduced with permission from Düzgüneş et al.[32]

[31] E. Pretzer, D. Flasher, and N. Düzgüneş, *Antiviral Res.* **34**, 1 (1997).

the end of the experiment is estimated by counting nuclei after staining the cells with Naphthol Blue Black.[33]

To evaluate the inhibition of HIV replication by ODNs, macrophages are cultured in 48-well plates. On day 7 after isolation they are infected with HIV-1$_{BaL}$ at a multiplicity of infection (MOI) of about 0.1 by incubation with 140–165 μl of virus-containing culture medium for 2 h at 37°. After infection, cells are washed with fresh DME-HG medium with 20% FBS, and treatments are added on the next day. The medium replacement on the following day is omitted, and further treatments are added concomitantly with the next two medium replacements. On day 8 after infection, fresh medium without treatments is added to the cells, and supernatants are saved for the subsequent analysis of p24 concentration by ELISA.[34] Viral p24 levels are then monitored in cell culture supernatants collected every 2–3 days.

Results demonstrate the effective delivery of antisense ODNs into cells by both pH-sensitive liposomes and sterically stabilized pH-sensitive liposomes.[32] In the free form the 15-mer anti-RRE phosphorothioate ODN was not active against HIV infection. When delivered in pH-sensitive liposomes, however, it inhibited p24 production by 42% at 1 μM and by 91% at 3 μM (Fig. 4). This relatively short ODN was also effective to a similar extent when encapsulated in sterically stabilized pH-sensitive liposomes (Fig. 4). Similar results were obtained with a 38-mer chimeric ribozyme complementary to HIV 5′-LTR.[32]

Concluding Remarks

Liposomes made of DOPE, CHEMS, and PEG-PE are stable *in vivo* and deliver their contents into cells in culture. Macromolecular antiviral agents encapsulated in the liposomes are highly effective in inhibiting HIV replication in human macrophages. Whether antisense ODN, ribozymes, or siRNA encapsulated in these liposomes will be effective in animal models or in the clinic will have to be determined in future experiments.

---

[32] N. Düzgüneş, S. Simões, V. Slepushkin, E. Pretzer, J. J. Rossi, E. De Clercq, V. P. Antao, M. L. Collins, and M. C. Pedroso de Lima, *Nucleosides Nucleotides Nucleic Acids* **20**, 515 (2001).

[33] A. Nakagawara and C. F. Nathan, *J. Immunol. Methods* **56**, 261 (1983).

[34] K. Konopka, B. R. Davis, C. E. Larsen, D. R. Alford, R. J. Debs, and N. Düzgüneş, *J. Gen. Virol.* **71**, 2899 (1990).

Acknowledgments

The studies from our laboratories described in this article were supported by Grants AI32399, AI33833, and AI35231 from the National Institute of Allergy and Infectious Diseases and by Grant FCT/BIO/36202/00 from the Foundation for Science and Technology of Portugal.

[9] Improved Preparation of PEG-Diortho Ester-Diacyl Glycerol Conjugates

By XIN GUO, ZHAOHUA HUANG, and FRANCIS C. SZOKA

Introduction

Low pH-sensitive liposomes[1–3] have been proposed as a triggered-release, drug delivery system capable of destabilizing biomembranes. Such systems could be useful for delivery into the cytoplasm of encapsulated compounds after the liposome is internalized into an endosome or lysosome. A recent advance in the design of pH-sensitive liposomes features the synthesis of cleavable surfactants whose hydrolysis is enhanced by elevated proton concentration.[4] Our group has reported the synthesis of a diortho ester conjugate of monomethyl poly(ethyleneglycol)2000 (mPEG-2000) and distearoyl glycerol (PODS2000).[5] At neutral pH, PODS2000 stabilizes the liposome during circulation in blood almost as much as conventional PEG-grafted lipids. At mildly acidic pHs, PODS2000 hydrolyzes rapidly, triggering a phase change of the phosphatidylethanolamine bilayer[5] and mediating the intracellular delivery of a cargo gene.[6] The biophysical mechanism of such a triggering event can be described by a "minimum surface shielding" model.[7]

These initial observations demonstrate the potential of PEG-diortho ester-diacyl glycerol (POD) conjugates for drug and gene delivery and

[1] N. Düzgüneş, R. M. Straubinger, P. A. Baldwin, and D. Papahadjopoulos, in "Membrane Fusion" (J. Wilschut and D. Hoekstra, eds.), p. 713. Dekker, New York, 1991.
[2] V. P. Torchilin, F. Zhou, and L. Huang, J. Liposome Res. 3, 201 (1993).
[3] D. C. Drummond, M. Zignani, and J. Leroux, Prog. Lipid Res. 39, 409 (2000).
[4] X. Guo and F. C. Szoka, Jr., Acc. Chem. Res. 36, 335 (2003).
[5] X. Guo and F. C. Szoka, Bioconjug. Chem. 12, 291 (2001).
[6] J. S. Choi, J. A. MacKay, and F. C. Szoka, Jr., Bioconjug. Chem. 14, 420 (2003).
[7] X. Guo, J. A. MacKay, and F. C. Szoka, Jr., Biophys. J. 84, 1784 (2003).

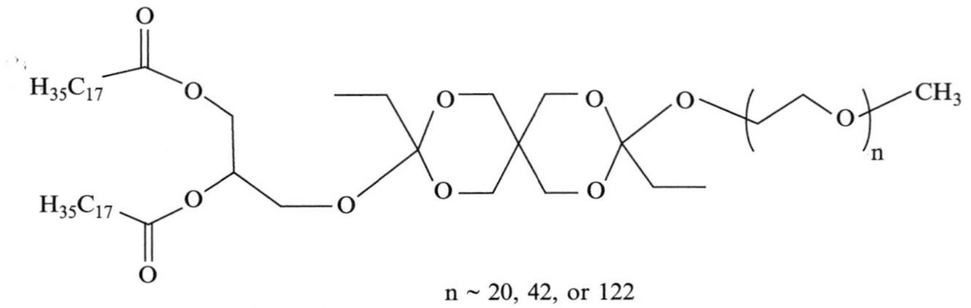

n ~ 20, 42, or 122

Fig. 1. Chemical structure of PODS.

prompted us to initiate a systematic structure–activity relationship on this category of surfactants. This article reports on the improved preparation of three POD conjugates Fig. 1, namely the mPEG2000-diortho ester-distearoyl glycerol conjugate (PODS2000, n ~ 42), the mPEG750-diortho ester-distearoyl glycerol conjugate (PODS750, n ~ 20), and the mPEG5000-diortho ester-distearoyl glycerol conjugate (PODS5000, n ~ 122).

Materials

The source of the chemical reagents and the general instrumentation procedures have been described previously[5] except for the following: the monomethyl ethers of PEG750 (mPEG750) and PEG5000 (mPEG5000) are from Sigma (St. Louis, MO) and ESIMS data of PODS750 and PODS5000 are obtained from an ion trap instrument (LCQDeca) from Thermo-Finnigan (Austin, TX) and calibrated according to the manufacturer's instructions.

The proper handling of the starting materials to exclude water is critical for the success of the syntheses. The diketene acetal, 3,9-diethylidene-2,4,8,10-tetraoxa-spiro[5,5]undecane, is stored under dry Ar in sealed round-bottom flasks at $-20°$. It is highly recommended that the diketene acetal be stored in small batches to avoid repetitive exposure of the diketene acetal to the atmosphere during removal of the material from the vial. Residual water in the PEG monomethyl ethers needs to be removed, most effectively by same-pot azeotropic distillation with anhydrous toluene, immediately before the reactions. Other drying methods used led to either lower yields (PODS2000) or failed reactions (PODS5000). Distearoyl glycerol is dried under high vacuum overnight in the presence of P_2O_5 before use. Distearoyl glycerol is used immediately after dissolution to avoid positional rearrangement of the acyl chains. Triethylamine is redistilled under Ar before use.

Synthesis of PEG-Diortho Ester-Diacyl Glycerol Conjugates

3,9-Diethyl-3-(2,3-distearoyloxypropyloxy)-9-[ω-methoxypoly(ethylene glycol)2000-1-yl]-2,4,8,10-tetraoxaspiro[5,5]undecane (PODS2000)

Polyethyleneglycol monomethyl ether (mPEG2000, MW 2000, 4 g, 2 mmol) is dried by azeotropic distillation with anhydrous toluene (50 ml) under Ar. A small amount of toluene is left with the residue, which is then cooled to 50°, followed by the addition of a solution of distearoyl glycerol (1.25 g, 2 mmol) in anhydrous THF (20 ml). The compound 3,9-bisethylidene-2,4,8,10-tetraoxaspiro[5,5]undecane is melted by heating gently with a heat gun, and 400 μl (ca. 424 mg, 2 mmol) of the melted compound is injected into the THF solution with a dry syringe. p-Toluenesulfonic acid in anhydrous THF (50 μl, 0.6 mg/ml) is added and the reaction mixture is stirred at 40° under Ar for 2 h. The reaction is stopped by adding 0.5 ml triethylamine and 20 ml methanol, and the reaction mixture is concentrated by rotary evaporation under reduced pressure. The residue is separated with a silica gel flash column (acetone/chloroform/triethylamine = 10/10/0.2). Fractions corresponding to the product are pooled, evaporated, and dried in high vacuum to give 1.83 g purified product. With the azeotropic distillation of mPEG2000 and the modified silica gel chromatography, the yield of purified PODS2000 is improved from 20[5] to 32.3%. TLC R_f 0.18 (acetone/chloroform/triethylamine = 10/10/0.2); FTIR 2910 cm^{-1} (CH$_2$ and CH$_3$), 2850 cm^{-1} (CH$_2$ and CH$_3$), 1743 cm^{-1} (ester C=O), 1109 cm^{-1} (PEG and orthoester C-O); ^{1}H NMR (400 MHz, CDCl$_3$, chemical shifts relative to TMS signal) δ 5.22 (1H, m, glycerol methine), 3.7–4.2 (4H, m, glycerol methylene), 3.2–3.7 ($\sim$170H, m, OCH$_2$ and OCH$_3$), 2.2–2.4 (4H, m, CH$_2$COO), 1.65–1.76 (4H, m, CH$_2$CH$_2$COO), 1.54–1.65 (4H, m, CH$_2$CH$_3$ on spiro rings), 1.04–1.36 [56H, m, CH$_3$(CH$_2$)$_{14}$CH$_2$CH$_2$COO], 0.82–0.97 (12H, m, CH$_2$CH$_3$); ESIMS, m/z calculated for [M+Na]$^+$ with 36–48 CH$_2$CH$_2$O units from PEG: C$_{123}$H$_{240}$O$_{46}$Na 2476.6, C$_{125}$H$_{244}$O$_{47}$Na 2520.7, C$_{127}$H$_{248}$O$_{48}$Na 2564.7, C$_{129}$H$_{252}$O$_{49}$Na 2608.7, C$_{131}$H$_{256}$O$_{50}$Na 2652.7, C$_{133}$H$_{260}$O$_{51}$Na 2696.8, C$_{135}$H$_{264}$O$_{52}$Na 2740.8, C$_{137}$H$_{268}$O$_{53}$Na 2784.8, C$_{139}$H$_{272}$O$_{54}$Na 2828.8, C$_{141}$H$_{276}$O$_{55}$Na 2872.9, C$_{143}$H$_{280}$O$_{56}$Na 2916.9, C$_{145}$H$_{284}$O$_{57}$Na 2960.9, C$_{147}$H$_{288}$O$_{58}$Na 3004.9, found 2478.3 (33%), 2521.1 (47%), 2565.3 (62%), 2609.3 (77%), 2654.2 (94%), 2697.3 (100%), 2741.4 (98%), 2785.3 (93%), 2830.4 (85%), 2874.4 (60%), 2917.5 (46%), 2962.4 (37%), 3006.4 (26%). Anal. calculate for C$_{135}$H$_{264}$O$_{52}$, C 59.62, H 9.78; HC$_{137}$H$_{268}$O$_{53}$, C 59.54, H 9.77; C$_{139}$H$_{272}$O$_{54}$,

C 59.46, H 9.76; $C_{141}H_{276}O_{55}$, C 59.39, H 9.76; $C_{143}H_{280}O_{56}$, C 59.31, H 9.75; $C_{145}H_{284}O_{57}$, C 59.24, H 9.74; $C_{147}H_{288}O_{58}$, C 59.17, H 9.73; found C 59.16, H 9.57, N < 0.2.

3,9-Diethyl-3-(2,3-distearoyloxypropyloxy)-9-[ω-methoxypoly(ethylene glycol)750-1-yl]-2,4,8,10-tetraoxaspiro[5,5]undecane (PODS750)

PODS750 is synthesized according to the procedure for PODS2000; yield 33%. TLC R_f 0.30 (acetone/chloroform/triethylamine = 10/10/0.2). ^{1}H NMR (400 MHz, $CDCl_3$, chemical shifts relative to TMS signal) δ 5.23 (1H, m, glycerol methine), 3.7–4.2 (4H, m, glycerol methylene), 3.5–3.75 (~90H, m, OCH_2 and OCH_3), 2.3–2.4 (4H, m, CH_2COO), 1.65–1.76 (4H, m, CH_2CH_2COO), 1.54–1.65 (4H, m, CH_2CH_3 on spiro rings), 1.2–1.3 [56H, m, $CH_3(CH_2)_{14}CH_2CH_2COO$], 0.82–0.97 (12H, m, CH_2CH_3). ESIMS, m/z calculated for $[M+NH_4]^+$ with 16–22 CH_2CH_2O units from PEG: $C_{83}H_{165}O_{26}N$ 1592.16, $C_{85}H_{169}O_{27}N$ 1636.18, $C_{87}H_{173}O_{28}N$ 1680.21, $C_{89}H_{177}O_{29}N$ 1724.24, $C_{91}H_{181}O_{30}N$ 1768.27, $C_{93}H_{185}O_{31}N$ 1812.29, $C_{95}H_{189}O_{32}N$ 1856.31, found (see Fig. 2) 1590.87 (47%), 1636.07 (71%), 1678.93 (86%), 1722.93 (100%), 1768.07 (74%), 1811.87 (52%), 1855.67 (35%). It is interesting to note a strong peak (m/z = 838) corresponding to the stabilized dialkoxycarbon cation fragment of PODS750, confirming the proposed mechanism of POD hydrolysis[5] (Fig. 2).

3,9-Diethyl-3-(2,3-distearoyloxypropyloxy)-9-[ω-methoxypoly(ethylene glycol)5000-1-yl]-2,4,8,10-tetraoxaspiro[5,5]undecane (PODS5000)

PODS5000 is synthesized according to the procedure for PODS2000 with slight modification on the purification method. The reaction mixture is first resolved partially by flash silica gel chromatography (acetone/chloroform/triethylamine = 10/10/0.2). Fractions corresponding to the pure PODS5000 are pooled, evaporated, and dried under high vacuum to give the purified product (12% yield). Additional pure PODS5000 can be obtained by further purification of column fractions corresponding to a mixture of PODS5000 and mPEG5000. Such fractions are pooled and evaporated, and the residue is dissolved in a minimum volume of 5 mM $(NH_4)_2CO_3$ aqueous solution (pH 9.0). The solution is applied to a Sephadex G-75 size-exclusion column (resins from Sigma) using 5 mM $(NH_4)_2CO_3$ aqueous solution (pH ~ 9.0) as the mobile phase. PODS5000 self-assembles into micelles in the aqueous mobile phase[8] and is eluted in

---

[8] K. Sou, T. Endo, S. Takeoka, and E. Tsuchida, *Bioconjug. Chem.* **11**, 372 (2000).

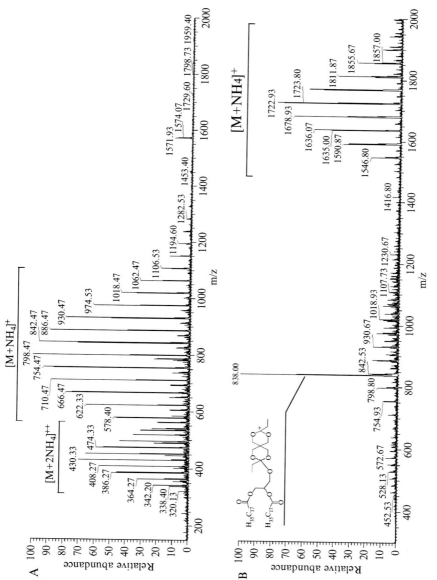

FIG. 2. Mass spectra of mPEG750 (A), PODS750 (B).

the void volume. mPEG5000 passes through the column more slowly and is removed readily from the product. Fractions of the resultant pure PODS5000 are pooled and lyophilized. Combined yield: 15%. TLC R_f 0.10 (acetone/chloroform/triethylamine = 10/10/0.2); ^{1}H NMR (400 MHz, CDCl$_3$, chemical shifts relative to TMS signal) δ 5.58 (1H, m, glycerol methine), 3.7–4.2 (4H, m, glycerol methylene), 3.5–3.6 (480H, m, OCH$_2$ and OCH$_3$), 2.3–2.45 (4H, m, CH$_2$COO), 1.65–1.76 (4H, m, CH$_2$CH$_2$COO), 1.39–1.45 (4H, m, CH$_2$CH$_3$ on spiro rings), 1.2–1.3 [56H, m, CH$_3$(CH$_2$)$_{14}$CH$_2$ CH$_2$COO], 0.80–0.95 (12H, m, CH$_2$CH$_3$); ESIMS, m/z calculated for [M+2H]$^{++}$ with 119–128 CH$_2$CH$_2$O units from PEG: [C$_{289}$H$_{574}$O$_{129}$]$^{++}$ 3057.1, [C$_{291}$H$_{578}$O$_{130}$]$^{++}$ 3079.1, [C$_{293}$H$_{582}$O$_{131}$]$^{++}$ 3101.1, [C$_{295}$H$_{586}$O$_{132}$]$^{++}$ 3123.2, [C$_{297}$H$_{590}$O$_{133}$]$^{++}$ 3045.2, [C$_{299}$H$_{594}$O$_{134}$]$^{++}$ 3167.3, [C$_{301}$H$_{598}$O$_{135}$]$^{++}$ 3189.3, [C$_{303}$H$_{602}$O$_{136}$]$^{++}$ 3211.3, [C$_{305}$H$_{606}$O$_{137}$]$^{++}$ 3233.3, [C$_{307}$H$_{610}$O$_{138}$]$^{++}$ 3255.3, [C$_{309}$H$_{614}$O$_{139}$]$^{++}$ 3277.3, found 3056.1 (37%), 3079.8 (71%), 3100.9 (70%), 3123.8 (100%), 3045.1 (82%), 3168.8 (94%), 3189.4 (68%), 3211.8 (86%), 3234.4 (78%), 3256.8 (85%), 3278.4 (55%).

Concluding Remarks

We are carrying out physical chemistry studies on these PODS conjugates, and preliminary data suggest that the size of the PEG head group plays a pivotal role in the phase behavior of the conjugates and their colloidal formulations with other lipids. This is consistent with the literature on conventional PEG–lipid conjugates.[9]

Acknowledgments

We gratefully acknowledge the generous gift of the diketene acetal from Dr. Jorge Heller of A. P. Pharma (Redwood City, CA). Support was provided by NIH grant GM61851 (FCS).

[9] F. K. Bedu-Addo, P. Tang, Y. Xu, and L. Huang, *Pharm. Res.* **13**, 710 (1996).

[10] Preparation of Plasmenylcholine Lipids and Plasmenyl-Type Liposome Dispersions

By David H. Thompson, Junhwa Shin,
Jeremy Boomer, and Jong-Mok Kim

Introduction

Plasmalogens are naturally occurring phospholipids containing *sn*-1-*Z*-1′-*O*-alkenyl chains of varying lengths and degrees of unsaturation that comprise as much as 18% of the total phospholipid mass in humans.[1] Modification of the *sn*-3-glycerol phosphate with choline or ethanolamine produces the corresponding plasmalogen variants known as plasmenylcholine or plasmenylethanolamine, respectively. These electron-rich vinyl ether phospholipids are found in the electrically active tissues of mammals, most prominently in brain, myelin, and heart.[2] Plasmenylcholines (PlsC) are known to be an important depot for arachidonic acid in heart tissue.[3,4] Human nervous system tissues, especially brain and myelin sheath, contain high concentrations of plasmenylethanolamine[5] and polyunsaturated fatty acids (PUFA). A biosynthetic pathway separate from that responsible for diacylglycerophosphocholine production exists for plasmalogen formation.[6] Genetic defects in this pathway are known to be the basis for peroxisomal disorders such as Zellweger's syndrome.[7]

Despite their extensive distributions in mammalian tissues, the biological role of plasmalogens remains unclear. Three hypotheses have been proposed to rationalize their existence. The first is based on the role of plasmalogens in arachidonate storage and signal transduction.[8–10] Gross and co-workers[11,12] have proposed, based on two-dimensional nuclear

---

[1] T.-C. Lee, *Biochim. Biophys. Acta* **1394**, 129 (1998).
[2] T. Sugiura and K. Waku, *in* "Platelet Activating Factor and Related Lipid Mediators" (F. Snyder, ed.), p. 55. Plenum Press, New York, 1987.
[3] H. W. Mueller, A. D. Purdon, J. B. Smith, and R. L. Wykle, *Lipids* **18**, 814 (1983).
[4] R. W. Gross, *Biochemistry* **23**, 158 (1984).
[5] L. A. Horrocks and M. Sharma, *in* "Phospholipids 4" (J. N. Hawthorne and G. B. Ansell, eds.), p. 51. Elsevier Biomedical Press, Amsterdam, 1982.
[6] N. Nagan and R. A. Zoeller, *Prog. Lipid Res.* **40**, 199 (2001).
[7] A. Poulos, A. Bankier, K. Beckman, D. Johnson, E. F. Robertson, P. Sharp, L. Sheffield, H. Singh, S. Usher, and G. Wise, *Clin. Genet.* **39**, 13 (1991).
[8] R. A. Wolf and R. W. Gross, *J. Biol. Chem.* **260**, 7295 (1985).
[9] A. N. Fonteh and F. H. Chilton, *J. Immunol.* **148**, 1784 (1992).
[10] F. Snyder, T.-C. Lee, and M. L. Blank, *in* "Advances in Lipobiology" (R. W. Gross, ed.), Vol. 2, p. 261. JAI Press, 1997.

magnetic resonance (NMR) evidence showing that plasmenylcholine (PlsC) lipids have a different glycerol backbone conformation with respect to the membrane interface than conventional diacyl-modified phospholipids that the Z-1-alkenyl ether bond provides a unique conformational motif at the membrane surface. This unique conformation motif is selectively recognized by phospholipase A_2 enzymes that are responsible for liberating arachidonic acid and initiating the arachidonate cascade. The second hypothesis—supported by mounting evidence from several laboratories,[13–23]—suggests that the alkenyl ether bond is present to serve as a sacrificial trap for reactive oxygen species that would otherwise attack the PUFA residues present at the nearby sn-2 position. Unique membrane protein activities observed in the presence of plasmalogens have led to the third proposal involving plasmalogen-specific membrane protein interactions.[1]

Plasmenylethanolamine[24] and plasmenylcholine[25,26] vesicles are also known to undergo lamellar-to-hexagonal II and lamellar-to-hexagonal I phase transitions, respectively, upon exposure to acidic and oxidative environments, suggesting that they may also be involved in membrane–membrane fusion processes. These attractive properties have led to the application of 2-palmitoyl-sn-glycero-3-phosphocholine (PPlsC), 1,2-di-O-(1′-Z-hexadecenyl)-sn-glycero-3-phosphocholine (DPPlsC), and a variety of plasmenyl-type cationic or polyethylene glycol (PEG)-modified lipids in the delivery of low molecular weight drugs,[27,28] proteins,[29] and

[11] X. Han and R. W. Gross, *Biochemistry* **29**, 4992 (1990).

[12] X. Chen, X. Han, and R. W. Gross, *Biochim. Biophys. Acta* **1149**, 241 (1993).

[13] R. A. Zoeller, O. H. Morand, and C. R. H. Raetz, *J. Biol. Chem.* **263**, 11590 (1988).

[14] G. Jürgens, A. Fell, G. Ledinski, Q. Chen, and F. Paltauf, *Chem. Phys. Lipids* **77**, 25 (1995).

[15] W. Jira and G. Spiteller, *Chem. Phys. Lipids* **79**, 95 (1996).

[16] G. Hofer, D. Lichtenberg, G. M. Kostner, and A. Hermetter, *Clin. Biochem.* **29**, 445 (1996).

[17] T. Brosche and D. Platt, *Exp. Gerontol.* **33**, 363 (1998).

[18] R. A. Zoeller, A. C. Lake, N. Nagan, D. P. Gaposchkin, M. A. Legner, and W. Lieberthal, *Biochem. J.* **338**, 769 (1999).

[19] P. J. Sindelar, Z. Guan, G. Dallner, and L. Ernster, *Free Radic. Biol. Med.* **26**, 318 (1999).

[20] N. Khaselev and R. C. Murphy, *Free Radic. Biol. Med.* **26**, 275 (1999).

[21] R. C. Murphy, *Chem. Res. Toxicol.* **14**, 463 (2001).

[22] R. A. Zoeller, T. J. Grazia, P. LaCamera, J. Park, D. P. Gaposchkin, and H. W. Farber, *Am. J. Physiol.* **283**, H671 (2002).

[23] D. H. Thompson, H. D. Inerowicz, J. Groves, and T. Sarna, *Photochem. Photobiol.* **78**, 323 (2003).

[24] P. E. Glaser and R. W. Gross, *Biochemistry* **33**, 5805 (1994).

[25] D. H. Thompson, O. V. Gerasimov, J. J. Wheeler, Y. Rui, and V. C. Anderson, *Biochim. Biophys. Acta* **1279**, 25 (1996).

[26] O. Gerasimov, A. Schwan, and D. H. Thompson, *Biochim. Biophys. Acta* **1324**, 200 (1997).

[27] Y. Rui, S. Wang, P. S. Low, and D. H. Thompson, *J. Am. Chem. Soc.* **120**, 11213 (1998).

[28] M. M. Qualls and D. H. Thompson, *Int. J. Cancer* **93**, 384 (2001).

[29] M. M. Qualls, Ph.D. Thesis. Purdue University, 2001.

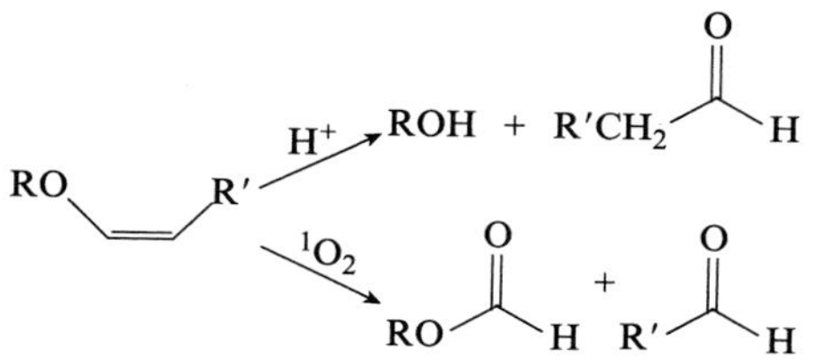

FIG. 1. Photooxidative and acid-catalyzed cleavage reactions of vinyl ether-containing lipids.[31]

genes[30] using triggering mechanisms that are activated by acidic or oxidative environments (Fig. 1).[31–34] The consequences of these degradative triggering reactions must be recognized in the handling of plasmenylcholine and diplasmenylcholine vesicle dispersions and precautions taken to avoid their premature decomposition due to adventitious exposure to oxidants or acidic media (i.e., <pH 6.5). In particular, given their role as endogenous antioxidants, plasmalogen derivatives must be viewed as especially prone to oxidation reactions that can produce fatty aldehyde, fatty acid, lysolipid, and allylic hydroperoxide degradation products (Fig. 2).[23] DPPlsC liposomes demonstrate good plasma stability at 37° relative to other pH-sensitive liposome formulations[27,33]; however, their practical utility has been limited by the difficulties encountered in their large-scale synthesis.

Naturally occurring plasmalogen mixtures are generally isolated from phosphocholine extracts derived from a number of animal sources. Because these extracts contain mixtures of alkyl chain lengths at the sn-1 and sn-2 positions, they are difficult to obtain as discrete molecular species without extensive HPLC purification, greatly limiting their utility in biophysical studies and drug delivery applications. This situation can be remedied partially by the preparation of semisynthetic plasmenylcholine using a base-catalyzed hydrolysis, lysolipid isolation, and reacylation procedure.[35] Advances have led to the discovery of several preparative routes for the total synthesis of plasmalogens wherein each element of the lipid structure can be controlled. Rui and Thompson developed the first synthetic

[30] J. A. Boomer, D. H. Thompson, and S. Sullivan, *Pharm. Res.* **19**, 1289 (2002).
[31] D. H. Thompson, Y. Rui, and O. V. Gerasimov, *in* "Vesicles" (M. Rosoff, ed.), p. 679. Dekker, New York, 1996.
[32] J.-M. Kim and D. H. Thompson, *in* "Reactions and Synthesis in Surfactant Systems" (J. Texter, ed.), p. 145. Dekker, New York, 2001.
[33] O. Gerasimov, J. A. Boomer, M. M. Qualls, and D. H. Thompson, *Adv. Drug. Del. Rev.* **38**, 317 (1999).
[34] P. Shum, J.-M. Kim, and D. H. Thompson, *Adv. Drug Del. Rev.* **53**, 273 (2001).
[35] V. C. Anderson and D. H. Thompson, *Biochim. Biophys. Acta* **1109**, 33 (1992).

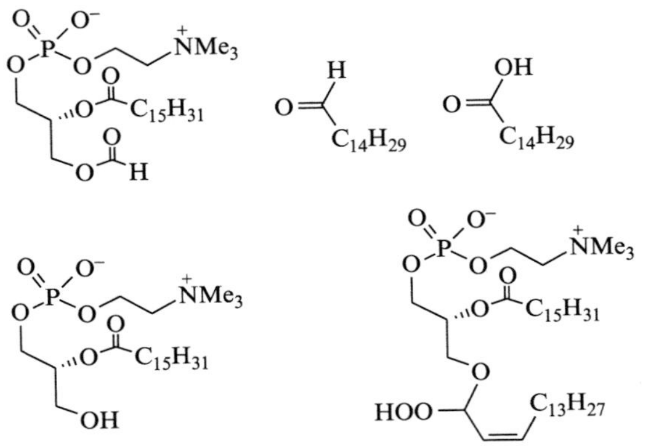

FIG. 2. Formate ester, fatty aldehyde, fatty acid, lysolipid, and allylic hydroperoxide products (left to right, top to bottom) formed during the singlet oxygen-mediated oxidation of palmitoyl plasmenylcholine (PPlsC).[23]

pathway to pure plasmenylcholines with (Z)-vinyl ether stereospecificity[36] via the transformation of acylglycerols to the corresponding vinyl phosphates, followed by reductive cleavage.[37] Qin and co-workers[38] subsequently reported a multistep synthesis of plasmenylcholine using a Lindlar catalyst reduction of alkynyl ethers as the key step in vinyl ether formation. Unfortunately, both of these pathways are tedious and produce low overall yields of plasmalogens. We have overcome these limitations by developing new, expedient synthetic methods for the preparation of PPlsC, DPPlsC, and other plasmenyl-type lipids. The strategies employed and detailed experimental procedures are outlined.

Synthesis of Plasmenyl-Type Lipids

Plasmenylcholine

Vinyl Phosphate Route. Our initial plasmenylcholine liposomes studies used semisynthetic lipid.[25,26,31,35] Batch-to-batch variations in the properties of these samples due to the variations in *sn*-1′-alkenyl chain composition, however, led to the development of the first total synthesis of stereochemically pure PPlsC (Fig. 3).[36] This was later extended to the

[36] Y. Rui and D. H. Thompson, *Chem. Eur. J.* **2**, 1505 (1996).
[37] Y. Rui and D. H. Thompson, *J. Org. Chem.* **59**, 5758 (1994).
[38] D. H. Qin, H. S. Byun, and R. Bittman, *J. Am. Chem. Soc.* **121**, 662 (1999).

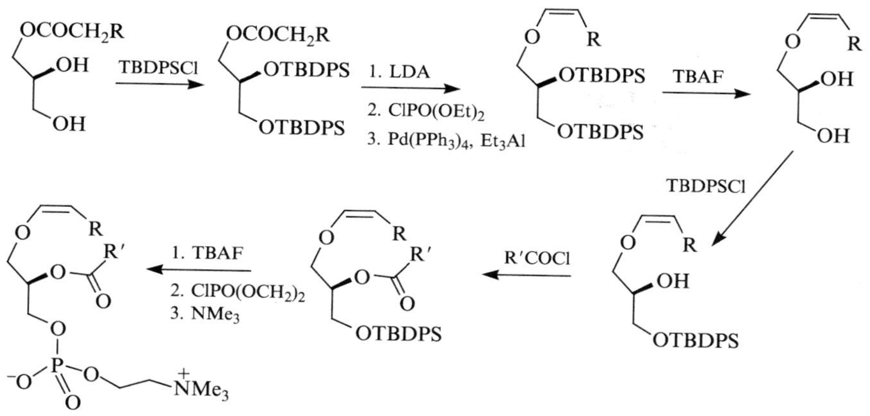

FIG. 3. Vinyl phosphate-mediated synthesis of PPlsC.[36]

preparation of DPPlsC[39] (see later). These pathways utilize a critical vinyl phosphate formation and reduction reaction sequence to install the Z-vinyl ether bond. Although effective at controlling vinyl ether stereochemistry, the capriciousness of this reaction pair motivated the development of more efficient synthetic procedures for the formation of plasmenyl-type compounds.[*]

The preparation of plasmenylcholines presents several synthetic challenges, including the (1) stereoselective formation of the naturally occurring (Z)-vinyl ether bond at the sn-1 position to avoid tedious purification of geometric isomers; (2) instability of the vinyl ether bond toward acidic or oxidative conditions, which limits the choice of reagents and isolation conditions; and (3) an sn-2 acyl group that is prone to rapid migration to the primary sn-1/sn-3 sites under basic or mildly acidic conditions (including silica gel chromatography with aprotic solvents). Two new synthetic routes, based on either a vinyl acetal reduction process or an allylic ether transformation, have been developed that accommodate these constraints.

---

[39] J. A. Boomer and D. H. Thompson, *Chem. Phys. Lipids* **99,** 145 (1999).

[*] Problems encountered with this transformation include (i) scrambling of vinyl ether stereochemistry as a result of the exotherms generated during lithium diisopropylamide treatment, vinyl phosphate trapping with ClPO(OEt)$_2$, and/or vinyl phosphate reduction; (ii) low catalytic efficiency of the Pd-catalyzed vinyl phosphate reduction; (iii) ethyl transfer—rather than hydride transfer—during the Et$_3$Al/Pd(PPh$_3$)$_4$ vinyl phosphate reduction reaction; and (iv) reformation of the glycerol ester if the vinyl phosphate intermediate is exposed to moisture and adventitious acid.

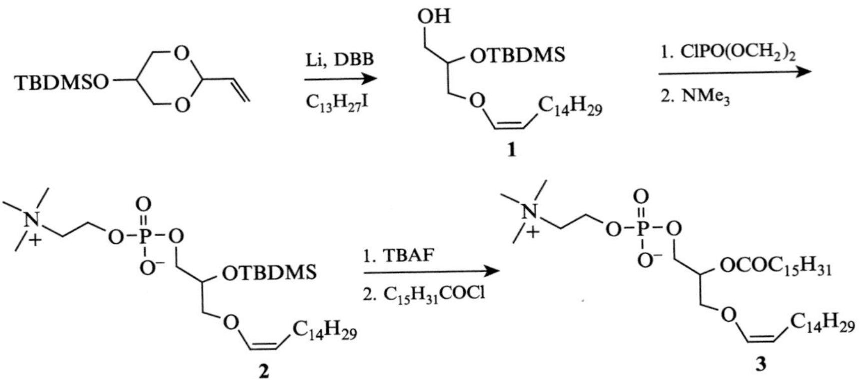

FIG. 4. PPlsC synthesis using the vinyl acetal reduction route under Barbier conditions.[40]

Vinyl Acetal Route.[40] Two important precursors for plasmenylcholine synthesis, a monosilylprotected *sn*-1 (*Z*)-vinyl ether glycerol (3-TBDMS-1-*O*-1'-(*Z*)-hexadecenyl glycerol and 2-TBDMS-1-*O*-1'-(*Z*)-hexadecenyl glycerol, **1**, Fig. 4), can be prepared in a one-step Barbier-type reaction of TBDMS-protected vinyl acetals and 1-iodoalkanes in moderate yields and excellent stereoselectivity (*Z*:*E* > 95:5).[†] Rui and Thompson[36] first synthesized 3-TBDPS-protected 1-*O*-1'-(*Z*)-hexadecenyl glycerol using a multistep protection/deprotection strategy. A key step in that sequence is the successful removal of the *sn*-3 TBDPS group with TBAF in the presence of imidazole at low temperature to minimize the *sn*-2 → *sn*-3 acyl migration problem that occurs typically with acyl-substituted lysolipids. The success of this pathway, therefore, is dependent critically on the attention given to the details of reaction and column separation conditions. The route employing **1**, a precursor of lysoplasmenylcholine, benefits from greater efficiency, as it avoids the acyl migration problem by introducing the acyl substituent in the last step. This approach also enables synthesis of a family of racemic plasmenylcholine analogs with varying chain composition at the glycerol 2-position. The formation of plasmenylcholines with stereocontrol at the *sn*-2 site, however, requires the use of a slightly less efficient alternate pathway involving the chiral intermediate, 3-TBDMS-1-*O*-1'-(*Z*)-hexadecenyl-*sn*-glycerol. Nonetheless, depending

---

[40] J. Shin, O. V. Gerasimov, and D. H. Thompson, *J. Org. Chem.* **67**, 6503 (2002).
[†] Barbier conditions involve the addition of a mixture of vinyl acetal and alkyl halide mixture to the lithium 4,4-di-*t*-butylbiphenyl (LiDBB) reductant solution.

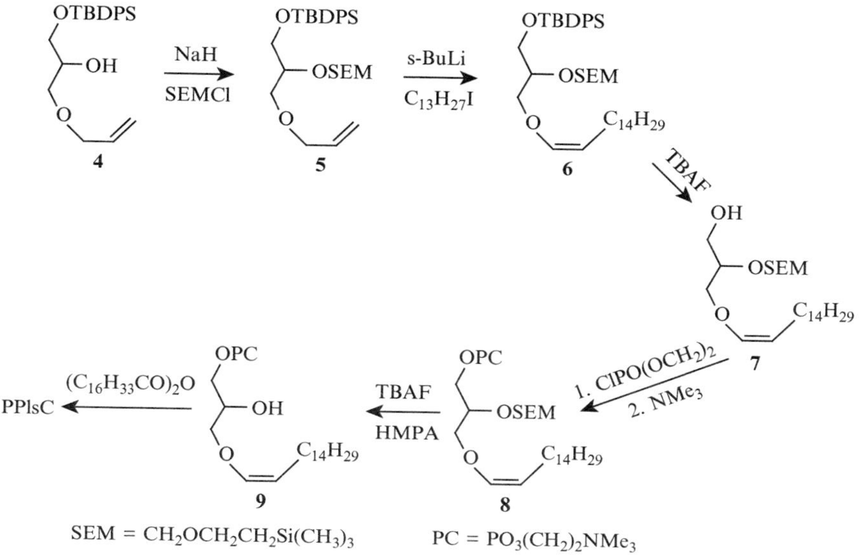

Fig. 5. Synthesis of PPlsC via the allyl ether route.[41]

on the requirements of the application, effective plasmenylcholine syntheses using either intermediate are possible and preferred over the prior
vinyl phosphate or acetylenic ether routes.

Allyl Ether Route.[41] Commercially available (R)-$(-)$-2,2-dimethyl-1,3-
dioxolane-4-methanol is utilized as starting material for the synthesis of
naturally occurring chiral plasmenylcholines via this pathway (Fig. 5).
The dioxolane starting material is protected readily with allyl bromide, hy-
drolyzed in the presence of solid acid catalysts, and protected with
TBDPSCl to give **4** in good yield. Chiral PPlsC is prepared subsequently
using 2-trimethylsilanyl ethoxymethyl (SEM) protection, TBDPS deprotection, phosphocholine installation, SEM deprotection, and palmitoyla-
tion.[36] A major advantage of this route is that *sn*-2 lysoplasmenylcholine
(**9**) serves as the key precursor for a general plasmenylcholine synthesis,
allowing many different types of biologically important plasmenylcholines
with different *sn*-2 acyl chains (*e.g.*, acetyl, arachidonyl) to be prepared
readily via a single acylation step.

[41] J. Shin and D. H. Thompson, *J. Org. Chem.* **68,** 6760 (2003).

Diplasmenylcholine

Previous methods for the preparation of DPPlsC have used the vinyl phosphate route (Fig. 3). Treatment of diacylglycerol with 2.8 equivalents of lithium diisopropylamide, followed by *O*-trapping of the bis-enolate with 4 equivalents of diethylchlorophosphate in HMPA, generates a bisvinyl phosphate intermediate.[39] Reduction with $(CH_3CH_2)_3Al$ catalyzed by $Pd(PPh_3)_4$ produces 1,2-di-*O*-(1*Z'*-dialkenyl)-3-*O*-TBDPS-*sn*-glycerol in modest overall yield. The efficiency of this reaction is highly dependent on catalyst quality; Pd(II) contaminants in the catalyst lead to the production of ethyl-transfer side products and the starting diester (arising from vinyl phosphate rearrangement). Efforts in our laboratory have focused on obviating these limitations by adapting the allyl ether reaction to the preparation of DPPlsC.[42]

Relative Merits of Different Plasmenylcholine Synthesis Pathways

The allyl ether route provides the best control over the synthesis of plasmenylcholines and is preferred for the preparation of chiral PPlsC, despite the multiple protection and deprotection steps required. This route is also the most widely applicable to the preparation of plasmenyl-type lipid analogs, including vinyl ether-linked PEG lipids with variable electron demand.[43] Racemic PPlsC, however, is prepared more efficiently using the vinyl acetal reduction process described earlier due to the symmetry of the key intermediate and limited use of protecting groups. Both of these methods are preferred over the vinyl phosphate route.

Poly(ethylene glycol)-Modified Plasmenyl-Type Lipids

The allyl ether coupling reaction has been used to prepare several cleavable vinyl ether PEG lipid conjugates, including ST302, as shown in Fig. 6.[44] In this case, 3-allyl-1,2-di-*O*-oleyl-*rac*-glycerol is converted to **10** after deprotonation and alkylation with TBDMS-protected 2-iodoethanol. This reaction gives the desired vinyl ether intermediate **10** in 24% yield after separation from the undesired α-coupled product. Deprotection of **10** with TBAF, followed by alumina chromatography, activation with 4-nitrophenyl chloroformate, and subsequent coupling with mPEG 2000 amine, provides the desired PEG lipid ST302.

[42] J. Van den Bossche, J. Shin, and D. H. Thompson, unpublished results.
[43] J. Shin, J. A. Boomer, A. Patwardhan, J. Nash, O. V. Gerasimov, and D. H. Thompson, unpublished results.
[44] J. Shin, P. Shum, and D. H. Thompson, *J. Control. Rel.* **91**, 187 (2003).

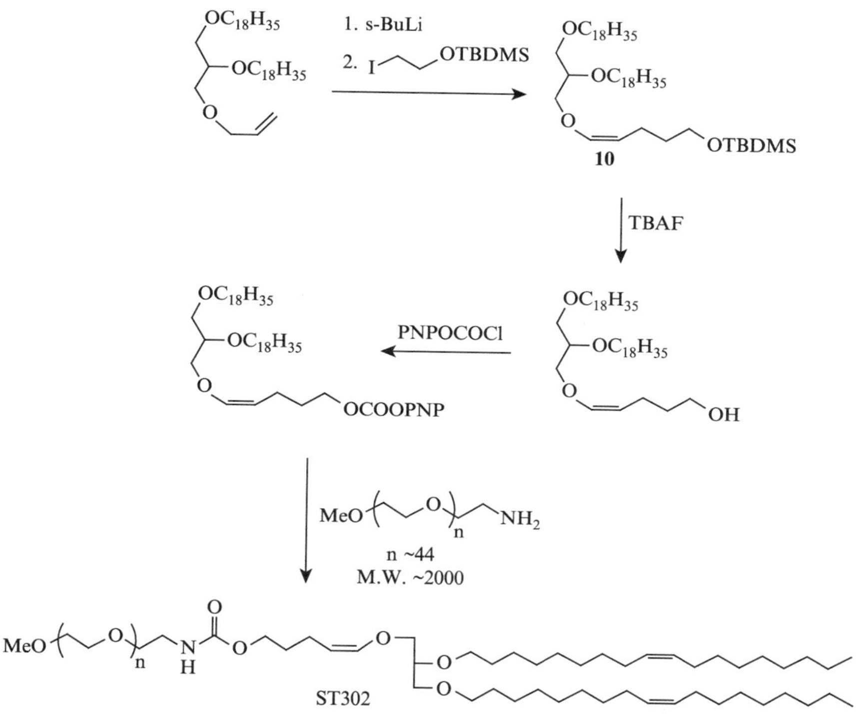

FIG. 6. Synthesis of ST302, a vinyl ether linked PEG lipid, via the allyl ether route.[44]

Preparation of Aqueous Dispersions of Plasmenyl-Type Lipids

The lability of the vinyl ether bond requires that special precautions be taken while handling plasmenyl-type lipids to avoid their premature decomposition. Exposure of these lipids to oxidative conditions, including prolonged contact with the atmosphere, can lead to their degradation via vinyl ether cleavage and radical-trapping reactions. Contact with acidic media, even as weakly acidic as pH 6.5, should also be avoided. The rate of vinyl ether hydrolysis displays a pseudo-first-order dependence on the H^+ concentration with an observed rate constant of 8.1×10^{-7} s^{-1} for PPlsC at pH 5.3 (38°, aqueous liposome dispersion); the reaction rate accelerates approximately an order of magnitude for each unit pH reduction.[26] For these reasons, the typical precautions observed in the handling of plasmenyl-type lipids include (1) maintenance of an inert atmosphere (Ar or N_2) above the lipid samples (i.e., during storage as a solid or as an organic or aqueous solution), (2) pretreating all solvents used to dissolve the lipid with a powerful basic dehydrating agent to remove adventitious moisture and acid, and (3) minimizing the time and temperature of acid exposure if acid-mediated liposome remote loading procedures are used. Solvent pretreatment by percolation through solid sodium carbonate is especially critical if chloroform is used to dissolve the lipid, as it often contains acidic impurities that become concentrated and catalyze the decomposition of the lipid during solvent evaporation. Observation of these simple handling procedures makes the processing of plasmenyl-type lipids straightforward and similar to standard liposome production techniques.

Experimental Methods

General Procedures

All chemicals are obtained from Sigma (St. Louis, MO) or Aldrich (Milwaukee, WI), unless specified otherwise. ^{1}H and ^{13}C NMR spectra are recorded at 200 MHz. Chemical shifts are reported in ppm relative to the residual solvent peaks as the internal standard. MS (EI/CI/ESI) is performed by the Purdue University MCMP mass spectrometry service. Liquid chromatography is performed typically on 230–400 mesh silica gel, using high-grade solvents as eluents. THF is distilled from Na. Benzene, acetonitrile, dimethylformamide, triethylamine, and pyridine are distilled from CaH$_2$. All other chemicals are used without further purification unless otherwise stated. Solid anhydrous MgSO$_4$ (Fisher, Pittsburg, PA) is used typically as a drying agent for acid-insensitive compounds; solid anhydrous Na$_2$CO$_3$ (Fisher) is used to pretreat solvents (especially CHCl$_3$!) that had not been freshly distilled before they were used to dissolve acid-sensitive compounds. CDCl$_3$ (Cambridge Isotope Labs, Andover, MA) is filtered through solid anhydrous Na$_2$CO$_3$ to remove traces of adventitious acid and moisture before preparation of NMR samples with acid-sensitive compounds.

Syntheses

2-tert-Butyldimethylsilyl-1-O-1'-(Z)-hexadecenyl glycerol (1). Imidazole (4.216 g, 61.93 mmol) is added to a flask containing a mixture of *trans/cis*-2-vinyl-4-hydroxymethyl-1,3-dioxane and *trans/cis*-2-vinyl-4-hydroxy-1,3-dioxane (4.03 g, 30.96 mmol) in dimethyl formamide (DMF) (100 ml); TBDMSCl (7.00 g, 46.4 mmol) is then added at 0°. The mixture is stirred at 23° overnight before diluting with Et$_2$O (500 ml) and washing with H$_2$O (3 × 50 ml). The organic layer is dried over anhydrous MgSO$_4$, filtered, and the solvent removed by evaporation under reduced pressure.

The residue is purified by silica gel chromatography (hexane:acetone, 20:1) and the silyl ether product is used in the subsequent vinyl acetal (Barbier) transformation.

Li (224 mg, 30 wt% in mineral oil, 9.66 mmol; MCB East Rutherford, NJ) is added quickly under Ar to an air-tight flask containing a glass-covered magnetic stirring bar. Hexane (30 ml) is added, stirred for 20 min, and then removed by cannula to wash out the mineral oil. This procedure is repeated one more time before the addition of 4,4-di-t-butylbiphenyl (DBB) (36 mg, 0.138 mmol) and tetrahydrofuran (THF) (30 ml) at 23° under Ar. The dark blue color of the radical anion appears within 10 s. The reaction mixture is cooled to 0°, a mixture of 1-iodotridecane (856 mg, 2.76 mmol) and vinyl dioxolane (336 mg, 1.38 mmol) in THF (3 ml) is added all at once at 0°, and the reaction mixture is stirred at 0° for 30 min. Hexane (10 ml) is then added, and the resulting mixture is quenched slowly with H_2O (5 ml) at 0°. The organic layer is washed with water (2 × 10 ml), separated, dried over anhydrous Na_2CO_3, and filtered. The solvent is removed by evaporation under reduced pressure, and the residue is purified by silica gel chromatography (hexane:Et_2O, 8:1) to give **1** in 47% yield after silica gel purification. ^1H NMR (C_6D_6): δ 0.04 (s, 3H), 0.08 (s, 3H), 0.92 (s, 12H), 1.20–1.50 (m, 25H), 2.30 (m, 2H), 3.42 (m, 2H), 3.55 (d, 2H, $J = 6$ Hz), 3.73 (m, 1H), 4.40 (1H, $J = 6$ Hz), 5.83 (d, 1H, $J = 6$ Hz); ^{13}C NMR (C_6D_6): δ-5.2, 5.0, 13.8, 17.7, 22.6, 24.1, 25.5, 29.3, 29.6, 29.7, 29.8, 31.8, 63.7, 71.8, 73.5, 106.4, 145.2; Cl calculated $(M + H)^+$ 429, found 429.

2-Hexadecanoyl-1-O-1'-(Z)-hexadecenyl   glycero-3-phosphocholine (PPlsC, 3). A solution of alcohol **1** (0.237 mmol) in benzene (25 ml) is added to a flask under Ar. Pyridine (57 μl, 0.71 mmol) and 2-oxo-2-chloro-1,3,2-dioxaphospholane (33 μl, 0.36 mmol) are then added to a flask that has been cooled to 5°. After stirring at 5° overnight under Ar, the solvent is removed under vacuum. The residue is transferred to a pressure bottle with benzene (2 ml) and acetonitrile (5 ml). Trimethylamine (~3 ml, 33 mmol) is then distilled into the reaction vessel, the vessel is sealed, and the mixture is stirred at 70° for 24 h. After slow release of reactor pressure at 0°, the resulting solution is purified using a silica gel column (gradient elution with CH_2Cl_2:MeOH:H_2O, 100:0:0, 80:20:0, 65:35:6). Suspended silica gel from the chromatographic fractions is removed using PTFE syringe filters (0.45 μm) to give a white solid (115 mg, 0.160 mmol, 67.6%) after lyophilization from benzene. ^1H NMR ($CDCl_3$): δ 0.82 (t, 6H, $J = J$ Hz), 1.20 (s, 48H), 1.51 (m, 2H), 1.94 (m, 2H), 2.24 (t, 2H, $J = 7$ Hz), 3.33 (s, 9H), 3.70–4.35 (m, 9H), 5.09 (m, 1H), 5.84 (d, 1H, $J = 6$ Hz); ^{13}C NMR ($CDCl_3$): δ 14.0, 22.6, 23.9, 24.9, 29.1, 29.3, 29.5, 29.6, 29.7, 29.9, 31.9, 34.4, 54.3, 59.2, 63.1, 66.2, 70.5, 71.7, 107.6, 144.7, 173.2; ESI $(M + H)^+$ calculated 718, found 718.

Imidazole (50 mg, 0.732 mmol) is added to a flask containing phospho-choline intermediate **2** (124 mg, 0.209 mmol) in THF (3 ml). TBAF (0.626 ml, 0.626 mmol) is added and the mixture is stirred at 23° for 5 h. The solution is loaded directly onto a silica gel column and purified by step gradient elution (CH_2Cl_2:MeOH:H_2O, 80:20:0, 65:35:6). Suspended silica gel from the chromatographic fractions is removed using PTFE syringe filters (0.45 μm) to give the desired product (98 mg, 0.204 mmol, 98%) after lyophilization from benzene. ^{1}H NMR (CD_3OD): δ 0.89 (t, 3H, J = 6 Hz), 1.28 (s, 9H), 2.04 (m, 2H), 3.22 (s, 9H), 3.6–3.96 (m, 7H), 4.22–4.38 (m, 3H), 4.90 (s, 1H), 6.00 (d, 1H, J = 6 Hz); ^{13}C NMR (CD_3OD): δ 14.5, 23.7, 25.0, 30.5, 30.6, 30.7, 30.8, 31.0, 33.1, 54.7, 60.4, 67.5, 68.0, 70.8, 73.9, 107.9, 146.3; ESI calculated (M + H)$^{+}$ 480, found 480.

DMAP (46 mg, 0.37 mmol) is added to a flask containing the depro-tected glycerophosphocholine (90 mg, 0.19 mmol) in CH_2Cl_2 (10 ml). Pal-mitic anhydride (167 mg, 0.338 mmol) is added, and the mixture is stirred at 23° for 20 h. The solution is loaded directly onto a silica gel column and purified by step gradient elution (CH_2Cl_2:MeOH:H_2O, 80:20:0, 65:35:6). Suspended silica gel from the chromatographic fractions is removed using PTFE syringe filters (0.45 μm) to give PPlsC as a powder (71 mg, 0.099 mmol, 53%) after lyophilization from benzene.

(R)-*(+)-1-Allyl-3-tert-butyldiphenylsilyl-sn-glycerol* *(4)*. (R)-(−)-2,2-Dimethyl-1,3-dioxolane-4-methanol (5.00 g, 37.8 mmol) is added to a solu-tion containing NaH (1.43 g, 56.7 mmol) in THF (150 ml) at 0°, and the mixture is stirred at 23° until gas evolution ceases. Allyl bromide (4.9 ml, 56.7 mmol) is added slowly and the mixture is stirred under Ar at 23° for 2 h. Hexane (200 ml) is then added and the solution is extracted with H_2O (2 × 50 ml). The organic layer is dried over $MgSO_4$ and concentrated to give an oil. AG50W-X2 (1 g) and the THF:H_2O (4:1, 50 ml) solution are added to the oil. The reaction mixture is heated at reflux overnight before removing the solid resin by filtration. The solution is evaporated under re-duced pressure and CH_2Cl_2 (100 ml) is added. The organic layer is dried over $MgSO_4$ and concentrated to give an oil (4.80 g). The oil is dried fur-ther under high vacuum. Imidazole (5.44 g, 80 mmol) is added to the oil in DMF (80 ml). TBDPSCl (9.3 ml, 36.3 mmol) is added slowly at 0°, and the mixture is stirred under Ar at 23° for 1 h. Hexane (200 ml) is added and the reaction mixture is washed with water (2 × 50 ml). Silica gel chro-matographic purification (CH_2Cl_2) of the organic residue gives **4** as an oil (11.75 g, 31.7 mmol, 84% yield).

$$\left([\alpha]_D^{25} = +2.9°[c1.00, CHCl_3]\right).$$

1-Allyl-3-tert-*butylmethylsilyl-2-(2-trimethylsilanylethoxymethyl)-sn-glycerol (5)*. Compound 4 (1.48 g, 6.00 mmol) in THF (5 ml) is added slowly to a solution containing NaH (182 mg, 7.20 mmol) in THF (30 ml) and the reaction mixture is stirred under Ar at 23° for 30 min. SEMCl (0.90 ml, 5.08 mmol) is added and the reaction mixture is heated at reflux for 1 h. Hexane (200 ml) is added after cooling and the mixture extracted with water (2 × 20 ml). Silica gel chromatographic purification (hexane:Et$_2$O, 8:1) gives 5 as an oil (1.569 g, 4.16 mmol, 82% yield). ^{1}H NMR (CDCl$_3$): δ 0.00–0.07 (m, 15H), 0.88–0.95 (m, 11H), 3.35–3.68 (m, 6H), 3.78 (quintet, 1H, J = 5 Hz), 3.98 (d, 2H, J = 5 Hz), 4.76 (s, 2H), 5.14 (d, 1H, J = 10 Hz), 5.24 (d, 1H, J = 17 Hz), 5.87 (ddt, 1H, J = 10, 17, 5 Hz); ^{13}C NMR (CDCl$_3$): δ −5.4, −1.4, 18.0, 18.2, 25.9, 62.9, 65.0, 70.0, 72.3, 76.6, 94.5, 116.7, 134.8.

3-tert-*Butyldiphenylsilyl-2-(2-trimethylsilanylethoxymethyl)-1-O-1'-(Z)-hexadecenyl-sn-glycerol (6)*. *sec*-BuLi (0.61 ml, 1.3 M in cyclohexane) is added to THF (7 ml) at −70°. 5 (0.609 mmol) in THF (1 ml) is added slowly, and the reaction mixture is stirred under Ar at −70° for 2 min. 1-Iodotridecane (245 mg, 0.792 mmol) in THF (4 ml) is added slowly and stirred at −70° for 10 min before warming to 0°. Hexane (30 ml) is then added and the mixture is washed with H$_2$O (2 × 5 ml). Silica gel chromatography (hexane:CH$_2$Cl$_2$, 8:1) of the organic residue gives 6 as an oil (229 mg, 29.0 mmol, 48% yield). ^{1}H NMR (CDCl$_3$): δ 0.93 (t, 3H, J = 6 Hz), 1.06 (s, 9H), 1.10 (s, 9H), 1.3 (m, 24H), 2.04 (m, 2H), 3.66–4.18 (m, 5H), 4.28 (q, 1H, J = 6 Hz), 5.82 (d, 1H, J = 6 Hz), 7.30–7.45 (m, 12H), 7.61–7.74 (m, 8H); ^{13}C NMR (CDCl$_3$): δ 14.2, 19.3, 19.4, 22.8, 23.4, 24.1, 26.9, 27.0, 29.5, 29.7, 29.8, 30.0, 32.0, 64.5, 72.8, 106.5, 127.6, 127.7, 129.6, 129.7, 133.6, 134.0, 135.6, 135.7, 135.9, 136.0, 145.5.

1-O-1'-(Z)-Hexadecenyl-2-(2-trimethylsilanylethoxymethyl)-sn-glycerol (7). Imidazole (276 mg, 4.06 mmol) and TBAF (3.5 ml, 1.0 M in THF) are added to 6 (794 mg, 1.16 mmol) in THF (20 ml), and the reaction mixture is stirred at 23° for 2 h. The reaction mixture is filtered through a silica gel plug, and the plug is washed with Et$_2$O. The organic solution is concentrated and purified by silica gel chromatography (hexane:Et$_2$O, 1:1) to give 7 as an oil (500 mg, 1.13 mmol, 97% yield). ^{1}H NMR (CDCl$_3$): δ −0.02 (s, 9H), 0.84 (t, 3H, J = 7 Hz), 0.92 (t, 2H, J = 8 Hz), 1.22 (m, 24H), 2.00 (q, 2H, J = 7 Hz), 2.88 (dd, 1H, J = 5, 8 Hz), 3.53–3.78 (m, 7H), 4.31 (q, 1H, J = 6 Hz), 4.70 (d, 1H, J = 7 Hz), 4.77 (d, 1H, J = 7 Hz), 5.87 (d, 1H, J = 6 Hz); ^{13}C NMR (CDCl$_3$): δ −1.4, 14.1, 18.1, 22.7, 24.0, 29.3, 29.4, 29.6, 29.7, 29.8, 31.9, 63.0, 65.7, 71.7, 79.1, 95.2, 107.8, 144.7; Cl (M + H)$^+$ calculated 445, found 445.

1-O-1'-(Z)-Hexadecenyl-2-(2-trimethylsilanylethoxymethyl)-sn-glycerophosphocholine (8). Compound 8 is prepared from 7 as described for 3.

1-O-1'-(Z)-Hexadecenyl-sn-*glycero-3-phosphocholine (9)*. TBAF (2.4 ml, 1.0 M in THF) is added to **8** (260 mg, 0.426 mmol) in HMPA (5 ml), and the mixture is stirred at 90° for 18 h. The solution is loaded directly onto a silica gel column and purified via step gradient elution with CH_2Cl_2:MeOH:H_2O (80:20:0, then 65:35:6). The suspended silica gel from the chromatographic fractions is removed using a 0.45-μm PTFE syringe filter to give **9** (164 mg, 0.342 mmol, 80% yield) after lyophilization from benzene. ^1H NMR (CD$_3$OD): δ 0.89 (t, 3H, $J = 6$ Hz), 1.28 (s, 9H), 2.04 (m, 2H), 3.22 (s, 9H), 3.6–3.96 (m, 7H), 4.22–4.38 (m, 3H), 4.90 (s, 1H), 6.00 (d, 1H, $J = 6$ Hz); ^{13}C NMR (CD$_3$OD): δ 14.5, 23.7, 25.0, 30.5, 30.6, 30.7, 30.8, 31.0, 33.1, 54.7, 60.4, 67.5, 68.0, 70.8, 73.9, 107.9, 146.3; ^{31}P NMR (CDCl$_3$): δ 1.468; ESI calculated $(M + H)^+$ 480, found 480.

2-Hexadecanoyl-1-O-1'-(Z)-hexadecenyl-sn-*glycero-3-phosphocholine (PPlsC, 3)*. Compound **3** is prepared from **9** by the reaction of palmitic anhydride in the presence of DMAP.[40] ^{31}P NMR (CDCl$_3$): δ 0.156.

$$\left([\alpha]_D^{25} = -1.7°[c1.00, CHCl_3] \right).$$

5-tert-Butyldimethylsilanyloxy-1-(1',2'-di-O-oleyl-rac-glyceryloxy)-1-pentene (10). *sec*-BuLi (2.4 ml, 1.3 M in cyclohexane) is added slowly to 3-allyl-1,2-di-*O*-oleyl-*rac*-glycerol (1.525 g, 2.41 mmol) in THF (50 ml) at −70°, and the reaction mixture is stirred under Ar at −70° for 7 min. 1-(*tert*-Butyldimethylsilyloxy-2-iodoethane (680 mg, 2.09 mmol) is added and stirred for 10 min at −70° before warming to 0°. Hexane (50 ml) is added, and the reaction mixture is washed with water (2 × 10 ml). Silica gel chromatographic purification (1:1 hexane:CH_2Cl_2) of the organic residue gives **10** as an oil (460 mg, 0.581 mmol, 24% yield). ^1H NMR (CDCl$_3$): δ 0.03 (s, 6H), 0.87 (m, 15H), 1.25 (m, 44H), 1.54 (m, 6H), 1.96–2.12 (m, 10H), 3.39–3.83 (m, 11H), 4.30 (q, 1H, $J = 6$ Hz), 5.32 (m, 4H), 5.92 (d, 1H, $J = 6$ Hz); ^{13}C NMR (CDCl$_3$): δ −5.3, 14.1, 18.3, 20.3, 22.7, 26.0, 26.1, 27.2, 29.2, 29.3, 29.5, 29.7, 29.8, 30.1, 31.9, 32.6, 33.0, 62.9, 70.3, 70.8, 71.7, 72.2, 77.8, 106.1, 129.8, 129.9, 145.5; Cl calculated $(M + H)^+$ 791, found 791.

5-(1',2'-Di-O-oleyl-rac-glyceryloxy)-4-penten-1-ol. Imidazole (66 mg, 0.97 mmol) is added to a flask containing **10** (220 mg, 0.278 mmol) in THF (10 ml). TBAF (0.80 ml, 1.0 M in THF) is added, and the mixture is stirred at 23° for 5 h. The solution is concentrated, and the resulting residue is purified by alumina chromatography (5:1 CH_2Cl_2:Et_2O) to give the desired product (174 mg, 0.257 mmol, 93%). ^1H NMR (CDCl$_3$): δ 0.83 (t, 6H, $J = 6$ Hz), 1.23 (m, 44H), 1.52 (m, 6H), 1.95 (m, 8H), 2.13 (q, 2H, $J = 6$ Hz), 2.27 (t, 1H, $J = 6$ Hz), 3.36–3.84 (m, 11H), 4.30 (q, 1H, $J = 6$ Hz), 5.29 (m, 4H), 5.98 (d, 1H, $J = 6$ Hz); ^{13}C NMR (CDCl$_3$): δ 14.1, 19.7, 22.7, 26.1, 27.2, 29.1, 29.2, 29.3, 29.4, 29.5, 29.6, 29.7, 29.8, 30.0, 31.9,

32.6, 61.4, 69.9, 70.7, 71.7, 72.1, 77.7, 105.5, 129.8, 129.9, 146.3; ESI calculated $(M + Na)^+$ 699, found 699.

5-(1′,2′-Di-O-oleyl-rac-glyceryloxy)-4-pentenyl 4-nitrophenyl carbonate. Triethylamine (60 mg, 0.59 mmol) is added to 5-(1′,2′-di-O-oleyl-*rac*-glyceryloxy)-4-penten-1-ol (150 mg, 0.222 mmol) in THF (5 ml). 4-Nitrophenyl chloroformate (PNPOCOCl, 67 mg, 0.332 mmol) is added, and the mixture is stirred for 1 h under Ar. The reaction mixture is concentrated, and the resulting residue is purified by silica gel chromatography (4:1 hexane:Et$_2$O) to give **5** as an oil (184 mg, 0.218 mmol, 98% yield). ^{1}H NMR (CDCl$_3$): δ 0.86 (t, 6H, $J = 6$ Hz), 1.25 (m, 44H), 1.54 (m, 4H), 1.80 (quintet, 2H, $J = 7$ Hz), 1.97 (m, 8H), 2.19 (q, 2H, $J = 7$ Hz), 3.38–3.86 (m, 9H), 4.30 (m, 3H), 5.32 (m, 4H), 6.01 (d, 1H, $J = 6$ Hz), 7.36 (d, 2H, $J = 9$ Hz), 8.25 (d, 2H, $J = 9$ Hz); ^{13}C NMR (CDCl$_3$): δ 14.1, 20.1, 22.7, 26.1, 27.2, 28.5, 29.3, 29.4, 29.5, 29.6, 29.7, 29.8, 30.1, 31.9, 32.6, 69.2, 70.1, 70.7, 71.7, 72.4, 77.8, 104.2, 121.7, 125.2, 129.9, 130.0, 145.3, 146.7, 152.4, 155.7; ESI calculated $(M + Na)^+$ 863, found 863.

ST302. Triethylamine (90 mg, 0.89 mmol), 4-nitrophenyl carbonate (324 mg, 0.385 mmol), and mPEG-NH$_2$ 2000 (592 mg, 0.296 mmol) are combined in DMF (5 ml) and stirred at 23° for 18 h. The reaction mixture is concentrated and purified by silica gel chromatography (10:1 CH$_2$Cl$_2$:MeOH) to give ST302 (758 mg, 0.281 mmol, 95% yield). In general, once the mPEG is coupled to the vinyl ether linker, the monitoring of reaction progress and product identification are best performed using ^{1}H NMR, as the TLC mobility of PEG compounds in PEGylation reactions may be dominated by the PEG component. ^{1}H NMR (CDCl$_3$): δ 0.86 (t, 6H, $J = 6$ Hz), 1.25 (m, 44H), 1.60 (m, 6H), 1.99 (m, 8H), 2.10 (q, 2H, $J = 7$ Hz), 3.3–3.9 (m, ca. 200H), 4.03 (t, 2H, $J = 7$ Hz), 4.29 (q, 1H, $J = 6$ Hz), 5.16 (s, 1H), 5.32 (m, 4H), 5.95 (d, 1H, $J = 6$ Hz).

Preparation of Aqueous Dispersions of Plasmenyl-Type Lipids

Liposomes are prepared by hydrating lipids films prepared by evaporation of lipid solutions in deacidified CHCl$_3$. After evaporating CHCl$_3$ under a gentle N$_2$ stream, the lipid residue is dried further under a 50-μm Hg vacuum for at least 8 h. The powder is then hydrated in neutral buffer solution via 10 freeze–thaw–vortex cycles, and the resulting multilamellar vesicle solution is extruded 10 times through two stacked 100-nm pore-diameter polycarbonate filters at 50°, 200 psi N$_2$. Liposomes in the 5–50 m*M* concentration range (~100 nm diameter) can be produced readily using these methods for PPlsC, 7:3 PPlsC:cholesterol, DPPlsC, 7:3 DPPlsC:cholesterol, and ST302:dioleoylphosphatidylethanolamine (1:99–1:9 molar ratios). Liposome sizes are determined by quasi-elastic light scattering (QLS) using a Coulter N4-Plus instrument and the manufacturer's supplied software (version 1.1).

Concluding Remarks

Three synthesis pathways for the preparation of plasmenylcholine and related plasmenyl-type lipid species have been developed. The relative merits of these routes and detailed experimental procedures have been described. A protocol for the preparation of plasmenyl-type liposome dispersions was also described.

The allyl ether route is the preferred method for the preparation of chiral PPlsC and other plasmenyl-type lipid analogs, including vinyl ether-linked PEG lipids. An alternative procedure, however, using vinyl acetal precursors, is more efficient for the production of racemic PPlsC. Either of these methods is preferred over the vinyl phosphate and acetylenic ether routes, which suffer from unpredictable product yields.

Acknowledgments

The authors thank the many students and postdoctoral associates that have been involved in these studies, particularly Oleg V. Gerasimov, Zhi-Yi Zhang, Pochi Shum, Nathan Wymer, Robert Haynes, and Jason Robarge. The authors gratefully acknowledge the NIH and Avanti Polar Lipids for their financial support.

Section III

Liposomal Oligonucleotides

[11] High Efficiency Entrapment of Antisense Oligonucleotides in Liposomes

By DARRIN D. STUART, SEAN C. SEMPLE, and THERESA M. ALLEN

Introduction

Antisense oligodeoxynucleotides (asODN) appear, at first glance, to be ideal candidates for targeted therapies due to their high degree of selectivity and low toxicities. Disease-related genes can be targeted specifically given knowledge of their unique mRNA sequence. Unfortunately, this approach has not met with widespread success in the clinic, despite more than 20 clinical trials and well over 10 years of development. The major growth in this technology has been in the genomics field, where asODN are being employed to assign function to newly discovered genes and to validate targets for pharmaceutical development.

Some of the efforts to advance this technology into useful therapies have focused on developing alternative chemistries to optimize the stability, specificity, affinity, and activity of nucleic acid oligomers.[1–6] For the most part, improvements in oligonucleotide chemistry have not improved their intracellular uptake, and asODN still require carrier-mediated delivery or permeabilization for activity.[7–10] Therefore, other efforts have revolved around the development of *in vitro* and *in vivo* carriers for oligonucleotides. Much of this activity has focused on the use of cationic lipids and polymers that interact electrostatically with anionic oligonucleotides to form complexes. Excess positive charge in these complexes allows them

[1] P. Wittung, J. Kajanus, K. Edwards, G. Haaima, P. Nielsen, B. Norden, and B. G. Malmstrom, *FEBS Lett.* **375,** 317 (1995).

[2] D. Stein, E. Foster, S. B. Huang, D. Weller, and J. Summerton, *Antisense Nucleic Acid Drug Dev.* **7,** 151 (1997).

[3] J. Summerton, D. Stein, S. B. Huang, P. Matthews, D. Weller, and M. Partridge, *Antisense Nucleic Acid Drug Dev.* **7,** 63 (1997).

[4] B. P. Monia, *Ciba Found. Symp.* **209,** 107 (1997).

[5] A. N. Elayadi, A. Demieville, E. V. Wancewicz, B. P. Monia, and D. R. Corey, *Nucleic Acids Res.* **29,** 1683 (2001).

[6] H. Wang, L. Nan, D. Yu, S. Agrawal, and R. Zhang, *Clin Cancer Res.* **7,** 3613 (2001).

[7] C. F. Bennett, M. Chiang, H. Chan, J. E. Shoemaker, and C. K. Mirabelli, *Mol. Pharmacol.* **41,** 1023 (1992).

[8] Q. Hu, C. R. Shew, M. B. Bally, and T. D. Madden, *Biochim. Biophys. Acta* **1514,** 1 (2001).

[9] P. A. Morcos, *Genesis* **30,** 94 (2001).

[10] C. Ghosh and P. L. Iversen, *Antisense Nucleic Acid Drug Dev.* **10,** 263 (2000).

to deliver antisense oligonucleotides intracellularly through mechanisms that may involve electrostatic binding to anionic cell membrane components.[11,12] These formulations can be quite efficient and simple to use for *in vitro* applications, with the only drawback being increased cell toxicities due, in large part, to the presence of the cationic lipids.[13,14]

In vivo, the requirement for a lipid carrier is much less clear. Several examples of therapeutic activity of asODN exist in animal models,[15–20] as well as in clinical trials in humans,[21,22] and in these examples lipid carriers have not been used. However, there is no evidence to suggest that the cellular uptake of free asODN is any more efficient *in vivo* than *in vitro.* Furthermore, asODN are cleared rapidly from circulation, with a large proportion being excreted via the kidneys.[23] Improvements in the intracellular delivery, pharmacokinetics (PK), and biodistribution (BD) of asODN may be possible with the use of well-designed lipid carriers. However, understanding the pharmacokinetics and biodistribution of liposomal asODN is an important prelude to determining appropriate therapeutic applications. Two previous studies on the biodistribution of cationic liposomal phosphorothioate asODN complexes found that, immediately following iv injection, the complexes accumulated primarily in lung and liver.[24,25]

[11] O. Zelphati and F. C. Szoka, *Pharm. Res.* **13,** 1367 (1996).

[12] O. Zelphati and F. C. Szoka, *Proc. Natl. Acad. Sci. USA* **93,** 11493 (1996).

[13] G. Lambert, E. Fattal, A. Brehier, J. Feger, and P. Couvreur, *Biochimie* **80,** 969 (1998).

[14] M. C. Filion and N. C. Phillips, *Biochim. Biophys. Acta* **1329,** 345 (1997).

[15] H. Wang, X. Zeng, P. Oliver, L. P. Le, J. Chen, L. Chen, W. Zhou, S. Agrawal, and R. Zhang, *Int. J. Oncol.* **15,** 653 (1999).

[16] H. Miyake, B. P. Monia, and M. E. Gleave, *Int. J. Cancer* **86,** 855 (2000).

[17] V. Arora, D. C. Knapp, B. L. Smith, M. L. Statdfield, D. A. Stein, M. T. Reddy, D. D. Weller, and P. L. Iversen, *J. Pharmacol. Exp. Ther.* **292,** 921 (2000).

[18] H. Zhang, J. Cook, J. Nickel, R. Yu, K. Stecker, K. Myers, and N. M. Dean, *Nature Biotech.* **18,** 862 (2000).

[19] D. E. Lopes de Menezes, N. Hudon, N. McIntosh, and L. D. Mayer, *Clin. Cancer Res.* **6,** 2891 (2000).

[20] H. Roh, D. W. Green, C. B. Boswell, J. A. Pippin, and J. A. Drebin, *Cancer Res.* **61,** 6563 (2001).

[21] B. R. Yacyshyn, M. B. Bowen-Yacyshyn, L. Jewel, J. A. Tami, C. F. Bennett, D. L. Kisner, and W. R. Shanahan, *Gastroenterology* **114,** 1133 (1998).

[22] B. Jansen, V. Wacheck, E. Heere-Ress, H. Schlagbauer-Wadl, C. Hoeller, T. Lucas, M. Hoermann, U. Hollenstein, K. Wolff, and H. Pehamberger, *Lancet* **356,** 1728 (2000).

[23] B. Tavitian, S. Terrazzino, B. Kuhnast, S. Marzabal, O. Stettler, F. Dolle, J. R. Deverre, A. Jobert, F. Hinnen, B. Bendriem, C. Crouzel, and L. Di Giamberardino, *Nature Med.* **4,** 467 (1998).

[24] C. F. Bennett, J. E. Zuckerman, D. Kornbrust, H. Sasmor, J. M. Leeds, and S. T. Crooke, *J. Control. Rel.* **41,** 121 (1996).

[25] D. C. Litzinger, J. M. Brown, I. Wala, S. A. Kaufman, G. Y. Van, C. L. Farrell, and D. Collins, *Biochim. Biophys. Acta* **1281,** 139 (1996).

The liposome formulations were similar to those used *in vitro:* a cationic lipid mixed with the fusogenic lipid dioleoylphosphatidylethanolamine (DOPE) and then complexed with asODN. However, the relatively static environment of a tissue culture dish is quite unlike the *in vivo* environment, and therefore some effort should be made to optimize liposomal asODN carriers for use *in vivo.*

Past studies using liposomes as drug carriers *in vivo* have shown that liposomes must be able to avoid sequestration by the liver and spleen in order to achieve passive (or ligand-mediated) targeting to other tissues.[26–28] The extensive liver and lung uptake and extremely short circulation times of cationic lipoplexes would hinder their targeting to diseased tissues *in vivo.* In addition, interaction of these complexes with serum proteins and nontarget cells, as a result of their excess positive charge, would interfere with ligand-mediated targeting to a specific cell type. Therefore, the goal of our group and many others has been to develop alternatives to cationic lipoplexes that may be more useful for systemic administration *in vivo.* In addition, we aimed to develop carriers for asODN that could be targeted specifically to disease cells and/or tissues *in vivo.*

Several studies have examined methods and formulations to increase the loading efficiency of asODN within neutral or anionic liposomes (for review, see Semple *et al.*[29]). Generally, less than 20% of added ODN can be entrapped within liposomes less than 200 nm in diameter. To date, the only way to efficiently entrap or complex asODN is to use positively charged lipids or polymers, and a few studies have described formulations with potential utility *in vivo.* These formulations can be assigned to two groups: formulations that resemble cationic lipoplexes with slight modifications and rationally designed systems that carry no excess surface positive charge and/or are shielded by polymer-grafted lipids such as polyethylene glycol (PEG).

Soni *et al.*[30] and Gokhale *et al.*[31] used cationic formulations in which dried lipid films were hydrated in the presence of asODN to form liposomes, which were then extruded or microfluidized to smaller diameters. These systems resemble cationic lipoplexes in that they are fairly simple

[26] K. K. Matthay, T. D. Heath, and D. Papahadjopoulos, *Cancer Res.* **44,** 1880 (1984).

[27] R. J. Debs, T. D. Heath, and D. Papahadjopoulos, *Biochim. Biophys. Acta* **901,** 183 (1987).

[28] A. Gabizon and D. Papahadjopoulos, *Proc. Natl. Acad. Sci. USA* **85,** 6949 (1988).

[29] S. C. Semple, S. K. Klimuk, T. O. Harasym, and M. J. Hope, *Methods Enzymol.* **313,** 322 (2000).

[30] P. N. Soni, D. Brown, R. Saffie, K. Savage, D. Moore, G. Gregoriadis, and G. M. Dusheiko, *Hepatology* **28,** 1402 (1998).

[31] P. C. Gokhale, V. Soldatenkov, F. H. Wang, A. Rahman, A. Dritschilo, and U. Kasid, *Gene Ther.* **4,** 1289 (1997).

to produce, have excess positive charge and therefore reasonable loading efficiencies, and have demonstrated some level of activity *in vivo*. The disadvantage of these formulations is that their PK and BD are still largely governed by the cationic surface charge and there is very little opportunity for delivery to tissues outside of the mononuclear phagocyte system (MPS, e.g., liver Kupffer cells) and the lungs. Indeed, the liver was the target organ in the studies by Soni *et al.*[30] as they were attempting to deliver hepatitis B virus (HBV) antisense oligonucleotides to infected hepatocytes. Gokhale *et al.*[32] observed significant antitumor activity of a *c-raf-1* asODN formulated in cationic lipoplexes [dimethyldioctadecyl ammonium bromide, phosphatidylcholine (PC), cholesterol; 1:3.2:1.6 molar ratio] in SQ-20B tumor xenografts. This study did not compare the efficacy of lipoplexes with free asODN, and biodistribution data indicated that lipoplexes did not increase intratumoral delivery compared to free asODN, but it did produce dramatic increases in disposition to the liver and spleen.

Steric stabilization of liposomes through the use of surface-bound hydrophilic polymers substantially increases the circulation time of liposomes and their entrapped drugs (for review, see Allen[33]). Several groups have examined liposomal asODN formulations that incorporate PEG to shield positive or negative charges. Li and Huang[34] formulated asODN into particles, termed LPDII, by first complexing asODN with polylysine at excess positive charge, followed by complexation with anionic liposomes containing a PEG-derivatized lipid, PEG-distearoylphosphatidylethanolamine (DSPE). The loading efficiency was reported to be 60–80% of added asODN, and the coupling of folate molecules to the PEG terminus as targeting ligands increased the activity of an asODN against the epidermal growth factor receptor in cell lines expressing the folate receptor. Unfortunately, there appears to be no *in vivo* data for this formulation. Meyer *et al.*[35] also demonstrated that PEG-lipids could be incorporated into liposomal asODN formulations, allowing ligand-mediated targeting via the attachment of anti-HER2 antibody fragments. Preformed cationic liposomes consisting of dioleoyltrimethylammonium propane (DOTAP), DOPE, and PEG-PE (1:1:0.12 molar ratio) were mixed with Bcl-2 asODN to form small (120 nm) particles that were effective in knocking out Bcl-2 protein levels in HER2-expressing cells. This formulation was stable over several

[32] R. C. Gokhale, D. McRae, B. P. Monia, A. Bagg, A. Rahman, A. Dritschilo, and U. Kasid, *Antisense Nucleic Acid Drug Dev.* **9,** 191 (1999).

[33] T. M. Allen, *Trends Pharmacol. Sci.* **15,** 215 (1994).

[34] S. Li and L. Huang, *J. Liposome Res.* **8,** 239 (1998).

[35] O. Meyer, D. Kirpotin, K. Hong, B. Sternberg, J. W. Park, M. C. Woodle, and D. Papahadjopoulos, *J. Biol. Chem.* **273,** 15621 (1998).

days at 4° and demonstrated some stability in the presence of plasma. Upon injection, however, most of the asODN dissociated from the liposomes (Kirpotin, personal communication), indicating that a great deal of the asODN is associated with the outside of the liposomes.

Stabilized antisense-lipid particles (SALP) were described by Semple *et al.*[29] and represent an important development toward the rational design of asODN formulations. This formulation takes advantage of the ionizable aminolipid dioleoyldimethylammonium propane (DODAP), which carries a positive charge at subphysiological pH (e.g., 4), but is neutral at pH 7.4. This should result in less binding of plasma proteins and fewer nonspecific cellular interactions as a result of loss of surface charge at physiological pH. Loading efficiencies of 65–80% are generally achieved, resulting in final asODN to lipid ratios of 0.15–0.2 (w/w). SALPs used to deliver c-myc asODN to subcutaneously growing melanoma cells in mice resulted in significant decreases in tumor growth and extended duration of action compared to nonformulated asODN.[36] The SALP formulation is described in greater detail in the following section.

The coated cationic liposome (CCL) formulation developed in the Allen laboratory represents another rationally designed liposomal asODN formulation.[37–40] The method optimizes the charge interaction between asODN and cationic lipids to produce particles that are close to charge neutral, as determined by zeta potential measurements. To overcome precipitation and aggregation that often occur in electrostatic systems at neutral charge ratios, this step is carried out in an organic solvent using a Bligh and Dyer monophase in a similar way to that described by Reimer *et al.*[41] for plasmid DNA and cationic lipids. Neutral lipids, as well as PEG-lipids and coupling lipids, are then added to the system and reverse phase evaporation vesicles (REV) are made. The CCL formulation meets the criteria for an ideal liposomal asODN formulation: it is efficient at loading asODN (80–100%), it can be extruded to diameters of less than 200 nm, it is stable in plasma, it demonstrates long-circulating pharmacokinetics, it can be targeted to specific cell types through the use of targeting ligands, and it is functionally active.

[36] C. Leonetti, A. Biroccio, B. Benassi, A. Stringaro, A. Stoppacciaro, S. C. Semple, and G. Zupi, *Cancer Gene Ther.* **8,** 459 (2001).

[37] G. Pagnan, D. Stuart, F. Pastorino, L. Raffaghello, P. G. Montaldo, T. M. Allen, B. Calabretta, and M. Ponzoni, *J. Natl. Cancer Inst.* **92,** 253 (1999).

[38] D. D. Stuart, G. Y. Kao, and T. M. Allen, *Cancer Gene Ther.* **7,** 466 (2000).

[39] D. D. Stuart and T. M. Allen, *Biochim. Biophys. Acta* **1463,** 219 (2000).

[40] F. Pastorino, D. Stuart, M. Ponzoni, and T. M. Allen, *J. Control. Rel.* **74,** 69 (2001).

[41] D. L. Reimer, Y. P. Zhang, S. Kong, J. J. Wheeler, R. W. Graham, and M. B. Bally, *Biochemistry* **34,** 12877 (1995).

The CCL and SALP formulations, while utilizing different methodologies, were each designed rationally to minimize residual surface charge and to achieve the ideal properties of a liposomal delivery system outlined earlier. The following sections describe the CCL and SALP formulations in detail with examples illustrating these characteristics.

Coated Cationic Liposomes

CCLs are formed through two sequential processes: (1) formation of hydrophobic cationic lipid-oligonucleotide particles, followed by (2) coating with neutral lipids through the formation of reverse-phase evaporation vesicles. Figure 1 illustrates the steps involved.

Bligh–Dyer Extraction

Reimer et al.[41] described the formation of hydrophobic plasmid DNA–cationic lipid particles through an organic extraction procedure. They demonstrated that cationic lipids could be used to extract plasmid DNA from an aqueous phase, into an organic phase, through a Bligh and Dyer monophase.[42] The extraction was shown to be mediated by the electrostatic interaction between the positively charged lipid and the negatively charged DNA. We hypothesised that this procedure would also work for negatively charged oligonucleotides and that the hydrophobic particles would serve as useful intermediates in the formation of REVs.[43]

A Bligh–Dyer monophase is formed by adding a slight excess of methanol to equal amounts of chloroform and aqueous (buffer or water) and was originally described as a method of extracting lipids from fish tissue.[42] The results in Fig. 2 show that asODN, like plasmid DNA,[41,44] can be extracted into chloroform using a cationic lipid.

In this experiment, 7×10-μg aliquots of an 18-mer phosphorothioate asODN complementary to the *MDR1* initiation codon (5'-GTCCCCTTC AAGATCCAT-3'), including a trace of $[^{32}\mathrm{P}]$asODN, are diluted into 0.25 ml distilled H_2O (dH$_2$O). In a separate set of microcentrifuge tubes, a dilution series of DOTAP (10 mM chloroform stock) is made by diluting 0 to 160 nmol DOTAP in 0.25 ml chloroform. Next, 0.51 ml methanol is added to each DOTAP tube, followed by 0.25 ml asODN. Each tube is mixed briefly and then inspected to ensure that a clear monophase is created. At higher DOTAP concentrations (>80 nmol), the monophase often

[42] E. G. Bligh and W. J. Dyer, *Can. J. Biochem. Physiol.* **37**, 911 (1959).
[43] F. Szoka and D. Papahadjopoulos, *Proc. Natl. Acad. Sci. USA* **75**, 4194 (1978).
[44] F. M. P. Wong, D. L. Reimer, and M. B. Bally, *Biochemistry* **35**, 5756 (1996).

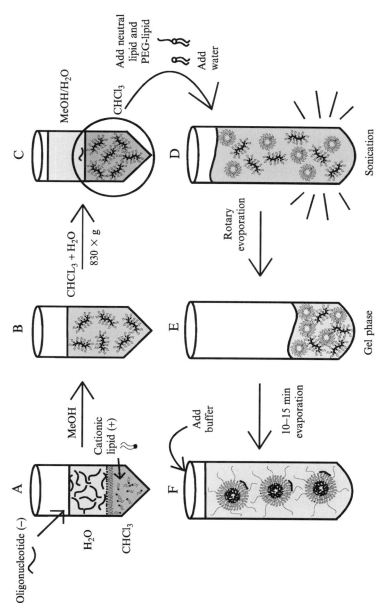

Fig. 1. Schematic demonstrating the steps and proposed intermediate structures involved in making coated cationic liposomes (CCL). (A–C) A Bligh–Dyer extraction carried out in a similar way to that described by Reimer et al.[41] for plasmid DNA. (D–F) Steps involved in the formation of reverse-phase evaporation vesicles (REVs) as described by Szoka and Papahadjopoulos.[43] Reproduced with permission from Stuart and Allen.[39]

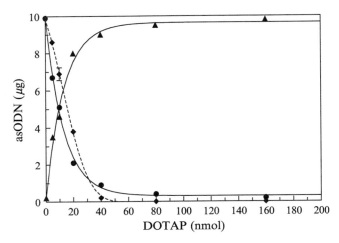

Fig. 2. Bligh–Dyer extractions of 10 μg 18-mer phosphorothioate asODN using the cationic lipid DOTAP. asODN in the organic phase (▲); asODN remaining in the aqueous phase as determined using a radiotracer, ^{32}P-asODN (●); and asODN remaining in the aqueous phase as determined by a spectrophotometric assay (◆). Mean of three experiments ± SD. Reproduced with permission from Stuart and Allen.[39]

appears slightly opaque. Samples are allowed to stand for approximately 30 min before 0.25 ml chloroform and then 0.25 ml dH$_2$O are added, and the tubes are vortexed briefly and then allowed to stand for 2–3 min. The additional chloroform and dH$_2$O cause the system to revert into two phases. The interphase can be defined more clearly by a brief (7 min) centrifugation at approximately 800g. The upper aqueous phase, which is composed of 0.5 ml of dH$_2$O and 0.51 ml methanol, is then removed and assayed for asODN. If a ^{32}P-labeled tracer is used, both aqueous and organic phases can be transferred to scintillation vials and the quantity of asODN in each phase can be determined using the specific activity. Alternatively, the absorbance of the aqueous phase at 260 nm can be determined using the appropriate blank. The amount of asODN extracted into the organic phase is assumed to be the difference between 10 μg and what was measured in the aqueous phase.

Figure 2 demonstrates that increasing the amount of cationic lipid increases the amount of asODN extracted, and 40 nmol of DOTAP will extract virtually all of the asODN into the organic phase. This corresponds to a charge ratio of approximately 1.4:1 (+/−). At a 1:1 charge ratio, approximately 90% of the asODN is extracted into the organic phase. The partitioning of the asODN into the organic phase is dependent on electrostatic interactions rather than hydrophobic forces, as neutral (zwitterionic) lipids (PC or PE) fail to extract the asODN. Furthermore, the extraction of asODN by polycationic lipids follows the same stoichiometry as for mono-cationic lipids, based on the charge ratio rather than the molar ratio of the lipid. Finally, the aqueous phase must be salt free, as ions will compete with the binding of lipid to asODN, interfering with the extraction.[39] When first attempting to make CCLs, or if a new cationic lipid is being used, it may be useful to carry out this exercise to confirm the stoichiometry of the extraction. The extraction has been carried out with up to 13 mg of asODN and 40 μmol DOTAP in volumes up to 7.5 ml. When larger volumes are used, longer and faster centrifugation may be required to produce a clearly defined interface between the phases.

There are several benefits to using the extraction procedure. In addition to forming an intermediate suitable for REV formation, the extraction optimizes the charge interaction between cationic lipid and asODN and allows for the titration of just enough cationic lipid to give a high loading efficiency without having to use excess positive charge. Figure 1C illustrates the most likely structure of the hydrophobic particles: the electrostatic interaction between the cationic lipid and the asODN produces an inverse micelle-like structure. The next step is to add the neutral "coating" lipids and form liposomes by performing a reverse-phase evaporation procedure similar to that first described by Szoka and Papahadjopoulos[43] (Fig. 1D and F).

Reverse-Phase Evaporation

Logically, to make CCLs, one begins with the amount of oligonucleotide that is required. For example, 10 mg of 18-mer oligonucleotide (M_r of approximately 6228 g/mol = 1.6 μmol) represents approximately (1.6 × 17 "+" charges/molecule) 27 μmol of "+" charges. Therefore, 27 μmol of DOTAP should be used in the extraction to give a charge ratio of 1:1 +/−. Following extraction, PC, cholesterol, and mPEG-DSPE are added to the organic phase. PC is added at a three-fold molar excess over DOTAP; so in this example 81 μmol PC would be added. Cholesterol is added at half molar ratio of DOTAP + PC; in this example, 59 μmol cholesterol would be added. mPEG-DSPE is added at 1/20 molar ratio (5 mol%) of DOTAP + PC; in this example, 5.9 μmol mPEG-DSPE would be added. The ratios of cholesterol and mPEG-DSPE are based on published observations for sterically stabilized liposomes that result in long-circulating pharmacokinetics.[45]

[45] T. M. Allen, C. B. Hansen, and D. E. Lopes de Menezes, *Adv. Drug Deliv. Rev.* **16,** 267 (1995).

The addition of neutral lipids and PEG-lipids to the organic phase does not cause aggregation or disruption of the DOTAP/asODN particles, because in organic media, where the neutral phospholipids are freely soluble, the hydrophobic interactions between the tails of the neutral phospholipids and the outward facing tails of the DOTAP/asODN inverse micelles should be minimal. Next, a volume of dH$_2$O is added such that the concentration of PC would be between 10 and 30 mM (e.g., for 10 μmol mol PC, between 0.3 and 1 ml dH$_2$O is added). The suspension is vortexed briefly and then sonicated in a high-power bath sonicator for 1 to 5 min. The emulsion should be stable enough that it will not phase separate over the course of the subsequent evaporation. It is also important in this step that a nonionic aqueous solution is used to prevent the dissociation of the DOTAP–asODN particles, which would reduce the loading efficiency.

The final steps in the formation of CCLs are predicted to be the same as for REVs,[43] which is supported by the following observations. After 5 to 15 min of rotary evaporation at 500 mmHg (the time depends on the type of container used; typically the process takes longer in a test tube than in a round-bottom flask), a gel is formed. At this stage it is predicted that, as the organic solvent is removed, the neutral lipids become less and less soluble and, through hydrophobic interactions, begin to fuse with ("coat") the DOTAP asODN inverse micellar particles. Subsequent evaporation will cause the system to revert to an aqueous solution. Care must be taken that the removal of the organic phase does not occur so rapidly that an explosive "bump" of the system occurs, resulting in a loss of material. For this reason, it is often useful to remove the sample from vacuum shortly after the gel state is reached and to agitate it gently. When the sample is reapplied to the vacuum it should begin to bubble and a small amount of aqueous solution should be observed. The vacuum can be increased to >500 mmHg to hasten the complete reversion from the gel state, taking care not to allow the gel to "bump." It is usually necessary to further agitate the mixture by hand, or even to vortex the gel, to complete the reversion. More water can also be added to maintain the volume, as a considerable amount of the initial volume seems to evaporate off with the chloroform. At this stage the system will bubble considerably under vacuum and care must always be taken to prevent "bumping." Strong intermittent vortexing is often required to solubilize completely any gel remaining on the sides of the container and particles of gel suspended in the mixture. Once bubbling stops and there are no aggregates apparent on visible inspection, CCLs are formed. Particle size measurements by dynamic light scattering indicate that the diameters range from around 300 to 600 nm.

To decrease the size following evaporation, CCLs can be extruded through polycarbonate filters using a Lipex extruder (Northern Lipids,

Vancouver, Canada). Typically, we use 200- or 100-nm pore-size filters. We have observed that if water is exchanged for buffer before extrusion, there is a dramatic decrease in the loading efficiency (<20%). Therefore, if the CCL sample requires dilution prior to extrusion, it should be diluted in dH$_2$O rather than buffers such as buffered physiological saline.

The final step in the formation of CCLs is to remove nonentrapped asODN and exchange the water for a physiological isotonic solution such as HEPES-buffered saline (HBS) (25 mM HEPES, 140 mM NaCl, pH 7.4). This may be accomplished by passing the CCLs down a Sepharose CL-4B column. The osmotic stress usually causes the diameter to decrease by approximately 10%.

Physical Characterization

By using a radio-labeled oligonucleotide as a tracer, the loading efficiency of CCLs can be determined by counting fractions collected from the Sepharose CL4B column. In our experience, 80–100% of the radioactivity elutes from the column with the lipid fractions, resulting in oligonucleotide-to-phospholipid ratios of approximately 17 nmol asODN/μmol phospholipid. On a total weight ratio, this represents approximately 74 μg asODN/mg total lipid. The final diameter of CCLs following extrusion through 200-nm filters is between approximately 150 and 200 nm, as determined by dynamic light scattering (Brookhaven B190 submicrometer particle sizer). CCLs are stable at 4° for several days without aggregating or losing entrapped asODN. Furthermore, overnight incubation in 50% human plasma results in no aggregation or loss of entrapped asODN.[38] These characteristics make CCLs good candidates for *in vivo* delivery of antisense oligonucleotides to diseased tissues such as solid tumors.

Ligand-Targeted CCLs

It is possible to produce ligand-targeted CCLs using the appropriate coupling lipids and procedures outlined previously.[46] Following extrusion (in water), CCLs are dialyzed against buffer at the appropriate pH required for the chemistry being employed. In our studies, we have included maleimide-PEG-DSPE at 1 mol% in the formulation. Following extrusion, the liposomes are dialyzed against 100 volumes of HBS, pH 7.4, for 1 h. During dialysis, free sulfhydryl groups can be generated in ligands containing amino groups, such as whole antibodies, using Traut's reagent and the –SH groups will then react with maleimide groups on the liposomes.

[46] C. B. Hansen, G. Y. Kao, E. H. Moase, S. Zalipsky, and T. M. Allen, *Biochim. Biophys. Acta* **1239**, 133 (1995).

Alternatively, Fab′ fragments or other ligands that contain free –SH group(s) can be used. A 10-mg/ml solution of the antibody in HBS, pH 8.0, is activated by adding Traut's reagent at a 15:1 molar ratio of Traut's: antibody and incubating for 1 h at room temperature. Following incubation, free Traut's reagent is removed on a Sephadex G-50 spin column, and the antibody is added to the liposomes at a 1:1000 ratio of antibody to lipid. Following overnight incubation, the noncoupled ligand is removed by separation on a Sepharose CL-4B column equilibrated in HBS, pH 7.4.

Stabilized Antisense-Lipid Particles

Like the CCL formulation, some of the primary criteria used in designing the SALP formulation are the reduction in net surface charge in the final particles and the maintenance of a high loading efficiency. To achieve this, ionizable aminolipids are used that would carry a net positive charge under slightly acidic conditions (pH $\sim$ 4.0). This strategy takes advantage of the interaction between the cationic lipid and anionic asODN under these conditions to facilitate loading, while allowing the final particles to be essentially free of surface charge when exposed to physiological pH.

SALP are prepared by a simple mixing process in which lipids are dissolved in 100% ethanol and added to asODN dissolved in an acidic buffer so that the final ethanol concentration is 40%. A typical SALP formulation consists of distearoylphosphatidylcholine (DSPC), cholesterol (CH), DODAP (Avanti Polar Lipids), and polyethylene glycol conjugated to a ceramide derivative (PEG-Cer; Northern Lipids). These lipids are dissolved in 100% ethanol and are mixed in a molar ratio of 20:45:25:10 (DSPC:CH:DODAP:PEG-Cer). A 0.4-ml aliquot of this mixture containing $\sim$10 mg of total lipid is then added slowly with continual mixing to 2.5 mg (0.6 ml) of asODN in 300 mM citrate buffer, pH 4.0. Visually, the sample will immediately begin to look milky white, but should remain translucent and be free of any aggregates. The order of addition is important: The ethanol solution of lipids should be added to the asODN. If performed in the reverse order, the high local concentration of ethanol will cause the added asODN to precipitate. Following mixing, size reduction and homogeneous particle sizes can be achieved by extrusion through 100-nm polycarbonate filters using a Lipex extruder. However, even in the absence of this step, particles have a mean diameter of approximately 150–200 nm and a reasonable size distribution. Ethanol and residual asODN are removed from the extruded particles by extensive dialysis (MW cutoff >10 kDa) or tangential flow diafiltration (MicroKros Ultrafiltration Module, >50-kDa pore size; Spectrum Microgen, Laguna Hills, CA) against PBS or HBS buffers, pH 7.4. Anion-exchange chromatography (DEAE-Sepharose

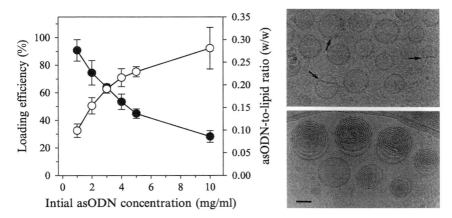

FIG. 3. Influence of asODN-to-lipid ratio on the loading and structure of SALP. (Left) The relationship between initial asODN-to-lipid ratio, loading efficiency (○) and final asODN-to-lipid ratios (●). (Right) Cryoelectron micrographs of SALP. The structure of SALP was evaluated at low (top right) and high (bottom right) initial ODN-to-lipid ratios [0.025 (w/w) and 0.25 (w/w), respectively]. Magnification in each panel was 50,000×. The size is represented by the bar (50 nm). Reproduced with permission from Semple *et al.*[47]

CL-6B) is also used sometimes to ensure complete removal of residual asODN. With this procedure, one should obtain particles with a mean diameter of approximately 110–120 nm, a loading efficacy of 50–70%, and a final asODN-to-lipid ratio of 0.17–0.19 (wt asODN/wt total lipid). The basic formulation described earlier can be scaled volumetrically as required, and it is suggested that formulations of at least 2–3 ml be prepared to minimize mechanical losses, particularly in the extrusion and dialysis steps.

An important consideration in controlling the final loading efficiencies and asODN-to-lipid ratios is the initial asODN-to-lipid ratio (Fig. 3A). Depending on the requirements of a given study, the process can be modified to increase or decrease loading efficacy at the expense of the final asODN-to-lipid ratio. Changing the initial ratio also has important implications for particle formation (Fig. 3B). At high asODN-to-lipid ratios (>0.10 wt/wt) the particles tend to form small multilamellar structures, whereas at initial ratios less than 0.05 (wt/wt) the particles tend to be predominantly unilamellar vesicles.[47] The ability to modulate readily the structure of the particles might have important consequences for the desired application. For example, the unilamellar structures would likely be more

---

[47] S. C. Semple, S. K. Klimuk, T. O. Harasym, N. Dos Santos, S. M. Ansell, K. F. Wong, N. Maurer, H. Stark, P. R. Cullis, M. J. Hope, and P. Scherrer, *Biochim. Biophys. Acta* **1510**, 152 (2001).

appropriate for antisense applications, whereas the multilamellar structures would likely have greater utility for the delivery of immunostimulatory oligonucleotides to antigen-presenting cells and macrophages.

Another important design feature of the SALP system is the inclusion of a PEG-lipid. In the method described previously, 10 mol% PEG-Cer is included in the formulation. This can be reduced to as low as 2–3 mol% of PEG-lipid relative to the other lipid components. The relatively high levels of ethanol and the fusogenic nature of the DODAP necessitate that some PEG-lipid, or other steric barrier lipid, be present in the formulation, as the particles will readily fuse and aggregate. Ethanol levels >30% (vol/vol) are also important to help fluidize the membrane and drive the loading process. Interestingly, significant loading can occur following the addition of asODN to the exterior of preformed vesicles, if sufficient ethanol is present in the mixture.[48]

One final, and very important, consideration for the SALP formulation involves the type of asODN chemistry used. The basic method described earlier is appropriate for phosphorothioate asODN, which are relatively insensitive to salt and buffer levels (Fig. 4). However, if native phosphodiester asODN are to be used, modifications to the formulation conditions must be made. We have found that phosphodiester ODN can be loaded

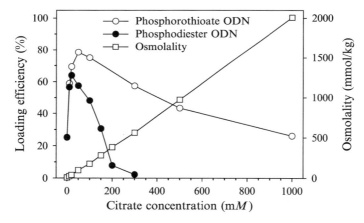

Fig. 4. Dependence of citrate concentration on the loading of phosphodiester asODN in SALP. The influence of citrate concentration in the initial ODN solution was evaluated over a range of citrate concentrations, pH 4.0. The osmolality of the buffers (0–1000 mM) was confirmed using a vapor pressure osmometer.

[48] N. Maurer, K. F. Wong, H. Stark, L. Louie, D. McIntosh, T. Wong, P. Scherrer, S. C. Semple, and P. R. Cullis, *Biophys. J.* **80,** 2310 (2001).

at high efficiency (>60%) using 10 mM citrate, pH 4.0. If 300 mM citrate is used, loading will be <3%. Basic methodologies for additional characterization of these particles and other liposomal delivery system for asODN are described in another volume of this series.[29]

Examples

The goal in designing CCLs or SALP was to produce an antisense carrier that is small in diameter (<200 nm), efficient at loading oligonucleotides, targetable, stable in plasma, displays long-circulating pharmacokinetics, and is functionally active. The following examples illustrate the progress made in these areas.

As described in previous sections, the CCL formulation can be extruded to less than 200 nm and still maintain >80% loading efficiency. Ligand-mediated targeting has been demonstrated with several cell lines with two different ligands. Disialoganglioside (GD_2) is expressed extensively on tumors of neuroectodermal origin.[49,50] CCL entrapping asODN against the c-myb oncogene were targeted to human neuroblastoma cell lines (ACN and GI-LI-N) using anti-GD_2 monoclonal antibodies.[51] There was a 10-fold increased uptake of c-myb asODN, delivered using anti-GD_2-CCL, compared to nontargeted or isotype-matched CCL in GI-LI-N cells. Uptake could be inhibited by excess free GD_2 antibody. Uptake of anti-GD_2-CCL by cells not expressing GD_2 (HL-60, HeLa) was no different than nontargeted CCLs. Anti-GD_2-CCL delivery of c-*myb* asODN resulted in greater downregulation of the c-myb protein and inhibition of cell proliferation than c-myb asODN delivered using nontargeted CCL.[51] Functional delivery of c-myc asODN has also been demonstrated with anti-GD_2-CCL in GD_2-positive melanoma cell lines (MZ2-MEL, COLO 853).[40] In these studies, delivery of c-myc asODN also correlated with a reduction in c-myc protein levels and decreases in cell proliferation.

The results in Fig. 5 illustrate another example of ligand-mediated targeting of CCL. The anti-CD19 monoclonal antibody was coupled to CCL to deliver an asODN against the mRNA of the multidrug resistance gene, *MDR1,* in a human B-cell line (Namalwa) that expresses the *MDR1* gene

[49] M. M. Uttenreuther-Fischer, C. S. Huang, and A. L. Yu, *Cancer Immunol. Immunother.* **41,** 331 (1995).

[50] J. D. Frost, J. A. Hank, G. H. Reaman, S. Frierdich, R. C. Seeger, J. Gan, P. M. Anderson, L. J. Ettinger, M. S. Cairo, B. R. Blazar, M. D. Krailo, K. K. Matthay, R. A. Reisfeld, and P. M. Sondel, *Cancer* **80,** 317 (1997).

[51] G. Pagan, D. D. Stuart, F. Pastorino, L. Raffaghello, P. G. Montaldo, T. M. Allen, B. Calabretta, and M. Ponzoni, *J. Natl. Cancer Inst.* **92,** 253 (2000).

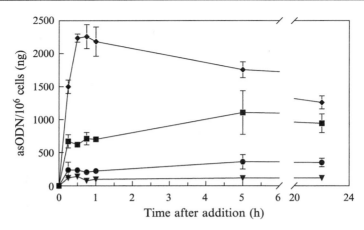

FIG. 5. Cellular association of *MDR1* phosphorothioate asODN using different delivery vehicles. Values were determined using a slot-blot assay hybridization assay similar to that described by Temsamani *et al.*[52] Free asODN (▼); anti-CD19-targeted CCL (■); nontargeted CCL (●); and asODN associated with LipofectAMINE (◆). Each point represents the mean ± SD of three different samples.

product P-glycoprotein (P-gp). This example serves to illustrate an important point. CCL or ligand-targeted CCL may not deliver as much asODN as traditional cationic lipoplexes (LipofectAMINE). Ligand-mediated delivery, however, provides specificity that is not available from cationic lipoplexes. In this experiment, LipofectAMINE increased dramatically the cellular uptake of asODN compared to free asODN; the net positive charge of these lipoplexes results in electrostatic binding with negatively charged cell surfaces. Conversely, CCLs are formulated to carry no excess surface positive charge and therefore they have no tendency to bind to cell membranes. Furthermore, they contain PEG-lipid, which provides a steric barrier that inhibits interactions with cell membranes and plasma proteins. The levels of asODN delivered with nontargeted CCLs are very close to those for free asODN. Delivery of asODN is increased significantly by the use of a targeting ligand. There is approximately a fourfold increase in the amount of asODN delivered using anti-CD19 CCL compared to nontargeted CCL. Delivery is still lower than for LipofectAMINE; however, it is specific and the long-circulating characteristics of CCLs (see later) should provide sufficient access to cell-associated target antigen to allow targeting to be effective *in vivo*.

[52] J. Temsamani, M. Kubert, and S. Agrawal, *Anal. Biochem.* **215,** 54 (1993).

The CCL and SALP formulations were designed to carry no excess positive charge, and the addition of PEG-lipid provides steric stabilization, which results in extended circulation times in the bloodstream. Indeed, this has been observed for both nontargeted CCL and anti-GD_2-CCL (unpublished results) as well as SALPs.

The pharmacokinetics and biodistribution of an 18-mer phosphorothioate asODN against the *MDR1* formulated in CCL was evaluated following iv administration into female CD_1/ICR mice. Figure 6 shows the blood clearance profile following a dose of 50 μg asODN. The clearance of the asODN followed a biphasic elimination curve, with the first phase representing approximately 30% of the injected dose. Twenty-four hours following injection, 10% of the injected dose was still circulating. The SALP formulation gave a similar long-circulating profile (separate study), with a slightly higher percentage in blood at earlier time points and a slightly lower percentage in blood at later time points compared to the CCL formulation. The blood clearance profiles of both formulations are quite similar to that of a typical, noncationic, stealth formulation.[38] Figure 6 also illustrates the blood clearance of cationic lipoplexes composed of DC-Chol:DOPE (1:1) mixed at a 4:1 +/− charge ratio with an 18-mer phosphorothioate asODN, including a trace of ^{125}I-labeled asODN (data from Litzinger et al.[25]). The DC-Chol cationic lipoplexes are cleared from the bloodstream, resulting in 10- to 40-fold lower blood levels of asODN compared to CCLs and SALPs.

The lung and liver account for the greatest proportion of the injected dose of the DC-Chol:DOPE lipoplexes, with some evidence for transient

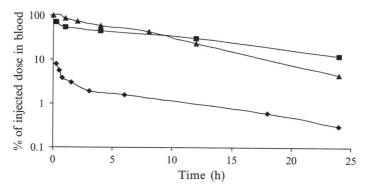

FIG. 6. Blood clearance of phosphorothioate asODN formulated in CCL (■), SALP (▲), and cationic lipoplexes composed of DC-Chol:DOPE, 1:1 molar ratio at a 4:1 +/− charge ratio (◆). CCL were dosed at 2 mg/kg asODN (GTCCCCTTCAAGATCCAT), SALP were dosed at 15 mg/kg asODN (TGCATCCCCCAGGCCACCAT), and lipoplexes were dosed at approximately 2 mg/kg asODN. Lipoplex data are from Litzinger et al.[25]

accumulation in the lung before redistribution to the liver. This would suggest that DC-Chol cationic lipoplexes would have very little utility in delivering asODN to other tissues, whereas CCLs or SALP should allow passive targeting to diseased tissues, such as solid tumors, by the mechanism of extravasation through leaky tumor vasculature. Passive accumulation into human melanoma xenografts in nude mice has been demonstrated for SALP formulation.[36] Lipoplexes would most likely result in greater levels of delivery and activity of asODN in cells *in vitro* due to their net positive charge. The important point here is that the optimal formulation depends on the circumstances: *in vitro,* polycationic particles may be most active (perhaps more toxic as well), whereas *in vivo,* small stable particles will be required to reach target tissues.

A c-myc asODN formulated in SALP resulted in c-myc protein down-regulation and significant inhibition of tumor growth compared to control groups in a human melanoma (NG) mouse xenograft model.[36] Free c-myc asODN also had significant impact on c-myc protein expression and primary tumor growth; however, the benefit of the SALP formulation was that it increased significantly the duration of action of the asODN and showed a dramatic enhancement in the reduction of lung metastases relative to the free c-myc ODN (76 and 26% reductions, respectively, unpublished data).

Summary

Current efforts are aimed at developing models in which to test both targeted and nontargeted CCLs *in vivo*. Until very recently it has not been possible to ask properly the question of whether a liposomal carrier will increase the activity of asODN, because when they are formulated into cationic lipoplexes they have a very poor PK/biodistribution profile. With the recent description of CCL formulation[37–39] and SALP formulation,[29,36,47] which have the capability of delivering asODN into diseased tissues such as solid tumors, we can begin to address this question and optimize and refine further the characteristics that will make active liposomal asODN formulations *in vivo*.

[12]   Cationic Liposomes/Lipids for Oligonucleotide Delivery: Application to the Inhibition of Tumorigenicity of Kaposi's Sarcoma by Vascular Endothelial Growth Factor Antisense Oligodeoxynucleotides

By Carole Lavigne, Yanto Lunardi-Iskandar,
Bernard Lebleu, and Alain R. Thierry

Introduction

Antisense oligodeoxynucleotides have been evaluated as therapeutic agents for the treatment of viral infections, cancer, and genetic disorders and to study the functional roles of various proteins. However, the efficacy of oligonucleotide (ODN) therapeutics appears to be limited in practice by several problems such as stability, cellular uptake, subcellular availability, and nonspecific effects. To overcome these problems, some chemically modified oligodeoxynucleotides have been designed. Phosphorothioate-modified oligonucleotides, in which an oxygen atom of the phosphodiester linkage has been replaced by a sulfur atom, are the most commonly used ODN analogs. Although the synthesis of nuclease-resistant ODN analogs has improved the application of antisense ODNs, modified ODNs do not show a significantly higher penetration rate into cells and have the same limitations as unmodified ODNs regarding cellular drug internalization. Therefore, effective clinical applications of these agents will most probably benefit from carriers that can improve their intracellular delivery. As illustrated by the few commercialized ODN therapeutic products, which are administrated by a local route (intraocular route for the anti-CMV ODN product from Isis, Carlsbad, CA), use of a carrier might broaden clinical application by making possible the systemic administration of ODN.

Nevertheless, these issues are not unique to ODN compounds; many conventional drugs also have problems with stability, cellular uptake, solubility, or toxicity. To deal with these pharmaceutical problems, several technologies have been developed since the early 1980s under the generic topic of "drug delivery systems."[1] The optimal delivery system for nucleic acid therapeutics should be nontoxic, protect the nucleic acid from the action of nucleases, deliver high levels of the therapeutic agent to the targeted cell types or tissues, and permit release of the nucleic acid once inside the cell to allow accumulation of these molecules at the intracellular site of

---

[1] M. Poznanski and R. L. Juliano, *Pharmacol. Rev.* **36,** 277 (1984).

action. One class of most promising delivery systems for nucleic acids are certainly the lipid-based carrier systems. The objective of this article is to provide an overview of progresses in the development of synthetic lipid-based systems and their therapeutic potential.

Advantages of Lipid-Based Carrier Systems

Liposomes were first characterized by Bangham[2] and have been utilized as a drug delivery system only fairly recently. Liposomes are generally defined as microscopic closed vesicles composed of phospholipid bilayer membranes surrounding an internal aqueous compartment in which various drugs, antibiotics, antitumor agents, hormones, and viruses can be entrapped. Liposomes are usually characterized in terms of their size, the number of concentric bilayers (lamellae), and the composition and physical properties of the lipids used.[3] According to the nomenclature for liposomes accepted at the New York Academy of Sciences meeting on "*Liposomes and Their Uses in Biology and Medicine*" in 1978, all types of lipid bilayers surrounding an aqueous space are considered to be in the general category of liposomes. Multilamellar vesicles (MLV) are very heterogeneous in size, ranging up to several micrometers in diameter. Unilamellar vesicles are classified into two categories: small unilamellar vesicles (SUV) that tend to be very uniform with particle diameters usually under 1000 Å and large unilamellar vesicles (LUV) with a greater diameter.

The advantages of liposomal carriers are numerous. Liposomes have been shown to protect the oligonucleotides against nucleases in biological fluids via encapsulation or complexation; increase cellular uptake and, therefore, the efficacy and therapeutic index; reduce toxicity of the encapsulated or complexed agent; have low immunogenicity; improve pharmacokinetics by reducing elimination and increasing circulation life times; allow repetitive administrations; and provide cellular targeting when coupled with antibodies or specific ligands.

Cationic lipid particles may have advantages over conventional encapsulating liposomes for the delivery of high levels of nucleic acids in cell culture and *in vivo*.[4–7] Cationic liposome-based delivery systems are being

[2] A. D. Bangham, *J. Mol. Biol.* **13**, 238 (1965).

[3] F. Szoka and D. Papahadjopoulos, *Annu. Rev. Biophys. Bioeng.* **9**, 467 (1980).

[4] A. R. Thierry, Y. Lunardi-Iskandar, J. L. Bryant, P. Rabinovich, R. C. Gallo, and L. C. Mahan, *Proc. Natl. Acad. Sci. USA* **92**, 9742 (1995).

[5] A. R. Thierry and L. C. Mahan, *in* "Advanced Gene Delivery" (A. Rolland, ed.), p. 123. Harwood Academic, Amsterdam, 1998.

[6] A. R. Thierry, P. Rabinovich, B. Peng, I. C. Mahan, J. L. Bryant, and R. C. Gallo, *Gene Ther.* **4**, 226 (1997).

evaluated in phase I and phase II clinical trials for the treatment of a variety of different types of human cancer and for the treatment of cystic fibrosis.[5,8,9]

Cationic Liposomes/Lipids

Negatively charged liposomes were the first liposomes used as vehicles for the gene transfer for *in vitro* and *in vivo* transfection processes. These liposome formulations presented some limitations associated with a low efficiency of nucleic acid encapsulation, DNA degradation induced by sonication, and the requirement to separate the DNA–liposome complexes from empty, "ghost" vesicles. In order to overcome these limitations and to increase the overall efficiency of nucleic acid delivery into the cells, positively charged liposomes were designed and used as delivery vectors.[10,11] Cationic liposomes form complexes with DNA through charge interactions. Cationic liposomes normally contain a cationic amphiphile and a neutral "helper" lipid such as dioleoylphosphatidylethanolamine (DOPE) to stabilize the lipid bilayer. All cationic lipids contain three different domains: a positively charged head group(s), a linker bond, and a hydrophobic anchor. The head group usually contains a simple or multiple amine groups and is responsible for the interactions between liposome and DNA and between the liposome–DNA complex and the cell membrane or other components inside the cell. The linker bond is responsible for the chemical stability and biodegradability of a cationic lipid, whereas the hydrophobic anchors help maintain the integrity of the bilayer membrane.[12]

Dioctadeayl amidoglycyl spermidine (DOGS) is a quaternary lipopolyamine, a class of cationic amphiphiles in which the cationic spermine group is connected to a double chain (C16:0) lipophilic group via an amidoglycyl spacer arm. Behr *et al.*[13] found that this compound was particularly efficient in compacting DNA and in transfecting a variety of cell types

[7] F. Pastorino, D. Stuart, M. Ponzoni, and T. M. Allen, *J. Control. Release* **74**, 69 (2001).

[8] N. J. Caplen, X. Gao, P. Hayes, R. Elaswarapu, G. Fisher, E. Kinrade, A. Chakera, J. Schorr, B. Hughes, J. R. Dorin, D. J. Porteous, E. W. F. W. Alton, D. M. Geddes, C. Coutelle, R. Williamson, L. Huang, and C. Gilchrist, *Gene Ther.* **1**, 139 (1994).

[9] N. J. Caplen, E. W. Alton, P. G. Middleton, J. R. Dorin, B. J. Stevenson, X. Gao, S. R. Durham, P. K. Jeffery, M. E. Hodson, C. Coutelle *et al.*, *Nature Med.* **1**, 39 (1995).

[10] P. L. Felgner, T. R. Gadek, M. Holm, R. Roman, H. W. Chan, M. Wenz, J. P. Northrop, G. M. Ringold, and M. Danielsen, *Proc. Natl. Acad. Sci. USA* **84**, 7413 (1987).

[11] J. P. Behr, B. Demeneix, J. P. Loeffler, and J. Perez-Mutul, *Proc. Natl. Acad. Sci. USA* **86**, 6982 (1989).

[12] X. Gao and L. Huang, *Gene Ther.* **2**, 710 (1995).

[13] J. P. Behr, *Tetrahedron Lett.* **27**, 5861 (1986).

without cellular toxicitiy.[11] In addition, DOGS offers a unique combination of properties such as cell membrane destabilization, endosome buffering capacity, and cellular kariophily (possibly nuclear tropism).[14] Previous studies reported a lipidic delivery system consisting of DOGS and DOPE, termed DLS, which is reproducible and stable, exhibits structural and particle mean size homogeneity, and results in high efficiency for *in vitro* and *in vivo* delivery of nucleic acids.[6,15–17] DLS liposomes were shown to destabilize cellular membranes and enhance transfection efficiency when compared to other cationic lipids.[18] The electrostatic interaction between negative charges on nucleic acids and positive charges on DOGS leads to DNA condensation and to DLS-DNA particle formation. Electron microscopic analysis showed the presence of lamellar vesicles when DLS-DNA was prepared at a 0.6 DOGS:DOPE molar ratio and a 0.008 to 0.16 DNA:lipid weight ratio.

Cationic Liposome–DNA Complexation and Lipoplexes

It was generally assumed that in contrast to negatively charged liposomes, cationic vesicles do not encapsulate or entrap the nucleic acids but bind them at their surface while maintaining their original size and shape. Gershon *et al.*[19] have proposed a model for cationic liposome–DNA complexation. At low ratios of liposomes to DNA, positive vesicles are adsorbed to nucleic acids to form aggregates that gradually surround larger segments of the DNA. As the amount of liposomes is increased, the aggregated liposomes along the DNA reach critical concentrations and charge densities at which membrane fusion and cooperative DNA collapse processes are initiated. Upon additional increase of the liposome concentration, the collapsed DNA structures are covered completely by the lipid bilayers. The size of the final product is dependent on the size of the initial lipid vesicles, the positive-to-negative charge ratio, the ionic strength of the medium, and the concentration of all reagents. DNA condensation in multilamellar complexes seems to be caused by mechanisms independent of the length of the individual DNA molecules[20,21] and the absolute concentration of the nucleic acids.[19] We have demonstrated[22] that a similar

[14] J. P. Behr, *Bioconj. Chem.* **5,** 382 (1994).

[15] C. Lavigne and A. R. Thierry, *Biochem. Biophys. Res. Commun.* **237,** 566 (1997).

[16] C. Lavigne, J. Yelle, G. Sauvé, and A. R. Thierry, *AAPS Pharmsci.* **3,** article 7 (2001).

[17] A. R. Thierry, *J. Liposome Res.* **7,** 143 (1997).

[18] P. Pinnaduwage and L. Huang, *Biochim. Biophys. Acta* **985,** 33 (1989).

[19] H. Gershon, R. Ghirlando, S. B. Guttman, and A. Minsky, *Biochemistry* **32,** 7143 (1993).

[20] Z. Reich, R. Ghirlando, and A. Minsky, *Biochemistry* **30,** 7828 (1991).

[21] V. A. Bloomfield, *Biopolymers* **31,** 1471 (1991).

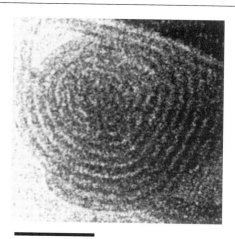

FIG. 1. Ultrastructure of a typical Neutraplex-ODN particle as observed by electron cryomicroscopy. Bar represents 50 nm. Image retrieved from Schmutz *et al.*[22]

structural morphology can be generated in the DNA–cationic lipid complex formulated with different structures and sizes of nucleic acids as with pDNA (circular supercoiled DNA of 10.4 kbp, MW ≈ 6,870,600) or with ODN (linear single-stranded DNA of 30 bases, MW = 9900). Using cryoelectron microscopy, a distinct concentric ring-like pattern with striated shells was observed when using linear double-stranded DNA, single-stranded DNA, and ODN (Fig. 1). DNA chains could be visualized in DNA/lipid complexes. In this study,[22] cationic lipid formulations were the result of meticulous optimization to enable the production of efficient gene delivery particles with a high level of reproduction and stability (Neutraplex particles). They were composed of a polycationic lipid (DOGS), a neutral lipid (DOPE), and an anionic phospholipid (cardiolipin) that interacted with DNA, a polyanionic macromolecule. Results indicated that both lamellar and hexagonal phases may coexist in the same lipoplex preparation or particle and that the transition between both phases may depend on equilibrum influenced by the type and the length of the DNA used.

Cationic lipid formulations such as lipofectin and lipofectamine (Gibco BRL), DOTAP (Boehringer-Mannheim, Wannheim, Germany), or DLS[15] made the formation of lipid/DNA complexes possible, which, in combination with a fusogenic lipid, showed high transfection efficacy in cell culture.

---

[22] M. Schmutz, D. Durand, A. Debin, Y. Palvadeau, A. Etienne, and A. R. Thierry, *Proc. Natl. Acad. Sci. USA* **96,** 12293 (1999).

Although often termed cationic liposomes, these complexes form clusters of lipid molecules and not a vesicularized membrane bilayer, and therefore lack an internal aqueous space (Fig. 1). It thus seems more appropriate to call them "nucleo-lipidic particles" or "lipoplexes" as proposed elsewhere.[23]

Cellular Uptake and Intracellular Distribution of Liposome–DNA Complex

Fusion between cationic liposomes and cell surfaces resulting in delivery of the entrapped nucleic acids directly across the membrane was the mechanism suggested a decade ago for delivery of nucleic acids.[10] The cellular uptake of liposomes is now generally considered to be mediated by the adsorption of liposomes onto the cell surface followed by endocytosis. Fusion of the liposome with endosomal membranes (or membrane destabilization) appears to be necessary for the effective delivery of the nucleic acids from the liposome to the cell. An understanding of the mechanism of liposome uptake by cells and the method of drug delivery by liposomes is now emerging.[6,12,24–26] Zelphati and Szoka[26] have proposed a model that describes oligonucleotide release from the endosome. After internalization of the DNA–liposome complexes by endocytosis, the endosomal membrane is destabilized, which causes a flip-flop of anionic lipids from the cytoplasmic face of the endosome into the internal face. The anionic lipids interact with the cationic lipids, displacing the nucleic acid from the cationic lipid and allowing the release of the anionic nucleic acid from the lipolex into the cytoplasm of the cell.

Cationic lipid vesicles can enhance the rate of ODN uptake into cells, change the subcellular distribution of ODN, and also enhance antisense activity by increasing the rate of ODN hybridization to its target mRNA.[27] Using autoradiography and fluorescent and laser-assisted confocal microscopy, we have observed increased cellular uptake and change in

[23] P. L. Felgner, Y. Barenholz, J. P. Behr, S. H. Cheng, P. Cullis, L. Huang, J. A. Jessee, L. Seymour, F. Szoka, A. R. Thierry, E. Wagner, and G. Wu, *Hum. Gene Ther.* **8,** 511 (1997).

[24] C. R. Miller, B. Bondurant, S. D. McLean, K. A. McGoven, and D. F. O'Brien, *Biochemistry* **37,** 12875 (1998).

[24a] P. Pires, S. Simões, N. Düzgüneş, and M. C. Pedross de Lima, *Biochim. Biophys. Acta* **1418,** 71 (1999).

[25] F. C. Szoka, Jr., Y. Xu, and O. Zelphati, *J. Liposome Res.* **6,** 567 (1996).

[25a] M. C. Pedroso de Lima, S. Simões, P. Pires, H. Faneca, and N. Düzgüneş, *Adv. Drug Deliv. Rev.* **47,** 277 (2001).

[26] O. Zelphati and F. C. Szoka, Jr., *Proc. Natl. Acad. Sci. USA* **93,** 11493 (1996).

[27] B. W. Pontius and P. Berg, *Proc. Natl. Acad. Sci. USA* **88,** 8237 (1991).

the intracellular distribution of plasmid DNA[6] or ODN[28,29] following lipo-some-mediated delivery. Fluorescein (FITC)-labeled ODN delivered using the DLS carrier system was taken up avidly by HeLa ovarian carcinoma cells and released rapidly from endocytotic vesicles, resulting in a high cytoplasmic and nuclear distribution of DNA (Fig. 2). In the nucleus, DNA was found particularly in the perinuclear and the chromosomal/nucleoli compartments. Use of the DLS vector allows uptake and intracel-lular localization of a peptide nucleic acid (PNA) ODN analog to KS-Y1 Kaposi's sarcoma cells (Fig. 3). Another study[28] investigated the cellular uptake and intracellular distribution of ^{125}I-labeled ODN delivered by the DLS lipid-based system in human fibrosarcoma HT-1080 cells. In this study, the use of cationic liposomes increased significantly and accele-rated the uptake of ^{125}I-labeled ODN. When free ^{125}I-labeled ODN were used, the radioactivity associated with the cells increased gradually and reached a plateau at 5% after 300 min of incubation. When ^{125}I-labeled ODN were delivered via liposomes, the maximum uptake was 35% and was reached after only 100 min. These observations suggest that complete release of the DNA from endocytotic vesicles can be achieved and sup-port the notion of the complete or partial release of the DNA from the liposomal carrier.

Use of Lipoplexes and Antisense Sequence Specificity

We have evaluated the specificity and activity of antisense ODN against HIV infection regarding a series of factors such as dose–response range, number and choice of experimental controls, backbone modifications of the oligonucleotides, type of cell infection, length of assays, and delivery by the DLS lipoplex system.[30] The highest level of inhibition was achieved in a long-term assay (30-day culture period) using lymphoma MOLT-3 cells infected acutely with HIV-1 (IIIB) and treated with free phosphorothiate (PS) oligonucleotides (ODN). The highest level of specificity was observed in our short-term assay (14-day culture period) using MOLT-3 cells in-fected acutely with HIV-1 (IIIB) and treated with free PS-ODNs. The highest potency (IC$_{50}$ level) was observed in a short-term chronic infection model with DLS-delivered ODNs where the DLS delivery improved the ODN activity up to 10[6] times compared to free ODNs. However, the near blocking of HIV replication obtained when using PS ODN appears due to

[28] O. A. Sedelnikova, I. G. Panyutin, A. R. Thierry, and R. D. Neumann, *J. Nuclear Med.* **39,** 1412 (1998).
[29] A. R. Thierry and A. Dritschilo, *Nucleic Acids Res.* **20,** 5691 (1992).
[30] C. Lavigne, J. Yelle, G. Sauvé, and A. R. Thierry, *AAPS Pharmsci.* **4,** article 9 (2002).

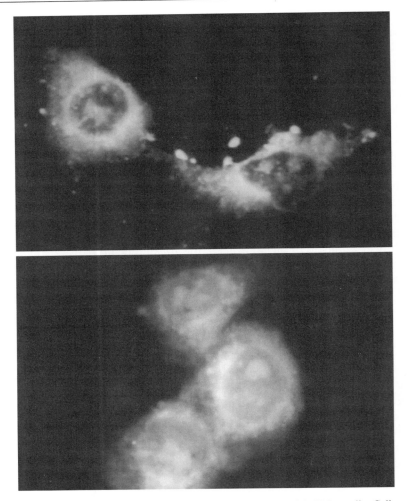

Fig. 2. Cellular localization of ODN following incubation with HeLa cells. Cells were treated with 1 μM FITC-labeled 16-mer ODN with the DLS system and observed following a 4 (top)- or 24 (bottom)-h incubation period. Each photograph represents images from fluorescent and laser-assisted confocal microscopy of live cells. Magnification, ×530.

the addition of extracellular and/or membrane effects. The higher efficacy of PS-ODNs, when compared to unmodified ODNs both delivered with the DLS system, was solely demonstrated in our short-term assay with MOLT-3 cells. Significant variations of the level of sequence specificity was observed on the type of control used and the type of cell assay employed.[30]

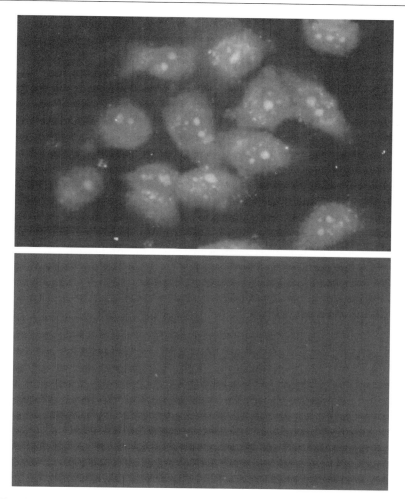

Fig. 3. Cellular localization of ODN following incubation with KS-Y1 cells. Cells were treated with 1 μM FITC-labeled 16-mer PNA with the DLS system (top) or in a free form (bottom). Each photograph represents images from fluorescent and laser-assisted confocal microscopy of live cells following 24 h of incubation. Magnification, ×320.

It seems that all control ODN tested (random, sense sequence and non-antisense T30177 ODN) have distinct activity and, consequently, different modes of action in inhibiting HIV replication. Those data buttress the notion that the contribution of the sequence-specific mode of action is minor compared to the other mechanisms involved in ODN antiviral activity.

Use of a delivery system highlights the potential of nonantisense specific activity and of ODN. Strategies such as RNAi or decoy ODN appear more promising for the development of pharmaceutical ODN toward HIV infection.

Targeting of Liposomes

Different approaches have been investigated to target liposomes by the attachment of antibodies[31–36] or cellular receptors ligands,[7,37–39] which may provide delivery via cellular receptor-mediated uptake mechanisms. The choice of the targeting ligand is important when designing targeted liposomes. The ligand should be relatively specific to the targeted cells, especially in contrast to cells readily accessible in the general circulation, and the bound epitope should result in internalization of the liposome.[40] Indeed, targeting to internalizing receptors can potentially increase the bioavailability of the drug, as demonstrated in studies with anti-HER2 immunoliposomes.[41,42] When anti-HER2-targeted immunoliposomes were prepared with an antibody that was not internalized, there was no increase in therapeutic efficacy compared with nontargeted liposomes.[40]

Using cationic lipid particles conjugated to ferric protoporphyrin IX (heme), we were able to increase oligoribonucleotide (ribozyme) uptake into HepG2 human hepatoma cells when compared with the same lipid

[31] J. P. Leonetti, P. Machy, G. Degols, B. Lebleu, and L. Leserman, *Proc. Natl. Acad. Sci. USA* **87**, 2448 (1990).

[32] C. B. Hansen, G. Y. Kao, E. H. Moase, S. Zalipsky, and T. M. Allen, *Biochim. Biophys. Acta* **1239**, 133 (1995).

[33] D. D. F. Ma and A.-Q. Wei, *Leukemia Res.* **20**, 925 (1996).

[34] J. W. Park, K. Hong, D. B. Kirpotin, O. Meyer, D. Papahadjopoulos, and C. C. Benz, *Cancer Lett.* **118**, 153 (1997).

[35] M. Mercadal, C. Carrion, J. C. Domingo, J. Petriz, J. Garcia, and M. A. de Madariaga, *Biochim. Biophys. Acta* **1371**, 17 (1998).

[36] I. Dufresne, A. Désormeaux, J. Bestman-Smith, P. Gourde, M. J. Tremblay, and M. G. Bergeron, *Biochim. Biophys. Acta* **1421**, 284 (1999).

[37] H. Schreier, P. Moran, and I. W. Caras, *J. Biol. Chem.* **269**, 9090 (1994).

[38] J.-S. Remy, A. Kichler, V. Mordvinov, F. Schuber, and J.-P. Behr, *Proc. Natl. Acad. Sci. USA* **92**, 1744 (1995).

[39] G. B. Takle, A. R. Thierry, S. M. Flynn, B. Peng, L. White, W. Devonish, R. A. Galbraith, A. R. Goldberg, and S. T. George, *Antisense Nucl. Acid Drug Dev.* **7**, 177 (1997).

[40] D. C. Drummond, O. Meyer, K. Hong, D. B. Kirpotin, and D. Papahadjopoulos, *Pharmacol. Rev.* **51**, 691 (1999).

[41] D. Goren, A. T. Horowitz, S. Zalipsky, M. C. Woodle, Y. Yarden, and A. Gabizon, *Br. J. Cancer* **74**, 1749 (1996).

[42] D. B. Kirpotin, J. W. Park, K. Hong, Y. Shao, R. Shalaby, G. Colbern, C. C. Benz, and D. Papahadjopoulos, *J. Liposome Res.* **7**, 391 (1997).

particles prepared without heme.[39] In this study, heme was conjugated to the amino lipid DOPE and used to form cationic lipid particles with dioleoyl trimethylammonium propane (DOTAP). Fluorescence microscopy showed that these targeted cationic lipids delivered oligoribonucleotides into both the cell cytoplasm and nucleus and that they may thus be potentially useful delivery vehicles for oligonucleotide-based therapeutics and transgenes, appropriate for use in liver diseases. Several questions involving *in vivo* applications of targeted DNA–liposome complexes remain to be answered, however. To mention a few, the stability of the liposome formulation in the circulation, the ability to make the nucleic acids bioavailable at the tumor site, the speed of release of the nucleic acids in the circulation and in the targeted tissue, and immunogenicity.

Pharmacological Considerations

Although a large number of studies have demonstrated that cationic lipids as synthetic carrier systems can deliver DNA to cells efficiently in tissue culture, their transfection efficiency in systemic administration is limited. Only a few reports have shown promising data with respect to efficient DNA delivery *in vivo*.[43–47] In light of the pharmacological considerations, technological development of lipid-based delivery systems has specific needs (Table I) that are broad in nature and thus difficult to combine in contrast to conventional drugs. (Several factors may influence the efficiency of *in vivo* DNA delivery.) The main problems encountered when cationic liposomes are used for DNA delivery *in vivo* are instability in the biological fluids and lack of target specificity. In addition, *in vivo* nucleic acid delivery is not only concerned with cellular barriers (which include the plasma membrane, the endosome, and the nuclear membrane), as discussed earlier, but also with extracellular barriers, such as posed by a specific tissue or the immune system. The blood–brain barrier, the multiple connective tissue layers found in muscle, epithelial cell linings, passive

[43] I. Aksentijevich, I. Pastan, Y. Lunardi-Iskandar, R. C. Gallo, M. M. Gottesman, and A. R. Thierry, *Hum. Gene Ther.* **7,** 1111 (1996).

[44] C. J. Wheeler, P. L. Felgner, Y. J. Tsai, J. Marshall, L. Sukhu, S. G. Doh, J. Hartikka, J. Nietupski, M. Manthorpe, M. Nichols, M. Plewe, X. Liang, J. Norman, A. Smith, and S. H. Cheng, *Proc. Natl. Acad. Sci. USA* **93,** 11454 (1996).

[45] S. Li and L. Huang, *J. Liposome Res.* **7,** 207 (1997).

[46] Y. K. Song, F. Liu, S. Chu, and D. Liu, *Hum. Gene Ther.* **8,** 1585 (1997).

[46a] Y. Liu, L. C. Mounkes, H. D. Liggitt, C. S. Brown, I. Solodin, T. D. Heath, and R. J. Debs, *Nature Biotechnol.* **15,** 167 (1997).

[47] N. Tomita, R. Morishita, J. Higaki, M. Aoki, Y. Nakamura, H. Mikami, A. Fukamizu, K. Murakami, Y. Kaneda, and T. Ogihara, *Hypertension* **26,** 131 (1995).

TABLE I
OPTIMIZATION OF LIPID-BASED SYSTEMS FOR ODN DELIVERY

Specific requirement	Focus of investigation
ODN compaction	Lipid composition
ODN protection	Formulation and lipid composition
Effective ODN transport to cells	Formulation and lipid composition
Stability and size homogeneity of ODN particle preparation	Formulation
Reproducibility of the preparation	Formulation
Intracellular release	Lipid composition
Cell targeting	Lipid composition/epitope addition
Noninflammatory and safe at clinically relevant effective dose	Lipid composition

trapping of liposomes in tumor tissues due to a discontinuous tumor microvasculature, and a high interstitial pressure are a few examples of potential tissue barriers.

Another limitation for the efficient delivery of DNA *in vivo* is the presence of interference substances in the body fluid, such as serum proteins, mucus, and surfactant. Blood contains plasma proteins carrying anionic charges that may bind nonspecifically to positively charged lipoplexes, removing them rapidly from circulation via the reticuloendothelial system. Plasma proteins may also alter greatly the lipid–DNA complex structure, leading to aggregation or deterioration of nucleic acids by exposure of DNA to nucleases. Furthermore, opsonization and activation of the complement system[48] by lipoplexes are additional physiological phenomena that participate in lowering the *in vivo* efficacy of cationic lipoplexes.

The pharmacokinetics and bioavailability following intravenous administration in baboons of an ODN delivered either in a free form or using a cationic or an anionic synthetic carrier system were compared.[49] Whole body distributions and metabolism were compared using positron emission tomography (PET) and an enzyme-based competitive hybridization assay. Free phosphodiester ODN showed typical pharmacokinetics for a phosphodiester oligonucleotide: high liver and kidney concentration, rapid degradation in plasma, and elimination from the body. The use of a cationic vector slightly protected ODN against degradation and enhanced uptake by the reticuloendothelial system but not in other organs. In contrast, the anionic vector enhanced dramatically the uptake in several organs, including

[48] C. Plank, K. Mechtler, F. Szoka, and E. Wagner, *Hum. Gene Ther.* **7,** 1441 (1996).
[49] B. Tavitian, S. Marzabal, V. Boutet, B. Kuhnast, S. Terrazino, M. Moynier, F. Dollé, J. R. Deverre, and A. R. Thierry, *Pharm. Res.* **19,** 367 (2002).

the lungs, spleen, and brain, with a prolonged accumulation of radioactivity in the brain. Using this vector, intact ODN were detected in the plasma for up to 2 h after injection, and the $T_{1/2\beta}$ (ODN half-life) and distribution volume increased by four- and seven-fold, respectively.[49] No evidence of toxicity was found at the administered dose (100 μg/kg) every week over a 4-week period. The anionic vector thus improved significantly the bio-availability and the pharmacokinetics profile and appears a promising delivery system for the *in vivo* administration of therapeutic ODN.

Future development of lipid-based gene delivery systems will focus on the design of lipoplex formulations regarding their intended tissue target, toxicity, pharmacological stability, and route of administration.

The following section provides methods for evaluating the *in vitro* and *in vivo* potential of a cationic lipid delivery system in a specific antisense approach.

Experimental Procedures

Kaposi's sarcoma occurs in nearly 34% of patients with AIDS and in individuals with organ transplants.[50] KS might be considered as a focal hyperplasia, leading eventually to tumorigenesis. Isolation of KS-derived cell lines with neoplastic characteristics was successful and is useful for studying this pathology at the tumor stage.[51] AIDS KS cells produce very high amounts of growth factors, such as the vascular endothelial growth factor (VEGF), which may play an important role in angiogenesis and then tumor growth. VEGF was also shown to be a crucial growth factor for other tumor types and has been considered a potential target in various therapeutic approaches. This section describes the use of antisense ODN directed against VEGF RNA. A screening of antisense sequences to six different regions of VEGF RNA and three different ODN lengths for their inhibitory effect on KS-Y1 tumorogenesis was performed previously.[52] The selected sequence (VEGF ODN) was delivered with a cationic liposome (lipoplex) system.

Oligodeoxynucleotides

Oligodeoxynucleotides are synthesized by Genset (Evry, France). Purification and quality control are ascertained by PAGE, anion-exchange HPLC, mass spectrometry, ^{1}H-NMR, total enzymatic digest, and pyrogenicity

[50] B. Ensoli, G. Barillari, and R. C. Gallo, *Hematol. Oncol. Clin. North Am.* **5,** 281 (1991).
[51] Y. Lunardi-Iskandar, J. L. Bryant, R. A. Zeman, V. H. Lam, F. Samaniego, J. M. Besnier, P. Hermans, A. R. Thierry, P. Gill, and R. C. Gallo, *Nature* **375,** 64 (1995).
[52] A. R. Thierry, R. A. Zeman, J. W. Hunh, J. L. Bryant, R. C. Gallo, and Y. Lunardi-Iskandar, *Proc. Am. Assoc. Cancer Res.* **36,** 413 (1995).

analyses. The VEGF ODN sequence (5′GTTCTCTGCGCAGAGT-CTCCTCATCCTTTC3′) is complementary to the human VEGF coding region 660–690 binding site. Random ODN corresponds to the scrambled VEGF ODN sequence and is used as a negative control.

Cells

The KS-Y1 cell line was established from pleural effusion of an AIDS patient. KS Y-1 cells are maintained in RPMI 1640 culture medium supplemented with 10% fetal bovine serum (FBS), 1% glutamine, sodium pyruvate, essential and nonessential amino acids, 1% nutridoma (Boehringer Mannheim, Germany). All reagents for cell culture are from Invitro-Gen (San Diego, CA). Cells are checked routinely for mycoplasma.

Lipoplex Formation

DLS liposomes consist of small unilamellar vesicles of approximately 50 nm diameter, which can complex with ODNs. DLS liposomes are formed by mixing 1 mg of dioctadecylamidoglycylspermidine (DOGS; Promega, Madison, WI) and 1 mg of dioleoylphosphatidylethanolamine (DOPE; Sigma, St. Louis, MO) as described previously.[15] ODN are mixed rapidly with DLS liposomes (0.26 mg lipid/ml) in sterile deionized water. After the addition of ODN to DLS liposomes, a completely different multilamellar structure is formed with a particle size ranging from 100 to 150 nm and which is very stable. The final lipoplex preparation contains 10 μg of ODNs for 38 μl of DLS liposomes. DLS–ODN complexes are stored at 4° for up to 1 month. The DLS lipoplex preparations are incubated at room temperature for at least 30 min prior to addition to the cells.

ODN Stability in a Biological Medium

DLS liposomes are prepared at a 0.6 DOGS/DOPE molar ratio and are mixed with P^{32}-labeled VEGF ODN at a 0.08 ODN/lipid weight ratio. Lipoplexes using Lipofectin (Gibco BRL, Gaithersburg, MD) and Transfectam [DOGS (Promega)] are formed under manufacturer's conditions at a 0.2 and 0.16 ODN/lipid weight ratio, respectively.

Ten micrograms of ODN with carrier is placed in an Eppendorf tube containing 150 mM NaCl in the presence or absence of 50% human serum (v/v). The reaction mixture (400 μl) is incubated for 5 h at 37° with rotation. Following incubation, the reaction mixture with serum is treated with 100 units proteinase K (Boehringer-Mannheim) in the presence of 0.05 M Tris–HCl, pH 8.0, 0.5% sodium dodecyl sulfate (SDS) for 1 h at 60°. Then

ODN are isolated by phenol–chloroform extraction, precipitated twice in the presence of EtOH and 3 mM sodium acetate, and resuspended in 30 μl TE buffer. ODN integrity is examined by polyacrylamide gel electrophoresis analysis.

Laser-Assisted Confocal Microscopy

KS-Y1 cells are grown on glass chamber slides (Nunc Inc., Naperville, IL) up to 50% confluency. Cells are incubated with 1 μM FITC-labeled VEGF ODN complexed with the DLS system for 4 h, with or without a 24-h postincubation period. They are then rinsed three times and mounted with SlowFade reagent without fixing agent (Molecular Probes, Inc., Eugene, OR) and coverslips.[53] Live cells are analyzed for confocal microscopy within 2 h following slide mounting. Optical sections of 1 μm are obtained for each sample using an MRC-600 (Bio-Rad, Hercules, CA) laser-scanning confocal system equipped with an inverted microscope. Images are analyzed with the variable numerical aperture set NA 1.1 and observed with the GeoLut output system (Bio-Rad).

Cellular Uptake

KS cells are grown in 6-well culture plates. Logarithmically growing cells are incubated with P^{32} end-labeled VEGF antisense ODN. Following incubation, cells are scraped and washed twice in phosphate-buffered saline (PBS). The pellet is suspended and centrifuged in 0.2 M glycine (pH 2.8) and then washed again in PBS. This treatment strips off membrane-bound ODN, and the remaining radioactivity represents intracellular ODN.[29] Then, the pellet is lysed in 10 mM Tris–HCl, 200 mM NaCl, 1% (SDS), 200 g/ml proteinase K, pH 7.4, for 2 h at 37°. Samples are then extracted with phenol. Aqueous fractions are analyzed by scintillation counting to assess cell-associated radioactivity. Cellular ODN uptake is expressed as micrograms per 10^6 cells.

Colony Assay

Cells (5 $\times$ 10^5/ml) are seeded in 0.8% (v/v) methyl cellulose (Fluka AG, Buchs, Switzerland) in the aforementioned culture medium but supplemented with 1.5% (v/v) FBS.[51] ODN are added to the medium, and the mixture is homogenized. One tenth milliliter of the methyl cellulose-containing cell preparation is seeded per well in 96-well flat-bottomed

[53] J. Lisziewicz, D. Sun, F. F. Weichold, A. R. Thierry, P. Lusso, J. Tang, R. C. Gallo, and S. Agrawal, *Proc. Natl. Acad. Sci. USA* **91**, 7942 (1994).

microtest plates. Cultures are incubated for 10–12 days and aggregates containing more than 50 cells are scored as colonies under an inverted microscope. The mean number of colonies (±SD) is determined from counting at least 4 wells. The SD is less than 10% of the mean values. The specific inhibition of cell colony formation is expressed as the percentage generally age of cells treated with randomly synthesized sequences.

In Vivo *Assay*

Bg nude mice are injected subcutaneously with 10^6 KS-Y1 cells/0.1 ml as described previously.[51] One week later, 100 μg of ODN in a free form or delivered with the DLS system is injected intratumorally daily for 11 days. Tumors are then removed and embedded for storage. Thin tissue sections of KS Y-1 tumors are fixed on slides and stained with eosin.[51]

Oligonucleotide Stability in DLS Lipoplexes

Study of the DLS system for the delivery of anti-HIV antisense ODN[15] demonstrated its capacity to completely encapsulate or integrate the entire quantity of loaded ODN as determined by the use of 5' ^{32}P end-labeled ODN. Results of the stability study presented in Fig. 4 demonstrated that phosphodiester ODN preserve their length integrity and are from enzymatic degradation protected in the presence of human serum. This level of protection is much higher than that observed when using the Transfectam (Promega) or Lipofectin (Gibco) commercial delivery reagents. Thus, the DLS system makes possible the *in vivo* use of highly unstable phosphodiester ODN. The ultrastructure of the DLS-ODN particulates showing ordered and high compaction of ODN[22] leads to more stable particles when compared to other cationic lipid delivery systems. In addition, this specific ultrastructure may be conducive to the high reproducibility and homogeneity of lipoplex particles as described in Schmutz *et al.*[22] It is remarkable to note that nucleic acids of very different length or form (from single-stranded ODN to double-stranded bacteriophage DNA) exhibit the same DNA packing with concentric winding within lipid layers. The mechanism underlying the acquisition of this specific ultrastructure is unresolved to date and under active investigation in our laboratory.

Uptake of Oligonucleotides by Cultured Cells

Analysis by confocal microscopy demonstrated that KS cells internalized efficiently FITC-labeled ODN delivered by the DLS system (Fig. 5). Four hours following DLS-ODN incubation, cells exhibited a

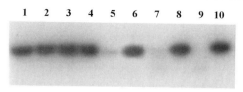

FIG. 4. Biological stability of ODN when presented in free form or entrapped in DLS lipoplexes. VEGF ODN were presented in free form (lanes 9 and 10) or as a complex with a synthetic vector such as DLS-1 (lanes 1 and 2), DLS-2 (lanes 3 and 4), Lipofectin (lanes 5 and 6), or Transfectam (lanes 7 and 8). ODN preparations were incubated with (lanes 1, 3, 5, 7, and 9) or without (lanes 2, 4, 6, 8, and 10) human serum. The resulting mixtures were subjected to PAGE as described in the text.

punctate pattern of fluorescence in cells, suggesting accumulation in endo-cytotic vesicles (Fig. 5A). Nevertheless, diffusion in the cytoplasm already occurred, demonstrating the fusogenic potential of DLS particles with en-docytotic vesicles. The presence of fluorescence in cell nuclei is already ob-servable following 4 h of incubation, illustrating the rapid translocation of free ODN from the cytoplasm to the cell nucleus. Nuclear accumulation, perinuclear and nucleolar accumulation in particular, is higher in cells incubated for 24 h following the removal of lipoplexes (Fig. 5B and C). This observation is consistent with results obtained from other efficient ODN delivery systems. Similar intracellular patterns were observed in HepG2 (a human hepatoma cell line) and in primary cultures of human macrophages obtained from peripheral blood treated with DLS-labeled-ODN. In nonadherent cells such as MOLT-3 human lymphoma cells and peripheral blood monocytes,[53] there is no punctuate pattern in the cyto-plasm, pointing out the poor endocytotic activity in these cells and the presence of another cell uptake mechanism, perhaps membrane fusion. It is crucial to analyze live cells instead of fixed cells, as fixation seems to perturb intracellular trafficking and is not reliable for the observation of intracellular fluorescence, especially when using fusogenic compounds. Thus, the use of a fluorescence aqueous stabilizer, such as the SlowFade re-agent, is imperative. Nevertheless, use of the DLS system makes possible the significant accumulation of ODN in the cell cytoplasm, and, consequently the therapeutic use of unmodified ODN, which appear less toxic and more specific compared to modified ODN such as phosphorothioated ODN.

The intracellular uptake of VEGF ODN in KS-Y1 cells was detectable at submicromolar (0.05 and 0.10 μM) concentrations and was saturable at concentrations higher than 3 μM (Fig. 6A). Kinetics of intracellular accu-mulation also showed a saturable uptake following a 10-h incubation (Fig. 6B). Free ODN accumulated quickly into cells but at a 14 times lower

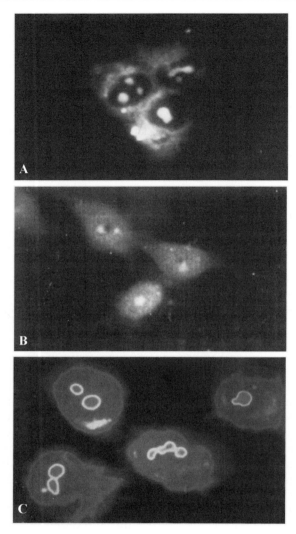

FIG. 5. Cellular distribution of VEGF ODN in KS-Y1 cells. Cells were treated with DLS-VEGF ODN for 4 (A) or 24 (B, C) under conditions described in the text. Each photograph represents images from laser-assisted confocal microscopy. (C) Images of cells observed with the GeoLut output system (Bio-Rad). Magnification, ×320 (A and B) and ×610 (C). (See color insert.)

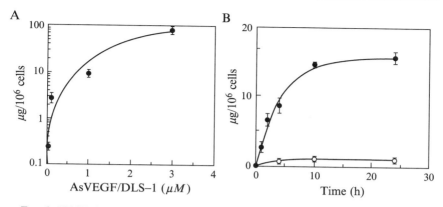

Fig. 6. VEGF ODN intracellular uptake in KS-Y1 cells. Dose curve (A) and kinetic curve (B) studies were carried out as described in the text. Full symbols, DLS-VEGF ODN; empty symbols, free VEGF ODN.

level. The dose curve and kinetic studies showed a typical pattern of cellular uptake. The highest uptake value reached 85 $\mu g/10^6$ cells, which corresponds to nearly 10^9 molecules/cell. This value appears very high compared to those found in the literature with free or delivered ODN, confirming the high efficiency of the DLS system to deliver ODN to cells. However, it highlights the relatively poor activity of ODN antisense molecules compared with other therapeutic drug compounds.

Inhibition of Colony Formation

Colony formation in methyl cellulose cell culture medium is increasingly inhibited with increasing concentrations of DLS-VEGF ODN up to 93% at 1 μM (Table II). The level of specific inhibition of 0.1 μM VEGF ODN using the DLS delivery system reached 60%, whereas the inhibition obtained with free VEGF ODN or DLS-random ODN was low (5 and 4%, respectively). DLS-random ODN treatment at led 1.0 μM to a 24% inhibition, whereas no activity was observed when cells were treated with DLS liposomes at an equivalent concentration. This suggests that ODN may exhibit nonsequence-specific activity in this assay, as demonstrated previously.[49] Nevertheless, significant activity (39% inhibition) was observed at the nanomolar range (0.010 μM). This result confirms those obtained previously in HIV culture models showing activity even at subnanomolar concentrations. This level of ODN activity is unprecedented, illustrating the potential of the DLS system for ODN delivery.

TABLE II
INHIBITORY ACTIVITY OF VEGF ODN ON KS-Y1 CELL TUMORIGENICITY

Assay	Dose (μM)	Colony number[a]	Total inhibition	Specific inhibition[b]
Untreated control		187 ± 24		
Free VEGF ODN	0.100	178 ± 7	5%	
DLS-random ODN	0.010	168 ± 8	10%	
	0.025	ND	ND	
	0.100	180 ± 30	4%	
	1.000	142 ± 33	24%	
DLS-VEGF ODN	0.010	115 ± 22	39%	32%
	0.025	79 ± 32	58%	ND
	0.100	72 ± 9	61%	60%
	1.000	14 ± 4	93%	90%

[a] The mean number of colonies ($\pm$SD) was determined from counting of at least four wells.
[b] The specific inhibition of cell colony formation was expressed as the percentage of cells treated with randomly synthesized sequences. ND, nondetermined.

Oligonucleotide Delivery *In Vivo*

Injection of KS-Y1 cells in immunodeficient mice led to aggressive tumor growth (Fig. 7A). Daily intratumoral administration of DLS-random ODN did not result in any marked change in tumor growth, in cell proliferation, or in the number of mitotic figures as observed in thin tissue sections (Fig. 7C) compared to the control (Fig. 7B). In contrast, DLS-VEGF ODN treatment led to a large heterogeneity of cell morphology, and necrosis was observed in nearly half of the tumor mass (Fig. 7D). No change was observed in tissue sections of control tumors treated with DLS liposomes (only containing lipids). Nevertheless, this control is not the most appropriate, as the structure and size of DLS liposomes (SUV) are completely different than that of highly autoassembled DLS-ODN particles (Fig. 1). These preliminary data demonstrated strong inhibition of tumor growth. More investigations are needed to evaluate whether ODN treatment prolongs life or prevents the reappearance of the tumor.

Concluding Remarks

Physicochemical studies of lipoplexes, such as the determination of the ODN encapsulation yield or ODN stability in a biological environment, were performed in order to control ODN dose and integrity. Analysis of ODN intracellular distribution and of cellular uptake demonstrated advantages in using the DLS lipoplex system compared to free ODN. Biological

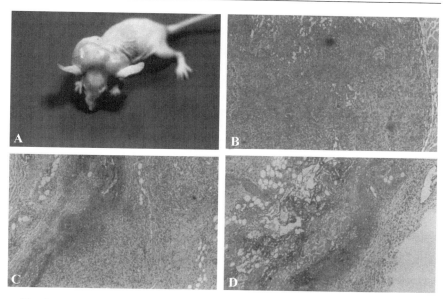

Fɪɢ. 7. Effect on KS-Y1 tumorigenesis in nude mice. Control mice exhibit a high tumor mass at the site of cell injection (A). Tumor tissue sections of control mice (B), scrambled sequence ODN (C)-, or DLS-VEGF ODN (D)-treated mice are shown. (See color insert.)

in vitro and *in vivo* assays showed that the selected sequence of VEGF ODN has easily detectable and specific activity. Experiments did not elucidate whether this sequence-specific activity is solely due to an antisense mechanism through ODN hybridization with VEGF mRNA. Increasing evidence indicates that a particular sequence might have a nonantisense mechanism that might originate from intracellular event such as specific protein binding.[30] The studies given as examples earlier contribute mainly to the development of an ODN delivery system, rather than show VEGF ODN as a therapeutic compound. Further studies are needed to evaluate the VEGF ODN mechanism of action. Nevertheless, this article points out the potential in targeting tumor growth factors such as the angiogenic VEGF. This strategy is under active investigation as a new approach to treat cancer.[54,55] We postulated and demonstrated previously that multiple targeting of different tumor autocrine growth factors of KS-Y1 cells,

[54] R. Masood, J. Cai, D. L. Smith, Y. Naidu, and P. S. Gill, *Proc. Natl. Acad. Sci. USA* **94,** 979 (1997).

[55] K. G. Carrasquillo, J. A. Ricker, I. K. Rigas, J. W. Miller, E. S. Gragoudas, and A. P. Adamis, *Invest. Ophtalmol. Vis. Sci.* **44,** 290 (2003).

interleukins such as IL-6 and IL-8 in addition to VEGF, might be a powerful treatment modality.[52] Moreover, ODN targeting VEGF may have applications in antiangiogenic therapies.[56,57]

Although the results obtained with the DL nucleic acid delivery system are promising, the safety of the carrier, especially of its cationic components, have to be demonstrated in toxicity experiments. A potential drawback of the DLS system is the pharmacological instability and clearance when systemic routes of administration are envisaged. In this regard, lipoplexes with a low negative charge appear to be pharmacologically more valuable.[22,49] Local administration of the DLS ODN delivery system is another area of potential application.

Acknowledgments

The authors thank Genset SA (Evry, France) and Marta Blumenfeld for kindly providing high-quality ODN. We thank Dr. Joseph Bryant (IHV, University of Maryland) for his expertise in animals models, Dr. Peter Rabinovich for his help in the ODN stability study, and Anuja Rastogi for her excellent technical assistance.

[56] K. L. Garrett, W. Y. Shen, and P. E. Rakoczy, *J. Gene Med.* **3,** 373 (2001).
[57] L. Maillard, P. Peycher, O. Fichaux, and D. Vincentelli-Djeffal, *Presse Med.* **29,** 1731 (2000).

[13] Requirements for Delivery of Active Antisense Oligonucleotides into Cells with Lipid Carriers

By Ilpo Jääskeläinen, Katriina Lappalainen,
Paavo Honkakoski, and Arto Urtti

Introduction

Antisense oligonucleotides (ODNs) bind specifically to a complementary sequence in mRNA of the target gene and subsequently inhibit target protein synthesis. The inhibition is dependent on the recruitment of RNase H that cleaves the target mRNA. Possible therapeutic targets are diseases that result from gene overexpression, expression of mutant genes, and viral infections. Additionally, ODNs can bind to pre-mRNA in the nucleus to correct aberrant splicing that causes, e.g., thalassemia.[1] Importantly, these

[1] H. Sierakowska, M. J. Sambade, S. Agrawal, and R. Kole, *Proc. Natl. Acad. Sci. USA* **93,** 12840 (1996).

targets are located intracellularly in the cytoplasm and nucleus, and ODNs must be able to penetrate into those intracellular sites for activity. The molecular weight and charge of ODNs do not allow simple diffusion across the cell membranes. Therefore, the delivery of antisense ODNs is problematic, both as experimental molecular biology tools and as therapeutic agents.

Most common ODN variants in the laboratory experiments are phoshodiesters (single-stranded DNA, PO-ODNs) and phosphorothioates (one oxygen in the phosphate bridge of DNA is replaced with sulfur, PS-ODNs). These ODNs are able to penetrate into the cells as such to some extent, but the efficacy of the endocytotic uptake mechanism is low.[2] The exact receptor protein(s) for ODN binding has not been identified yet.[3] Cationic liposomes can be used to improve the uptake of ODNs in cell culture[4] and to augment their stability against nucleases.[5] The delivery of ODNs from endosomes into the cytosol can be increased, at least for phosphorothioate modifications (PS-ODNs), by adding membrane active components such as the fusogenic lipid dioleoylphosphatidylethanolamine (DOPE), or peptides, to the liposomes. *In vivo* simple intravenous complexes of ODNs accumulate predominantly into the lung and subsequently into the liver,[6,7] thus restricting their use for other targets. In many local routes of administration the situation may be different.

This article presents experimental protocols of ODN delivery into cells with cationic lipids. It presents the structure–activity relations of cationic complexes in ODN delivery, some experimental factors that affect delivery, and techniques to study the intracellular kinetics of ODNs.

Delivery Systems for Antisense Oligonucleotides

Efficient delivery systems must be able to increase the uptake of ODNs into the cells and result also in the right intracellular distribution of ODN. The uptake of the delivery systems takes place by endocytosis, but ODNs must be able to escape from the endosomes to the cytoplasm and nucleus to hybridize with the target RNA. Without endosomal escape,

[2] A. Thierry and A. Dritschilo, *Nucleic Acids Res.* **20,** 5691 (1992).
[3] S. Wu-Pong, *Adv. Drug. Deliv. Rev.* **44,** 59 (2000).
[4] C. F. Bennett, M.-Y. Chiang, H. Chan, J. E. E. Shoemaker, and C. K. Mirabelli, *Mol. Pharmacol.* **41,** 1023 (1992).
[5] K. Lappalainen, A. Urtti, E. Söderling, I. Jääskeläinen, K. Syrjänen, and S. Syrjänen, *Biochim. Biophys. Acta* **1196,** 201 (1994).
[6] C. F. Bennett, J. E. Zuckerman, D. Kornbrust, H. Sasmor, J. M. Leeds, and S. T. Crooke, *J. Control. Rel.* **41,** 121 (1996).
[7] D. C. Litzinger, J. M. Brown, I. Wala, S. A. Kaufman, G. Y. Van, C. L. Farrell, and D. Collins, *Biochim. Biophys. Acta* **1281,** 139 (1996).

ODNs will be trapped in the endosomes and shuttled to lysosomes for degradation. Although ODNs are too large for simple diffusion across the lipid bilayers, they are small enough to diffuse in the cytoplasm and enter into the nucleus via nuclear pores.[8]

Background

Polymeric Systems. Polymeric delivery systems for antisense oligonucleotides include cationic polymers, such as poly-L-lysine[9], polyamidoamine dendrimers, nanoparticles, and polyethyleneimine.[9–12] These polymers bind negatively charged oligonucleotides and form complexes with them. This is analogous to plasmid DNA complexation by the cationic polymers. However, in this case the oligonucleotides are smaller than the cationic polymers and, therefore, no DNA condensation takes place in the same sense as with plasmid DNA. The complexation of ODNs and cationic polymers does not take place in 1:1 stochiometry. Instead, each complex may contain a large number of polymer and ODN molecules and the size distribution of the complexes is polydisperse. The cationic polymers augment the binding of the complexes on the cell surface due to their cationic charges and hence the intracellular delivery.

Liposomes. Classical liposomal encapsulation with neutral and negatively charged lipids have been used to formulate ODNs, but these liposomes are not effective delivery systems for ODNs *in vitro,* with some exceptions.[12a] Therefore, this section discusses delivery techniques that are based on cationic lipids.

Cationic lipids form cationic micelles or cationic liposomes in water. The architecture of these self-associated systems depends on the lipid structure: large head groups (cone-shaped molecules) favor micelle formation, whereas rod-shaped lipids tend to form liposomes.[13] Cationic liposomes and micelles are able to bind ODNs by the cationic charges on the

---

[8] D. J. Chin, G. A. Green, G. Zon, F. C. Szoka, Jr., and R. M. Straubinger, *New Biol.* **2,** 1091 (1990).

[9] A. J. Stewart, C. Pichon, L. Meunier, P. Midoux, M. Monsigny, and A. C. Roche, *Mol. Pharmacol.* **50,** 1487 (1996).

[10] G. Lambert, E. Fattal, and P. Couvreur, *Adv. Drug Deliv. Rev.* **47,** 99 (2001).

[11] A. Bielinska, J. F. Kukowska-Latallo, J. Johnson, D. A. Tomalia, and J. R. Baker, Jr., *Nucleic Acids Res.* **24,** 2176 (1996).

[12] B. Boussif, F. Lezoualc'h, M. A. Zanta, M. D. Mergny, D. Scherman, and J. P. Behr, *Proc. Natl. Acad. Sci. USA* **92,** 7297 (1995).

[12a] N. Düzgüneş, S. Simões, V. Slepushkin, E. Pretzer, J. J. Ross, E. De Clercq, V. P. Antao, M. L. Collins, and M. C. Pedroso de Lima, *Nucleosides Nucleotides Nucleic Acids* **20,** 515 (2001).

[13] J. Mönkkönen and A. Urtti, *Adv. Drug. Deliv. Rev.* **34,** 37 (1998).

polar head groups, but the liposomal structure is usually lost during the complexation.[14] ODNs are not encapsulated in the aqueous core of the cationic liposomes. Instead, sandwich-type structures with ODNs embedded between lipid bilayers or tubular structures with ODNs covered with lipids are formed, especially in the case of DOPE-containing liposomes.[14] There are many cationic lipid structures available and they are mostly used for the delivery of plasmid DNA. The basic principle of ODN and DNA complexation is electrostatic and both complexes are taken up into cells via endocytosis after binding of the positively charged complexes on the cell surface. However, the ODN molecules are usually about 10^4 in molecular weight, whereas plasmid DNA is 10^6–10^7. This causes some differences in the intracellular kinetics. Most importantly, ODNs can diffuse freely in the cytoplasm and through the nuclear pores, but the diffusion of plasmid DNA is restricted in both cases.

Fusogenic compounds. Simple complexation of ODNs with cationic lipids can enhance the cellular uptake of ODN and plasmid DNA, but normally this is not adequate for antisense activity. Most of ODN may be located in the endosomes and lysosomes, and only a fraction is delivered into the cytoplasm and nucleus. Therefore, liposomes should improve the escape from endosomes. This functionality can be added to the delivery system with fusogenic compounds. These agents enhance the fusion of the cationic lipid complexes with the lipid bilayers in the endosomal wall. The compounds are either fusogenic lipids or peptides. The lipids in this class (e.g., DOPE) usually have small head groups and unsaturated acyl chains, enabling them to adopt a tubular hexagonal architecture under certain conditions. Upon mixing with the lipid bilayer of the endosomes they may release ODN into the cytoplasm.[15] However, these liposomes may form less defined structures and their mean size is often in the micrometer range.

Fusogenic peptides are usually derived from viruses that use them for cellular penetration and endosomal escape. These peptides adopt helical or β-sheet conformations in contact with lipid bilayers, often in a pH-sensitive way. The lipid bilayer is ruptured as a result. A membrane-active, pH-sensitive peptide, JTS-1 (GLFEALLELLESLWELLLEA),[16–18] has been used to enhance cationic lipid–mediated DNA delivery. When JTS-1 is added to DOTAP/ODN complexes, it seems to be more effective

[14] I. Jääskeläinen, B. Sternberg, J. Mönkkönen, and A. Urtti, *Int. J. Pharm.* **167,** 191 (1998).
[15] O. Zelphati and F. C. Szoka, *Proc. Natl. Acad. Sci. USA* **93,** 11493 (1996).
[16] S. Gottschalk, J. T. Sparrow, J. Hauer, M. P. Mims, F. E. Leland, S. L. C. Woo, and L. C. Smith, *Gene Ther.* **3,** 448 (1996).
[17] J. G. Duguid, C. Li, M. Shi, M. J. Logan, H. Alila, A. Rolland, E. Tomlinson, J. T. Sparrow, and L. C. Smith, *Biophys. J.* **74,** 2802 (1998).
[18] V. W. Y. Lui, Y. He, and L. Huang, *Mol. Ther.* **3,** 169 (2001).

than DOPE in increasing PS-ODN activity.[19] In our studies, JTS-1 was, however, ineffective with relatively high molecular weight polyethyleneimine or poly-L-lysine. The JTS-1 peptide has been shown to have lytic activity on phosphatidylcholine liposomes and erythrocytes, and the activity is more pronounced at pH 5 than at pH 7, suggesting high membrane activity at the endosome level.[17] This peptide, together with DC-Chol and protamine, has been used also for transfecting ribozymes against the human protooncogene c-neu.[18] Additionally, other peptides, such as HA (GLFFEAIAEFIEGGWEGLIEGC), show similar effects, at least with plasmid DNA transfections.[19]

Testing Antisense Efficacy

Cationic Liposomes and Polymers. Cationic liposomes composed of DOTAP, DOTAP/DOPE (1:1 mole ratio), DOTAP/Cholesterol (chol) (1:1 mol ratio), and DOTAP/DOPE/Chol (2:1:1 mole ratio) are prepared in sterile water as 3.2 mM (cationic lipid) stock solutions by the thin lipid hydration method.[20] DOTAP, DOPE, and Chol can be obtained from Avanti Polar Lipids (Alabaster, AL). Cytofectin GS/DOPE vesicles are prepared according to the manufacturer's instructions (Glen Research; unfortunately, cytofectin GS is not available commercially anymore). The sizes of PS-ODN/cationic lipid complexes are determined by a Nicomp 380 ZLS Zeta potential/particle sizer (Particle Sizing Systems, Santa Barbara, CA). Polyethyleneimines with mean molecular masses of 25 and 800 kDa (PEIs 25 and 800) are from Aldrich and Fluka, respectively. Poly-(L-lysine) hydrobromides (PLLs) of mean molecular weights of 4000, 20,000, and 200,000 are from Sigma.

Charge Ratio Optimization of Complexes. For charge ratio optimization for use in cells expressing luciferase (D 407 cells, a retinal pigment epithelial cell line), the complexes are formed at +/− charge ratios of 16, 8, 4, 2, 1, and 0.5 for polymeric and 8, 4, 2, 1, and 0.5 for lipid-based carriers, respectively. The final concentration of ODN in the cell culture medium is 360 nM (20 μl in 400 μl of cell growth medium) in all cases. Antisense phosphorothioate ODN against luciferase (5'-TGG CGT CTT CCA TTT-3') and sense oligonucleotides (5'-ACC GCA GAA GGT AAA-3') as control are used. Carriers (in polystyrene tubes) and ODNs in sterile water are diluted and then complexed (in polystyrene tubes) in equal volumes in 75 mM NaCl, 50 mM MES, 50 mM HEPES, pH 7.2, and incubated for

[19] L. Vaysse, I. Burgelin, J. P. Merlio, and B. Arveiler, *Biochim. Biophys. Acta* **1475**, 369 (2000).

[20] I. Jääskeläinen, S. Peltola, P. Honkakoski, J. Mönkkönen, and A. Urtti, *Eur. J. Pharm. Sci.* **10**, 187 (2000).

approximately 15 min before transfection. JTS-1 (five negative charges/ molecule) is used with DOTAP, PEI, or PLL and is added to carrier/ ODN complexes at a charge ratio of 8 $(+/-)$ in amounts to yield final complexes at $+/-$ ratios of 4, 2, 1, and 0.5. The peptide in buffer is added to the complex after about 15 min (final DOTAP/ODN/JTS-1, 1:1:1 by vol) and is incubated for a further 15 min before addition to the cells.

Effects of Complexation Medium on Transfection Activity. ODNs and carriers are diluted and then complexed in equal volume with the carriers in sterile water, DMEM or 75 mM NaCl 50 mM MES, 50 mM HEPES, buffer, pH 7.2, as described earlier, before transfection (CV-1, a green African monkey kidney cell line) with 0–40 μl (0–360 nM) of the complexes.

Optimization of the Amount and Serum Resistance for JTS-1 Peptide. The amount of JTS-1 peptide (five negative charges) in the complexes is optimized by adding the peptide to the plain DOTAP/PS-ODN complexes with various $+/-$ charge ratios. To the plain complexes (in 75 mM NaCl 50 mM MES, 50 mM HEPES buffer, pH 7.2) prepared at $+/-$ charge ratios of 8, 6, 4, 3, 2, and 1, the peptide is added to yield DOTAP/ODN/ JTS-1 complexes with apparent charge ratios of 6, 4, 3, 2, 1, and 0.5.

Transfection. Cells are grown in DMEM with 5% (D 407) or 10% fetal bovine serum (FBS) (CV-1) containing 100 U/ml penicillin and 100 μg/ml streptomycin, plated into 24-well plates (about 1×10^5 cells/well), and grown overnight. Cells are treated with the complexes (20 μl, 180 nM ODN for CV-1 cells or 20 μl, 360 nM ODN for D 407 cells) for charge ratio optimization for 4 h in 400 μl of serum-free medium and, in the case of JTS-1, also in the presence of 10% FBS. The medium is replaced with normal growth medium and the cells are grown for a further 20 h. Cells are washed with phosphate-buffered saline (PBS) and lysed with lysis buffer (25 mM Gly-Gly, 15 mM MgSO$_4$, 4 mM EGTA, 1% Triton X-100, pH 7.8). Luciferase activity is measured with a luminometer by adding 100 μl of luciferase reagent (20 mM tricine, 1.07 mM magnesium hydroxycarbonate, 2.67 mM MgSO$_4$, 0.1 mM EDTA, 33.3 mM dithiothreitol, 0.53 mM MgATP, 270 μM coenzyme A, 470 μM luciferin) to 50 μl of cell lysate. Protein is measured with the BCA reagent (Pierce, Rockford, IL).

Results. None of the polymeric carriers tested had any enhancing effect as antisense carriers in D 407 cells.[20] In the case of PEI, the results are interesting, as with similar molecular weights it has been shown to be very efficient in plasmid transfections.[21] This is possibly due to too tight binding of PS-ODNs compared to plasmid DNA and PO-ODNs.[22]

[21] Z. Hyvönen, A. Plotniece, I. Reine, B. Chekavichus, G. Duburs, and A. Urtti, *Biochim. Biophys. Acta* **1509**, 451 (2000).

[22] S. Dheur, N. Dias, A. van Aerschot, P. Herdewijn, T. Bettinger, J.-S. Remy, C. Helene, and E. T. Saison-Behmoaras, *Antisense Nucleic Acid Drug Dev.* **9**, 515 (1999).

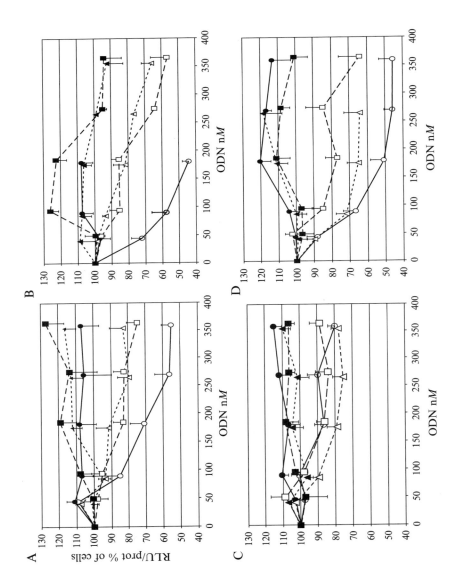

Therefore, the efficacy was tested only for lipid-based carriers on CV-1 cells. The antisense effects are shown in Fig. 1 with complexes prepared in different media and the subsequent size distributions in Fig. 2.[20] These indicate the relationship of complex size/fusogenic compounds on the transfection activity. Figure 3 shows the enhancing effect of JTS-1 when used with DOTAP, and also the relatively good serum resistance of DOTAP/ODN/JTS-1 complexes.

Conclusions. The activity of PS-ODNs in cell culture is dependent on the delivery system. PS-ODNs show an antisense effect only when delivered with liposomes that include a fusogenic component such as DOPE or the JTS-1 peptide. Complexes with neutral net charge are able to deliver antisense ODN if such fusogenic compounds are incorporated in the liposomes.

Methods for Intracellular Characterization of ODN

Importance of Intracellular Kinetics

Antisense ODNs must hybridize with their target RNA either in the cytoplasm or in the nucleus. Therefore, ODNs must reach these cellular compartments in a free form. Protein-bound ODNs may not hybridize and may cause nonspecific effects. Also, ODNs bound to cationic liposomes or polymers cannot hybridize; thus, ODN must be released from the carrier in the cells. Cationic complexes of ODNs and DNA are taken up into the cells via endocytosis and ODNs must be released from endosomes to the cytoplasm and nucleus for activity. Free ODNs in the cytoplasm also diffuse to the nucleus via nuclear pores.

The cellular uptake of ODNs (or DNA) can be measured using fluorescent or radiolabeled ODNs with subsequent analysis with fluorescence microscopy[23] or liquid scintillation counting,[5] respectively. It is important to

---

[23] E. G. Marcusson, B. Bhat, M. Manoharan, F. C. Bennett, and N. M. Dean, *Nucleic Acids Res.* **26,** 2016 (1998).

FIG. 1. Inhibition of luciferase expression by different carrier/PS–ODN complexes incubated for 15 min before adding to CV-1 cells. Changes in relative luciferase expression (LUC)/mg protein are shown as a percentage of the levels in untreated cells. Cells were incubated with PS-ODN concentrations of 0, 45, 90, 180, 270, and 360 nM. (A) ODN complexes with DOTAP/DOPE (+/− 2) (B) Cytofectin (+/− 4). (C) DOTAP (+/− 4). (D) DOTAP/JTS-1 (+/− 1). Water: ○, antisense; ●, sense). MES-HEPES: △, antisense; ▲, sense. DMEM: □, antisense; ■, sense. Error bars are expressed as ± SEM (*n* = 5–8). Reproduced, with permission, from Jääskeläinen *et al.*[20]

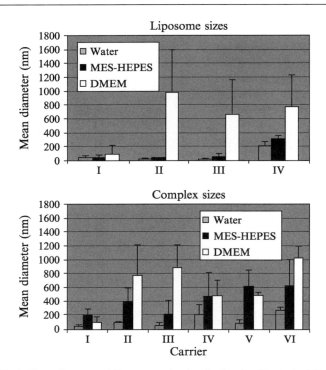

Fig. 2. (Top) Mean diameter of liposomes after incubation for 15 min in different media: (I) DOTAP, (II) DOTAP/DOPE (1/1), (III) DOTAP/DOPE/Chol (2/1/1), and (IV) cytofectin. (Bottom) Size distribution of liposome/PS-ODN complexes after incubation for 15 min in different media: (I) DOTAP at $+/-$ 4, (II) DOTAP/DOPE (1/1) at $+/-$ 2, (III) DOTAP/DOPE/Chol (2/1/1) at $+/-$ 2, (IV) DOTAP/DOPE/Chol (2/1/1) at $+/-$ 1, (V) DOTAP/JTS-1 at $+/-$ 1, and (VI) cytofectin at $+/-$ 4. Reproduced, with permission, from Jääskeläinen et al.[20]

analyze the degradation products of ODNs (i.e., shorter fragments) because cellular nucleases may cleave ODNs. This analysis can be done using capillary gel electrophoresis (CGE),[24] mass spectrometry,[24] or SDS–PAGE electrophoresis.[5,24] The endocytotic uptake mechanism is demonstrated usually by fluorescence microscopy (or confocal microscopy) in which lysosomes and the nucleus can be costained with appropriate markers. Also, metabolic inhibition experiments (with sodium azide or

[24] S. T. Crooke, M. J. Graham, J. E. Zuckerman, D. Brooks, B. S. Conklin, L. L. Cummins, M. J. Greig, C. J. Guinosso, D. Kornbrust, M. Manoharan, H. M. Sasmor, T. Schleich, K. L. Tivel, and R. H. Griffey, J. Pharmacol. Exp. Ther. 277, 923 (1996).

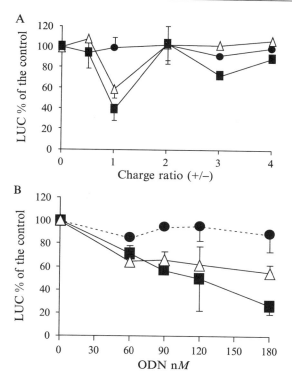

FIG. 3. Optimization of JTS-1 peptide-containing complexes. (A) DOTAP is complexed with a 15-mer PS-ODN at a $+/-$ 4 charge ratio (no JTS-1). Subsequently, JTS-1 is added to yield DOTAP/ODN/JTS-1 complexes with $+/-$ charge ratios of 3, 2, 1, and 0.5. Luciferase expression is compared to the control cells. (B) Effect of oligonucleotide concentration in optimized complexes ($+/-$ 1) on luciferase expression. ■, antisense (serum-free medium); △, antisense (10% FBS containing medium); ●, sense. Because the sense oligonucleotides showed similar results in both media, combined results are shown.

deoxyglucose) and uptake experiments at lower temperature can be used to elucidate these mechanisms.[25] As mentioned earlier, ODN distribution does not prove the antisense activity if the ODN is in the complexed form. Double-labeled ODN–lipid complexes can be used to elucidate the release of ODN from the liposomal complexes. This method utilizes fluorescence resonance energy transfer (FRET). For example, Jääskeläinen et al.[14] showed that DOTAP/DOPE liposomes release ODN in contact with endosome-mimicking liposomal bilayers, whereas DOTAP liposomes

[25] O. Zelphati and F. C. Szoka, *Pharm. Res.* **13**, 1367 (1996).

do not. FRET, combined with confocal imaging, was also used to show that cationic lipids and ODNs dissociate from each other upon contact with liposomes containing acidic lipids.[15]

This section describes in more detail some microscopic techniques and genetically engineered cells for intracellular ODN studies.

Regulated Cell Lines Controlling Mature mRNA Levels. The antisense ODN-mediated arrest of gene expression depends not only on the specificity and efficacy of the ODN, but also on the concentration of free ODN and its target mRNA. Therefore, in mechanistic studies, antisense ODN delivery it is advantageous to regulate the level of mRNA. Regulated gene expression systems can be used for this purpose.

Gene expression systems that can be controlled by extracellular signals allow detailed studies on gene function and the physiological effects of a given cellular protein. Undesirable pleiotropic effects and leakiness of the inactive state that plagued the early systems[26] were neatly avoided in the system based on the bacterial *tet* operon and its repressor.[27,28] A fusion protein (tTA, tetracycline-controlled transactivator) between the *tet* repressor and the activation domain of the viral VP16 protein can bind to its *tetO*-binding site and activate transcription only when tetracycline (or doxycycline) is not present. In a reverse system (rtTA), the opposite happens due to the site-directed mutagenesis of the DNA-binding domain of the *tet* repressor. The accumulation and, hence, poor kinetics of tetracycline may prevent rapid regulation; in addition, the VP16 moiety may be toxic.

A novel cell line expressing the luciferase gene under the control of a drug-responsive nuclear receptor constitutive androstane receptor (CAR) was developed.[29] The regulatory DNA sequences are based on phenobarbital-responsive enhancer elements (PBREM, Fig. 4), and the cell lines are based on HEK293 cells. CAR-dependent LUC can be suppressed by 3α-androstenol (ANDR), its 16α-reduced, and 3-keto derivatives but not by some other steroids, not by tetracycline. In addition, CAR-dependent LUC can be increased by structurally diverse drugs but not by TET. These data indicate that CAR can be regulated (either induced or suppressed) by a broad range of different ligands. The wide ligand specificity of CAR may be useful in giving a better choice of inducing chemicals that have desirable kinetics and a better profile of side effects.

[26] G. T. Yarranton, *Curr. Opin. Biotechnol.* **3,** 506 (1992).
[27] M. Gossen and H. Bujard, *Proc. Natl. Acad. Sci. USA* **89,** 5547 (1992).
[28] P. E. Shockett and D. G. Schatz, *Proc. Natl. Acad. Sci. USA* **93,** 5173 (1996).
[29] P. Honkakoski, I. Jääskeläinen, M. Kortelahti, and A. Urtti, *Pharm. Res.* **18,** 146 (2001).

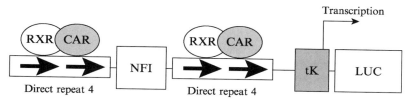

Cyp2b10 PBREM enhancer

FIG. 4. The CAR-responsive PBREM enhancer described in the text. The PBREM enhancer from the Cyp2b10 gene contains a nuclear factor I (NFI) site surrounded by two direct repeat 4 motifs for the nuclear receptor CAR/RXR heterodimer. Ligand binding by CAR modulates luciferase reporter (LUC) gene expression driven by the thymidine kinase (tk) promoter.

Splicing Correction. Correction of aberrant splicing may have therapeutic use, e.g., in the treatment of thalassemia[1,30] or cystic fibrosis.[31] A system utilizing the luciferase gene interrupted by a mutated human β-globin intron 2 was developed by Kang *et al.*[32] In this system, luciferase pre-mRNA shows aberrant splicing that can be corrected with targeted antisense ODN. As splicing takes place only in the nucleus, this system measures the access of ODN into the nucleus. The system has been used to compare the efficacy of various carrier/ODN complexes. It has been used to compare the nuclear delivery of 2′-*O*-methyl-oligoribonucleotide with different carriers.[33] Because RNase H-independent oligonucleotides are needed in these kinds of experiments, direct comparison of the results to those obtained with nonmodified (unmodified or only partly modified PO- or PS-ODNs) should be avoided. Also, the site of action (nucleus/cytoplasm) of RNase H-dependent and -independent ODNs may be different and the structural modifications may change the complex properties.

Methods for the Use of Genetically Engineered Cells in Antisense Testing

Chemicals. The synthesis of 1,4-bis[2-(3,5-dichloropyridyloxy)]benzene (TCPOBOP) is described in Honkakoski *et al.*[34] ANDR is from Steraloids (Newport, RI), and G418 and hygromycin B are from Calbiochem (La Jolla, CA).

[30] G. Lacerra, H. Sierakowska, C. Carestia, S. Fucharoen, J. Summerton, D. Weller, and R. Kole, *Proc. Natl. Acad. Sci. USA* **97**, 9591 (2000).
[31] F. J. Friedman, J. Kole, J. A. Cohn, M. R. Knowles, L. M. Silverman, and R. Kole, *J. Biol. Chem.* **274**, 36193 (1999).
[32] S.-H. Kang, M.-J. Cho, and R. Kole, *Biochemistry* **37**, 6235 (1998).
[33] S.-H. Kang, E. L. Zirbes, and R. Kole, *Antisense Nucleic Acid Drug Dev.* **9**, 497 (1999).
[34] P. Honkakoski, R. Moore, K. Washburn, and M. Negishi, *Mol. Pharmacol.* **53**, 597 (1998).

Plasmids. Plasmids coding for tTA (pTetOff), tTA-responsive luciferase (pTREluc), pTKhyg, and pCMVβ are from Clontech Inc. (Palo Alto, CA). LUC reporter plasmid pPBREMluc is constructed by *Bgl*II excision of the PBREM element containing CAR-binding sites and the thymidine kinase promoter from pPBREMtkCAT[34,35] and insertion into the *Bgl*II site of pGL3-Basic plasmid (Promega, Madison, WI). The CAR cDNA is released as a *Bam*HI (blunt)-*Xho*I fragment and is inserted in the *Eco*RI (blunt) and *Sal*I sites of the pCI-neo vector (Promega).[36] Plasmids are purified with Qiagen columns (Hilden, Germany) and verified by dideoxy sequencing and restriction mapping.

Generation of Cell Lines Expressing Regulatable Luciferase. Cells (HEK293, ATCC CRL-1573) are grown in DMEM with 10% FBS and 100 U/ml penicillin–100 μg/ml streptomycin (Gibco-BRL, Gaithersburg, MD). Cells are transfected with the pCI-neo/CAR plasmid by calcium phosphate, and transformed cells are selected with 0.4 mg/ml G418. Colonies are expanded and tested for expression of CAR by transient transfection with pCMVβ plus pPBREMluc plasmids. The transfected cells are grown in the presence of either 5 μM ANDR or 500 nM TCPOBOP to repress or activate CAR-regulated LUC activity, respectively.[34–37]

Colonies that give strong CAR-dependent responses are transfected with pTKhyg plus pPBREMluc (1:20 ratio) and selected with 0.1 mg/ml hygromycin B (Hyg B). Colonies resistant to both G418 and Hyg B are tested for ANDR-repressed and TCPOBOP-activated LUC. Colonies expressing both the tTA transactivator and the TET-responsive LUC are generated similarly using pTetOff and pTKhyg plus pTREluc plasmids, transient transfections, and TET (2 μg/ml) treatments, according to the manufacturer's instructions. The sublines HEK293:CAR/PBREMluc and HEK293:tTA/TREluc are expanded and characterized.

Treatments with Inducing Chemicals. HEK293:CAR/PBREMluc cells are grown in the presence of 5 μM ANDR for 24 h, and known CAR-activating chemicals[34–36] or the vehicles are added for up to 48 h. In dose and time course experiments, ANDR-treated cells are treated with 3.2–2000 nM TCPOBOP for up to 48 h or with 500 nM TCPOBOP for up to 48 h.

Treatments with Suppressing Chemicals. Untreated HEK293:CAR/PBREMluc cells are grown for up to 48 h with 5 μM ANDR and, in

[35] P. Honkakoski, I. Zelko, T. Sueyoshi, and M. Negishi, *Mol. Cell. Biol.* **18,** 5652 (1998).

[36] T. Sueyoshi, T. Kawamoto, I. Zelko, P. Honkakoski, and M. Negishi, *J. Biol. Chem.* **274,** 6043 (1999).

[37] B. M. Forman, I. Tzameli, H.-S. Choi, J. Chen, D. Simha, W. Seol, R. M. Evans, and D. D. Moore, *Nature* **395,** 612 (1998).

dose–response experiments, with 0.03 to 10 μM. The HEK293:tTA/TREluc cells are grown in the absence of TET and are then treated with 2 μg/ml TET for up to 48 h and with 0.01 to 1000 ng/ml in dose–response experiments. To test the depression of LUC, 31-4 cells are grown in the presence of 10 ng/ml TET, the cells are washed thoroughly with medium, and fresh medium without TET is added for up to 48 h.

Transfection with Antisense Oligodeoxynucleotides. The cells are transfected for 4 h in normal growth medium with 2.5 μg/ml GS2888:DOPE (Cytofectin)[38,39] and 37.5–300 nM antisense ODNs, 5′ TGGCGTCTTC CATGG for HEK293:CAR/PBREMluc (M29) and 5′ TGGCGTCTT CCATTT for HEK293:tTA/TREluc (31-4), spanning the translation initiation codon (underlined) in respective plasmids or control sense ODNs.[40] Fresh medium (400 μl) is added, and the cells are cultured for 24 h with 5 μM ANDR or 500 nM TCPOBOP for M29 cells, and in the presence or absence of 2 μg/ml TET for 31-4 cells. LUC is measured with a luminometer standardized with *Photinus pyralis* luciferase (19 $\times$ 10^6 RLU/mg protein) from Sigma (St. Louis, MO).

Antisense Effects in Engineered Cells. LUC expression can be reduced by 40–70% in both M29 and 31-4 cells (Fig. 5). The half-maximal values for inhibition are less than 100 nM. These results are comparable with previously reported half-maximal values of 150–1000 nM and 60–80% maximal antisense effects obtained with cationic lipids and unmodified phosphorothioate ODNs directed against various genes.[23,25,41,42] Most likely, cell-specific factors determine the extent of inhibition and the maximum amount of ODN/cationic lipid tolerated by the cells. At lower doses the results show no difference between high or low levels of mRNA on antisense potency and are similar to those obtained earlier with modified (C-5 propyne) oligonucleotides.[39] The residual uninhibited LUC activity is probably due to preexisting LUC protein because pretreatment of cells with LUC-suppressing ANDR or TET tended to improve the antisense effect at higher ODN doses. These studies suggest the utility of our LUC-expressing cell lines in evaluating various ODN carriers for antisense research.

[38] J. G. Lewis, K.-Y. Lin, A. Kothavale, W. M. Flanagan, M. D. Matteucci, R. B. DePrince, R. A. Mook, Jr., R. W. Hendren, and R. W. Wagner, *Proc. Natl. Acad. Sci. USA* **93,** 3176 (1996).

[39] W. M. Flanagan, A. Kothavale, and R. W. Wagner, *Nucleic Acids Res.* **24,** 2936 (1996).

[40] M. Antopolsky, E. Azhayeva, U. Tengvall, S. Auriola, I. Jääskeläinen, S. Rönkkö, P. Honkakoski, A. Urtti, H. Lönnberg, and A. Azhayev, *Bioconj. Chem.* **10,** 598 (1999).

[41] F. Cumin, F. Asselbergs, M. Lartigot, and E. Felder, *Eur. J. Biochem.* **212,** 347 (1993).

[42] C. F. Bennett, D. Mirejovsky, R. M. Crooke, Y. J. Tsai, J. Felgner, C. N. Sridhar, C. J. Wheeler, and P. L. Felgner, *J. Drug Target* **5,** 149 (1998).

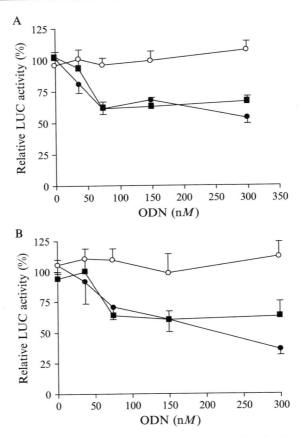

FIG. 5. HEK 293 M29 cells (A) and 31–4 cells (B) are transfected with sense (open) or antisense (closed) PS-ODNs and cultured in the presence of suppressing ANDR or TET (circles) or with TCPOBOP or in the absence of TET (squares). Because sense ODNs do not inhibit LUC, only results from one condition are shown. LUC and protein are assayed 24 h posttransfection. Data are expressed as relative LUC activity compared to untransfected cells (=100). Mean ± SEM of five to eight determinations are shown.

It should be noted that cytofectin is not readily available anymore, but can be replaced with DOTAP/DOPE or DOTAP/JTS-1 liposomes.

Figure 6 shows schematically tetracycline-regulated cell lines based on retinal pigment epithelial cells (D 407). D 407 6-2 cells (Fig. 6, top) can be used to study antisense effects of both RNase-dependent and -independent oligonucleotides, whereas D 407 6-27 mut cells (splicing correction) can be used to study the nuclear delivery of RNase-independent modified

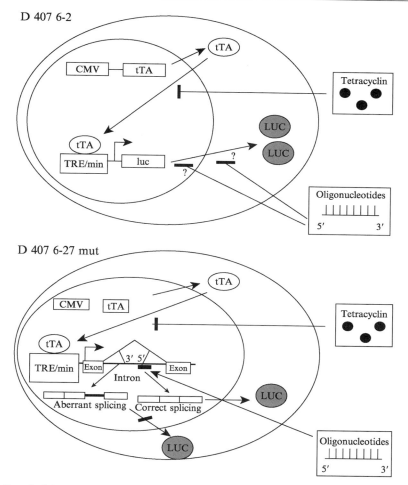

FIG. 6. Schematic presentation of tetracycline-regulated systems based on luciferase (LUC) that can be used to study antisense effects. (Top). Testing of antisense activity in the tetracycline-responsive retinal pigment epithelial cell line (D 407 6-2). Luciferase expression and the level of luciferase mRNA can be regulated with tetracycline. Antisense affects the expression of luciferase. Effects can be studied at different levels of target mRNA.[27,28] (Bottom) Testing of splicing correction with antisense. The retinal pigment epithelial mutant cell line (D 407 6-27 mut) produces inactive luciferase due to aberrant splicing (sites indicated as 3' and 5').[32] The short bar under the intron indicates the binding of oligonucleotide ON-705 to the aberrant splicing site (T to G mutation at nucleotide 705), leading to correct splicing.

oligonucleotides. Figure 7 shows the restoration of splicing and the effect of tetracycline after the transfection of 2'-*O*-methyl-oligoribonucleotide (ON-705) with cytofectin (top) and the potency of different carriers (liposomal/polymeric) on splicing correction (bottom). Results show a similar inability of polymeric carriers to transfect ONs in active form, as shown earlier with D 407 6-2 cells.[20]

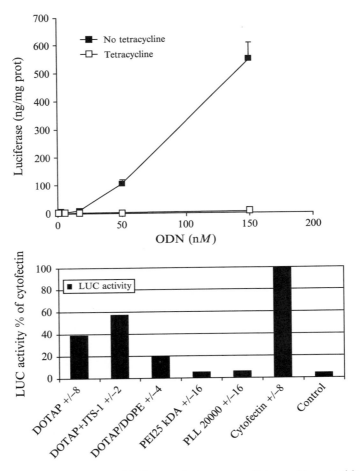

Fig. 7. (Top) Oligonucleotide binding results in correct splicing and increased luciferase activity in D 407 6-27 mut cells at increasing concentrations of oligonucleotide (ON-705). Transcription of the luciferase gene can be regulated with tetracycline. (Bottom) Mutant cell line with a splicing defect can be used to study the nuclear delivery of oligonucleotides with different delivery systems. Restoration of splicing in D 407 cells by ON-705 complexed with different carriers at optimized charge ratios is shown (cytofectin = 100%).

Microscopic Methods Used to Study Intracellular Distribution

Fluorescence microscopy has revealed some important features in the cellular delivery of ODNs with cationic lipids. However, fluorescence microscopy does not allow high-resolution images at the ultrastructural level. Electron microscopy (EM) provides that possibility, but like most microscopic methods, it is qualitative.

Microscopic Studies Using Digoxigenin-Labeled ODNs. 5′-Digoxigenin (DIG)-labeled PO-ODNs are used to study the intracellular fate of ODNs in CaSki cells (a human cervical cancer cell line). The sequence 5′-GTGTATCTCCATGCAT-3′, complementary to the initiation site of the *E7* oncogene of HPV 16 (human papilloma virus), and the random sequence 5′-CATCTGTATTGGATCG-3′ are used and complexed with either dimethyldioctadecylammonium bromide (DDAB)/DOPE (2/5 w/w) or LipofectAMINE (2,3-dioleoyloxy-*N*-[2(sperminecarboxamido)ethyl]-*N,N*-dimethyl-1-propanaminium trifluoroacetate (DOSPA)/ DOPE (3/1 w/w) and are incubated for 5 min (light microscopy), 20 min, 1.5 h, 4 h, and 24 h (for light and electron microscopy). Both of the cationic lipids are used as 10 μM final concentrations in the culture medium with DIG-ODN concentrations of 0.1, 0.2, and 1.0 μM.[43]

Materials. 5′ end-labeled ODNs are from Genosys Biotechnologies Inc. (Cambridge, UK), and horseradish peroxidase–labeled sheep polyclonal antibody against DIG (anti-DIG-POD, 1:500) is from Boehringer Mannheim (Pentzperg, Germany) LipofectAMINE and DDAB/DOPE liposomes (LipofectACE) can be obtained from Gibco-BRL (Paisley, UK).

Methods. Anti-DIG-POD is used against DIG. After incubation with DIG-ODNs or their lipid complexes, the cells are fixed in 3% paraformaldehyde and 0.5% glutaraldehyde and are treated with 1% NaBH$_4$ in order to reduce free aldehyde groups and double bonds, immersed in 25% sucrose, and freeze-thawed three times in liquid nitrogen in order to increase antisera accumulation.

The samples are washed with 0.05 *M* Tris-buffered saline (TBS, pH 7.4), including 0.2% Triton X-100 in the first washing. After incubation for 15 min in 10% normal goat serum (NGS) in TBS and washing with 1% NGS the, samples are incubated with anti-DIG-POD in 1% NGS in TBS overnight at 4°. 3,3′-Diaminobenzidine (DAB) is used as chromogen in the immunoperoxidase reaction.

Reaction products for light microscopy can be intensified by using OsO$_4$ (0.02%), and to visualize the reaction in EM 1% is needed.

---

[43] K. Lappalainen, R. Miettinen, J. Kellokoski, I. Jääskeläinen, and S. Syrjänen, *J. Histochem. Cytochem.* **45,** 265 (1997).

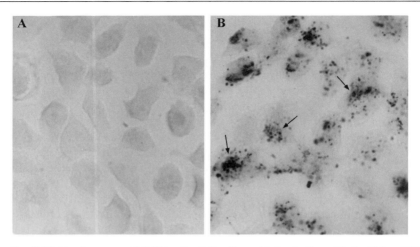

FIG. 8. Light microscopy of CaSki cells. Cells after a 4-h incubation with 1 μM free DIG-ODNs (A) or with net negatively charged ($-/+$ 1.6) DDAB/DOPE complexes (B). Some punctate immunoreactivity is indicated by arrows. Reproduced, with permission, from Lappalainen et al.[43]

For EM the cells are stained with 2% uranyl acetate, embedded in Epon, and ultrathin sections on copper grids are stained with uranyl acetate and lead citrate or left unstained to avoid nuclear chromatin staining by the agents.

Cellular Distribution of DIG-ODNs in CaSki Cells. Immunopositive signals with light microscopy are seen already after incubation with the complexes for 5 and 20 min, especially in the case of the highest (1.0 μM) ODN concentration. At incubation times of 5 min to 1.5 h, signals are more intense for DDAB/DOPE than DOSPA/DOPE complexes, but similar with longer incubation times. However, some cytoplasmic and perinuclear (or even nuclear) immunostainings are detected for DOSPA/DOPE complexes (20 min and 1.5 h). Figure 8 shows light microscopy of CaSki cells after a 4-h incubation with plain ODNs (A) or DDAB/DOPE complexes of ODNs (B).

Figure 9A and B show the electron microscopy of untreated CaSki cells with or without grid staining, respectively. In Fig. 9C, the signals after a 1.5-h incubation are located in intracellular vesicles (negatively charged complexes), whereas net positive complexes (without grid staining) show ODN localization in the nuclear envelope and nucleoplasm, in addition to cytosolic immunoreactivity after 4 h of incubation (Fig. 9D). In general, EM shows no drastic difference in cell association between net positively

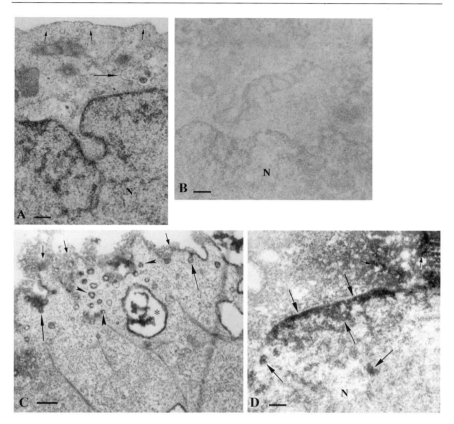

Fig. 9. Electron microscopy of CaSki cells. Control cells with (A) or without grid staining (B). The large arrow in A shows a multivesicular body and small arrows indicate plasmalemmal vesicles. (C) Cells after a 1.5 h incubation with 1 μM DIG-ODNs complexed with DDAB/DOPE ($-/+$ 1.6, net negative). Immunoreactivity is seen on the cell surface (small arrows), plasmalemmal vesicles (arrows), small intracellular vesicles (arrowheads), and a multivesicular body (asterisk). (D) Cells after a 4-h incubation (0.2 μM DIG-ODN, $-/+$ 0.32, net positive) without grid staining. Immunopositivity is seen in the nuclear envelope and the nucleoplasm (large arrows) and in the cytosol (small arrows). Reproduced, with permission, from Lappalainen et al.[43]

and negatively charged complexes. Also, the type of cationic lipid (DOSPA, five positive charges, or DDAB, one positive charge) does not affect immunopositive signals to any great extent, although the complex sizes differ significantly.[43]

Summary

Contrary to expectations, antisense ODNs are not "magic bullets," drugs with perfect selectivity. PS-ODNs bind to proteins and thereby exert nonspecific effects. Efficient delivery systems for ODNs would improve the therapeutic index of these drugs and improve their use as tools in cell culture experiments. Despite much research, the delivery of ODNs still remains a problem.

The intracellular delivery of ODNs involves many critical steps, and multiple methods are needed to elucidate these factors. Some requirements and study methods for efficient ODN delivery with liposomes have been presented here. Clearly, more mechanistic studies will be needed before these delivery processes can be understood thoroughly.

Acknowledgments

The authors thank the National Agency of Technology (TEKES), Graduate School of ESPOM, and the Academy of Finland for financial support.

[14] Mass Spectrometry and Enzyme-Linked Immunosorbent Assay Methods for the Quantitation of Liposomal Antisense Oligonucleotide (LE-rafAON) in Human Plasma

By Ateeq Ahmad, Sumsullah Khan, and Imran Ahmad

Introduction

Raf-1, a cytoplasmic serine and threonine kinase, acts as a downstream effector of Ras and as the activator of MEKI in the Ras-MAPK signaling transduction pathway.[1-3] This pathway is associated with cell proliferation and survival via the engagement of downstream transcription factors.[4-10]

[1] F. McCormick and A. Wittinghofer, *Curr. Opin. Biotechnol.* **7,** 440 (1996).
[2] J. M. Shields, K. Pruitt, A. McFall, A. Shaub, and C. J. De, *Trends Cell Biol.* **10,** 147 (2000).
[3] W. Kolch, G. Heidecker, P. Lloyd, and U. R. Rapp, *Nature (Lond.)* **349,** 426 (1991).
[4] U. Kasid, A. Pfiefer, T. Brennan, M. Beckett, R. R. Weichselbaum, A. Dritschilo, and G. E. Mark, *Science* **243,** 1354 (1989).
[5] U. Kasid, S. Suy, P. Dent, S. Ray, T. L. Whiteside, and T. W. Sturgill, *Nature (Lond.)* **382,** 813 (1996).
[6] D. K. Morrison and R. E. Cutler, Jr., *Curr. Opin. Biol.* **9,** 174 (1997).
[7] W. Kolch, *Biochem. J.* **351,** 289 (2000).

The importance of Raf-1 activity in cancer has been well established.[11–13] Gokhale and others have demonstrated that human tumor cells transfected with antisense c-raf-1 DNA show inhibition of Raf-1 expression and delayed tumor growth.[14–17] These observations validated the importance of Raf-1 as a critical target in cancer therapeutics.

In recent years, the usefulness of the antisense sequence–specific inhibition of gene expression and pharmaceutical application of antisense molecules has been shown.[18–25] Presently, many antisense oligonucleotides (AONs) are in clinical trials for the treatment of cancer. These AONs are directed against several cancer-related genes, including c-raf-1. Efforts are being made to develop reliable methods for the delivery of AONs, such as the use of liposomes as carrier systems.[16,17,26–28] Currently, a liposomal

[8] K. Podar, Y. Tai, F. E. Davies, S. Lentzsch, M. Sattler, T. Hideshima, B. K. Lin, D. Gupta, Y. Shima, D. Chauhan, C. Mitsiades, N. Raje, P. Richardson, and K. C. Anderson, *Blood* **98**, 428 (2001).

[9] U. Kasid and S. Suy, *in* "Apoptosis Genes" (C. S. Potten, C. Booth, and J. Wilson, eds.), p. 85. Kluwer Academic Publisher, Norwell, MA, 1998.

[10] J. Zhong, J. Troppmair, and U. R. Rapp, *Oncogene* **20**, 4807 (2001).

[11] J. Chen, K. Fujii, L. Zhang, T. Roberts, and H. Fu, *Proc. Natl. Acad. Sci. USA* **98**, 7783 (2001).

[12] C. G. Broustas, N. Grammatikakis, M. Eto, P. Dent, D. L. Brautigan, and U. Kasid, *J. Biol. Chem.* **277**, 3053 (2002).

[13] B. P. Monia, J. F. Johnston, T. Geiger, M. Muller, and D. Fabbro, *Nature Med.* **2**, 668 (1996).

[14] A. Pfeifer, G. Mark, S. Leung, M. Dougherty, E. Spillare, and U. Kasid, *Biochem. Biophys. Res. Commun.* **252**, 481 (1998).

[15] A. Rasouli-Nia, D. Liu, S. Perdue, and R. A. Britten, *Clin. Cancer Res.* **4**, 1111 (1998).

[16] P. C. Gokhale, B. Radhakrishnan, S. R. Husain, D. R. Abernethy, R. Sacher, A. Dritschilo, and A. Rahman, *Br. J. Cancer* **74**, 43 (1996).

[17] P. C. Gokhale, J. Pei, C. Zhang, L. Ahmad, A. Rahman, and U. Kasid, *Anticancer Res.* **21**, 3313 (2001).

[18] S. T. Crooke, *Annu. Rev. Pharmacol. Toxicol.* **32**, 329 (1992).

[19] S. Agrawal, *Antisense Nucleic Acid Drug Dev.* **9**, 371 (1999).

[20] E. Koller, W. A. Gaarde, and B. P. Monia, *Trends Pharmacol. Sci.* **21**, 142 (2000).

[21] B. P. Monia, J. Holmlund, and A. F. Door, *Cancer Invest.* **18**, 632 (2000).

[22] I. Lebedeva and C. A. Stein, *Annu. Rev. Pharmacol. Toxicol.* **41**, 403 (2001).

[23] C. A. Stein, *J. Clin. Invest.* **108**, 641 (2001).

[24] V. A. Soldatenkov, A. Dritschilo, F.-H. Wang, Z. Olah, W. B. Anderson, and U. Kasid, *Cancer J. Sci. Am.* **3**, 13 (1997).

[25] F. McPhillips, P. Mullen, B. P. Monia, A. A. Ritchie, F. A. Dorr, J. F. Smyth, and S. P. Langdon, *Br. J. Cancer* **85**, 1753 (2001).

[26] A. Rahman, G. White, N. More, and P. S. Schein, *Cancer Res.* **45**, 796 (1985).

[27] M. Fishman, L. Strauss, J. Pei, P. C. Gokhale, C. Zhnag, P. LoRusso, E. Kraut, C. Fleming, A. Ahmad, A. Zhang, S. Khan, and U. Kasid, *Proc. Am. Soc. Clin. Oncol. Annu. Meet.* **21**, 84b (2002).

[28] J. N. Moreira, C. B. Hansen, R. Gaspar, and T. M. Allen, *Biochim. Biophys. Acta* **1514**, 303 (2001).

formulation of antisense oligonucleotides (LE-rafAON) targeted toward c-raf-1 is in phase I clinical trials. To understand the pharmacokinetics of LE-rafAON, simple and sensitive methods are required to quantify AON in human plasma.

The two assays described in this article monitor the concentration of antisense oligonucleotides in human plasma using two different methodologies. Other assays have been used to quantitate antisense oligonucleotides.[29] However, LC/MS/MS and ELISA assays have not been examined thoroughly for their reliability. We have developed and validated LC/MS/MS bioanalytical and ELISA methods for a 15-mer rafAON in human plasma. These assays are robust and reproducible from 8 to 10,000 ng/ml for LC/MS/MS and from 1 to 50 ng/ml for ELISA. These methods are free from any interference of matrix or dilution effect and meet the sensitivity and reproducibility criteria needed for pharmacokinetic studies of LE-rafAON in human plasma.

LC/MS/MS Assay

The quantification of 15-mer rafAON in human plasma is based on protein precipitation followed by solid-phase extraction and analysis using reversed-phase HPLC with Z-spray electrospray ionization MS/MS detection. Negative ions for rafAON are monitored in multiple reaction monitoring mode (MRM). Samples containing rafAON are spiked with an internal standard (IS) before processing. Drug to IS peak area ratios for the standards are used to create a linear calibration curve.

Preparation of LE-rafAON

LE-rafAON formulation is prepared by the following procedure.[30] Briefly, lyophilized lipids (dimethyldioctadecyl ammonium bromide, egg phosphatidylcholine, and cholesterol) are reconstituted at room temperature with a rafAON solution in normal saline using a rafAON to lipid ratio of 1:15 (w/w). The mixture is vortexed vigorously for 2 min, followed by hydration at room temperature for 2 h and sonication for 10 min in a bath-type sonicator (Model XL 2020; Misonix, Inc., Farmingdale, NY).

[29] P. C. Gokhale, V. Soldatenkov, F.-H. Wang, A. Rahman, A. Dristchilo, and U. Kasid, *Gene Ther.* **4,** 1289 (1997).
[30] P. C. Gokhale, C. Zhang, J. T. Newsome, J. Pei, I. Ahmad, A. Rahman, A. Dritschilo, and U. Kasid, *Clin. Cancer Res.* **8,** 3611 (2002).

Standard and Quality Control Samples

Seven nonzero standards and five quality control human plasma samples containing LE-rafAON are prepared. These samples are used for assay validation parameters and are stored at $<-20°$.

Extraction of Samples

Plasma samples (1.0 ml) are treated with acetonitrile to precipitate proteins. The solid-phase extraction of rafAON and the internal standard (Fig. 1) are carried out using a Waters Oasis tC1812cc (2 g) cartridge. The extracts are evaporated to dryness and reconstituted with a solution of 5 mM ammonium acetate, pH 7.5.

Chromatographic Settings and Conditions

Samples are injected onto a Synergi Max-RP (50 × 2 mm, 4 μm) analytical column with a solvent delivery system (LC-10Ad vp, Shimadzu Corporation, Tokyo, Japan), vacuum degasser (DGU-14 Shimadzu Corporation), and autoinjector (PE Series 200 Injector, Perkin Elmer Mountain View, CA). The analytes are eluted with methanol, water, and ammonium acetate, pH 8.0, using a gradient system. The chromatographic run time is 7 min.

Chromatography Settings

Column type:	Synergi Max-RP, 4 μm, 50 × 2 mm, Phenomenex
Mobile phase composition:	(A) MeOH:H$_2$O:NH$_4$OAc (1 M, pH 8)/10:90:0.5 (v:v:v)
	(B) MeOH:H$_2$O:NH$_4$OAc (1 M, pH 8)/90:10:0.5 (v:v:v)
Program:	Gradient:(min) 0, 0.1, 1.0, 3.5, 3.6, 6.6, 6.7
	%B: 0, 0, 100, 60, 100, 100, 0
Flow rate:	180 μl/min
Analysis time:	10 min
Injection volume:	~8 μl
Retention time:	Antisense oligonucleotide = ~3.0 min
	5'-*GUGCUCCAUUGAUGC* (IS) = ~ 3.0 min

MS/MS Conditions

A Micromass Quattro Ultima triple quadrupole mass spectrometer with an electrospray ionization source at -25 V cone voltage and 30 eV collision energy is used to detect the analytes by MRM in negative ion

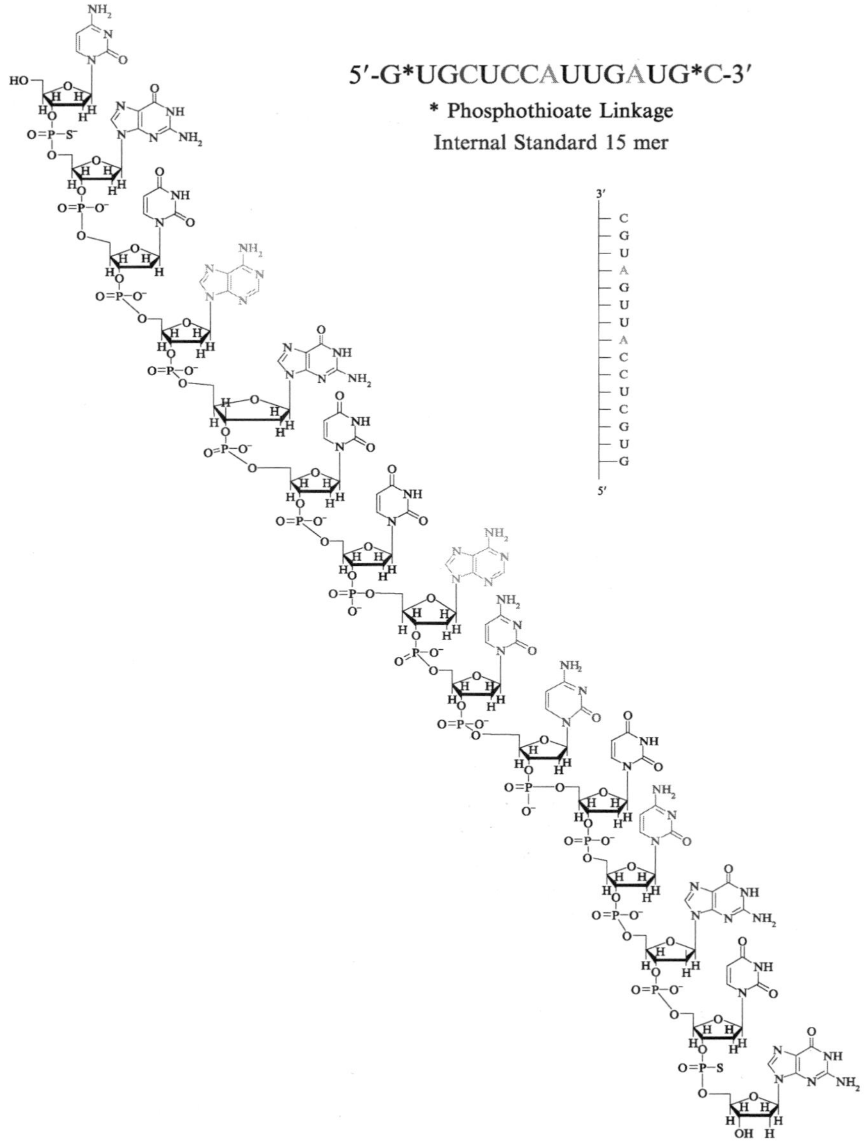
5′-G*UGCUCCAUUGAUG*C-3′
* Phosphothioate Linkage
Internal Standard 15 mer
3′
C
G
U
A
G
U
U
A
C
C
U
C
G
U
G
5′

FIG. 1. Structure of the internal standard.

mode. The mass transitions are monitored at m/z 1146.2 → 745.9 for AON and m/z 1128.72 → 731.9 for the internal standard. AON is quantified by the peak area ratio. The source temperature is 150°.

Compound	Ionization mode	Cone voltage (V)	Collision energy (eV)	Transition
Antisense Oligonucleotide	ESI-	25	30	1146.2 → 745.9
5'-GUGCUCCAUU GAUGC* (IS)	ESI-	25	30	1128.72 → 731.9

Chromatography Results

A typical chromatogram of a rafAON standard at 8 ng/ml and an internal standard at 1 μg/ml from the human plasma sample is shown in Fig. 2. Similarly, a typical quality control chromatogram with a rafAON concentration of 8000 ng/ml and an internal standard are shown in Fig. 3. To generate a calibration curve for LE-rafAON standards in human plasma, standards of 8, 20, 100, 300, 1000, 3000, and 10,000 ng/ml are prepared and analyzed. The peak area ratio of rafAON to internal standard are plotted with respect nominal concentration. A weighted ($1/x^2$) least-squares regression analysis is used to generate a calibration curve. A typical calibration curve

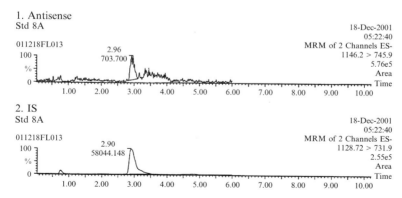

FIG. 2. Typical calibration standard chromatogram of antisense oligonucleotide (8 ng/ml) in human plasma and internal standard (bottom).

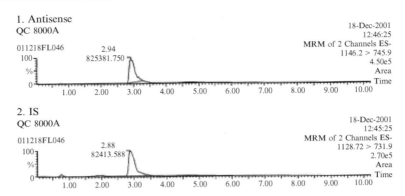

FIG. 3. Typical quality control sample chromatogram of antisense oligonucleotide (8000 ng/ml) in human plasma and internal standard (bottom).

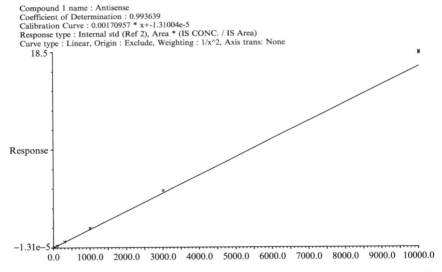

FIG. 4. Typical calibration curve for LE-rafAON in human plasma. Standards contain 8, 20, 100, 300, 1000, 3000, and 10,000 ng/ml LE-rafAON.

is shown in Fig. 4, and validation parameters are summarized in Table I. In human plasma, rafAON is stable at room temperature for 1.5 h, in the autosampler at room temperature for 34 h, in the autosampler at 4° for 51 h, freeze/thaw for three cycles, and long-term storage at −20° for 61 days.

TABLE I
SUMMARY OF VALIDATION PARAMETERS FOR LC/MS/MS ASSAY OF rafAON IN
HUMAN PLASMA[a]

Analyte	Assay range	Intrabatch precision (%CV)	Intrabatch accuracy (%difference)	Between-batch precision (%CV)	Between-batch accuracy (%difference)
Antisense oligonucleotide	8 to 10,000 ng/ml	2.3 to 14.0%	−8.8 to 7.8%	4.5 to 12.3%	−11.4 to −3.2%

[a] Sample volume: 1000 μl.

ELISA

The principle and stepwise procedure used for the LE-rafAON ELISA method is presented schematically in Fig. 5. Briefly, LE-rafAON is hybridized to its complementary oligonucleotide, which is biotinylated at the 3′ end and labeled at the 5′ end with digoxigenin. The hybridized species is bound to a neutravidin-coated plate, and S1 nuclease is added to digest any unhybridized oligonucleotides. Antidigoxigenin conjugated to alkaline phosphatase is added, which binds to the 5′ end of the complementary strand. The fluorescence of the BBT anion is measured after the addition of the Attophos substrate. This section describes the experimental details of this assay and its validation parameters.

Chemicals and Reagents

Raf-1 oligonucleotides are synthesized at Lofstrand Laboratories Limited (Gaithersburg, MD), and liposomes are prepared by the Pharmaceutics Department at NeoPharm, Inc. (Waukegan, IL). EDTA, glycerol, sodium chloride, sodium phosphate (dibasic), Tris hydrochloride, Tween 20, zinc acetate and complementary oligonucleotide, or cutting probe (5′-5GCATCAATGGAGCAC7) are obtained from Sigma-Aldrich (St. Louis, MO). Antidigoxigenin-AP is from Boehringer Mannheim (now Roche Diagnostics, Indianapolis, IN), S1 nuclease is from Invitrogen Life Technologies (Carlsbad, CA), the Attophos-AP fluorescent substrate system is from Promega (Madison, WI), 1 N hydrochloric acid is from VWR (West Chester, PA), and sodium acetate is from Baker (Phillipsburg, NJ). The superblock buffer in TBS and Reacti-Bind neutravidin-coated ELISA plates blocked with superblock blocking buffer are products of Pierce (Rockford, IL). Blank human plasma is from normal human volunteers. Deionized water from a Milli-Q system (Millipore, Milford, MA) is used throughout.

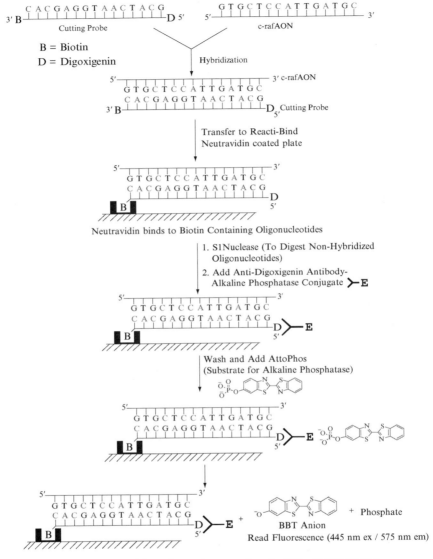

FIG. 5. The principle and stepwise procedure for LE-rafAON ELISA.

The cutting probe is reconstituted in Milli-Q water and stored at −20° until use. It is then diluted to working concentration in SSPE/Tween 20 buffer. The buffer consists of 14.2 g of sodium phosphate (dibasic), 87.66 g of sodium chloride, 10.0 ml of Tween 20, and 100 ml of 0.5 *M* EDTA in a total volume of 1.0 l. The SSPE/Tween 20 buffer is diluted 6:4 with Milli-Q water and is then used to prepare a 57.6-ng/ml cutting probe solution.

S1 nuclease buffer is prepared by dissolving 10.35 g of sodium acetate and 0.54 g of zinc acetate in Milli-Q water, adjusting the pH to 4.6 using acetic acid, and bringing the total volume to 125 ml. The solution is brought to 250 ml by adding 125 ml of glycerol, mixing well, and storing until use at 2–8°. The S1 nuclease working solution is prepared by diluting S1 nuclease to 50 units/ml in the buffer.

Drug Solutions

The LE-rafAON stock solution is prepared by first reconstituting the oligonucleotide with 25 ml of normal saline and shaking gently. This solution is added to a vial containing blank liposomes, shaken gently, placed in a sonicating water bath (Misonix), and sonicated at the highest setting for 10 min with occasional, gentle shaking. The stock solution is spiked into blank human plasma to make an intermediate standard, which is then diluted to prepare the calibration standards. A separate LE-rafAON solution is reconstituted from which quality control (QC) samples in human plasma are prepared. All solutions are aliquoted into polypropylene tubes and stored at −80° until use. For each analytical run, calibration standards of concentrations 1.0, 2.0, 4.0, 8.0, 16.0, 25.0, and 50.0 ng/ml are used to generate the standard curve, and a minimum of three QC samples (1.0, 2.0, 3.0, 9.0, and 27.0 ng/ml) are analyzed.

Sample Preparation

Samples are removed from storage at −80° and thawed on ice. Each thawed sample is vortexed; 250 μl is aliquoted into an appropriate tube and 250 μl of cutting probe working solution is added. These mixtures are vortexed, sonicated for 1 min at a low setting, and then incubated at ambient temperature on a shaker plate for 1 h. The Reacti-Bind neutravidin-coated plate is washed once with wash buffer and blotted on absorbent paper. Aliquots (250 μl) of the sample/cutting probe mixtures are transferred into the appropriate wells of the Reacti-Bind plate, which is then covered using a plate sealer and allowed to incubate at ambient temperature for 30 min. At that time, the plate is washed four times with wash buffer (10× Tris-buffered saline with 10% Tween 20) and blotted.

The S1 nuclease working solution (300 μl) is added to each well, the plate is again covered using a plate sealer and then incubated at ambient temperature. After 2 h, the plate is washed four times with wash buffer and blotted. Antidigoxigenin-AP (1:2000) is diluted 1:10 in superblock buffer, 200 μl is added to each well, and the plate is sealed, incubated at ambient temperature for 30 min and then washed four times and blotted. The Attophos substrate (reconstituted according to the kit instructions) is added (200 μl) to each well and is developed by incubating in a dark room at ambient temperature for a minimum of 15 min. Stop solution and 200 μl of 20% EDTA solution are added, and the fluorescence is read on a Victor 2 (Wallac, Turku, Finland) plate reader (445 nm excitation, 575 nm emission) within 15 min. The responses of the calibration standards are used to generate a four-parameter logistic function. Concentrations of QC and unknown samples are determined by interpolation from this standard curve.

Validation

Each validation run includes a blank, the seven calibration standards, at least three QC samples, all in duplicate, and any unknown samples to be quantified. A total of 13 validation runs are performed on 10 separate days. Interassay and intraassay precision and accuracy are assessed for the QC samples and are summarized in Table II. Precision is measured as %CV (standard deviation divided by the average observed concentration $\times$ 100), and accuracy is presented as %difference, defined as

$$\%\text{difference} = [(\text{observed concentration} - \text{nominal concentration})/ \text{nominal concentration}] \times 100$$

Acceptance criteria for validation are $\pm$ 20%, for precision and accuracy, at all concentrations except the lower limit of quantitation, which should be within 25%. The specificity, effect of hemolysis, and temperature stability of the assay are also assessed.

TABLE II
ELISA VALIDATION CHARACTERISTICS OF LE-rafAON IN HUMAN PLASMA

| Nominal (ng/ml) | Intraassay | | | | Interassay | | | |
	N	Observed (ng/ml)	% difference	%CV	N	Observed (ng/ml)	% difference	· %CV
1.00	6	1.05	4.64	6.91	20	1.03	3.25	20.6
2.00	6	1.89	−5.54	3.33	20	2.27	13.4	16.4
3.00	5	2.70	−9.95	5.79	27	2.59	−13.6	14.8
15.0	6	15.7	4.88	4.05	28	14.4	−4.20	14.9
40.0	5	37.4	−6.46	5.15	28	38.8	−2.90	14.6

ELISA Results

The precision and accuracy of this method are within the acceptable limits for a bioanalytical assay based on immunological techniques (Table II). All calibration curves had a correlation coefficient greater than 0.997. The linear range of 1.0 to 50.0 ng/ml allows for the analysis of samples from clinical trials. This assay was found to have the requisite specificity for LE-rafAON, as analyses of plasma blanks from six individuals were free of any interference.

Acknowledgments

We thank PPD Immunochemistry Laboratory (Richmond, VA) and Quest Pharmaceutical Services (Newark, DE) for experiments involving both studies.

[15] Liposomal Antisense Oligonucleotides for Cancer Therapy

By Doris R. Siwak, Ana M. Tari, and Gabriel Lopez-Berestein

Introduction

After bacterial cells and viruses were discovered to naturally regulate gene activity by antisense, scientists have developed and utilized antisense to regulate protein expression in eukaryotic cells. In 1984, Izant and Weintraub[1] demonstrated that full-length antisense RNA against chicken thymidine kinase (TK) inhibited TK expression in chicken cells. In 1985, Miller's group demonstrated that short, complementary strands of DNA (12 bases or less) could also inhibit protein expression by inhibiting rabbit globin expression with globin antisense oligonucleotides.[2,3]

Whereas oligonucleotides, rather than full-length genes, enabled a simpler preparation of antisense due to the automated synthesis of oligonucleotides, other problems, such as nucleic acid instability and difficulty in penetrating the cell membrane, also had to be overcome. One commonly used technique for increasing the resistance to nuclease degradation is to substitute the nonbridging oxygen atom in the phosphate backbone.

[1] J. G. Izant and H. Weintraub, *Cell* **36**, 1007 (1984).

[2] K. R. Blake, A. Murakami, and P. S. Miller, *Biochemistry* **24**, 6132 (1985).

[3] K. R. Blake, A. Murakami, S. A. Spitz, S. A. Glave, M. P. Reddy, P. O. Ts'o, and P. S. Miller, *Biochemistry* **24**, 6139 (1985).

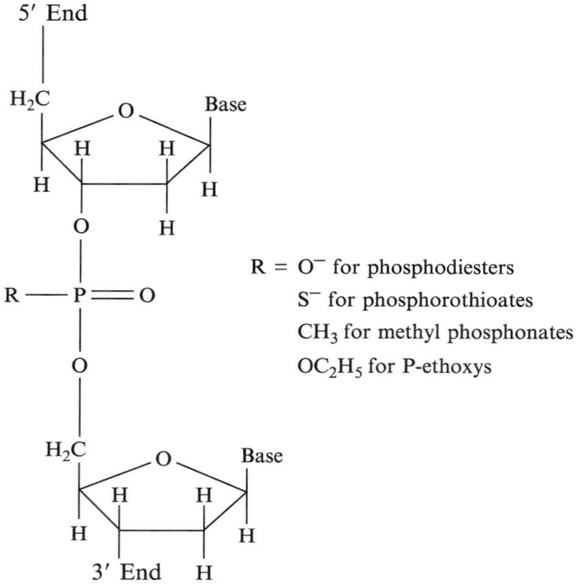

5' End

H_2C   O   Base

H   H

H

H

O   H

R —— P ══ O

O

H_2C   O   Base

H   H

H

H

3' End   H

$R = O^-$ for phosphodiesters
S^- for phosphorothioates
CH_3 for methyl phosphonates
OC_2H_5 for P-ethoxys

FIG. 1. Structure of phosphodiester and nuclease-resistant analogs used in the synthesis of antisense oligonucleotides.

Among the first-generation oligonucleotides, the most commonly used substitution is a sulfur atom, which produces a phosphorothioate backbone (Fig. 1). Other substitutions include methyl phosphonates and P-ethoxys. These substitutions are uncharged and lack ionizable groups. Thus, they are more hydrophobic and presumably better able to penetrate the cell membrane.

An increased effectiveness of delivering antisense oligonucleotides into cells was achieved by the incorporation of oligonucleotides into liposomes. One of the earliest reports on liposomal antisense oligonucleotides was by Stein and colleagues in 1988. Using phosphorothioate c-Myc oligonucleotides incorporated into phosphatidylserine liposomes, the authors induced decreased c-Myc expression and cell growth.[4] From the late 1980s to the early 1990s, various groups reported increased inhibitory effects of liposomal antisense oligonucleotides compared to unencapsulated oligonucleotides. The actual quantitation of liposomal antisense oligonucleotide

---

[4] S. L. Loke, C. Stein, X. Zhang, M. Avigan, J. Cohen, and L. M. Neckers, *Curr. Top. Microbiol. Immunol.* **141,** 282 (1988).

uptake has been examined more recently; for example, Wang *et al.*[5] determined that uptake of the epidermal growth factor receptor (EGFR) antisense oligonucleotide encapsulated in folate-PEG-liposomes was 16 times greater than that of the unencapsulated antisense oligonucleotide.

Carcinogenesis is aided by the overexpression of oncogenes; thus, the utility of liposomal antisense oligonucleotides as therapeutic agents has been apparent from the earliest stages of antisense development. As evident from the literature, various liposomal antisense oligonucleotides against various oncogenes in a variety of cancer cell types have been studied. This article presents some of our most recent work in utilizing liposomal antisense oligonucleotides to determine the effects of two different oncogenes, Grb2 and Bcl-2, on the biological and biochemical properties of breast cancer and leukemia cells, respectively.

Preparation of Liposomal Antisense Oligonucleotides

For simplicity of preparation, a single type of phospholipid, 1,2-dioleoyl-*sn*-glycero-3-phosphocholine (DOPC) (Avanti Polar-Lipids, Inc., Alabaster, AL) is used. DOPC (Fig. 2), a zwitterionic (neutral) phospholipid at physiological pH, has a fluid phase temperature (T_c) of approximately $-22°$.[6] This is due to its large phosphatidylcholine head group and its fluid, unsaturated hydrocarbon chains. The low T_c enables liposomal preparation as well as liposomal reconstitution to be performed at room temperature. DOPC is dissolved in sterile *tert*-butanol to produce a 20-mg/ml stock solution. *tert*-Butanol is the organic solvent of choice for human delivery, as much of the solvent can be removed by evaporation and it solubilizes liposomes effectively. DOPC is then added to a sterile glass vial.

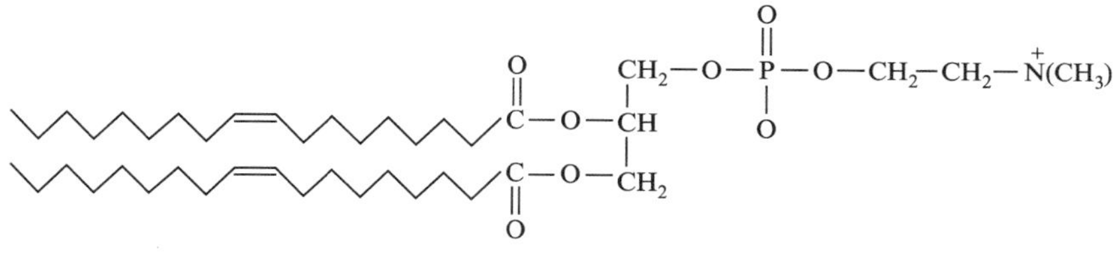

FIG. 2. 1,2-Dioleoyl-*sn*-glycero-3-phosphocholine (DOPC).

[5] S. Wang, R. J. Lee, G. Cauchon, D. G. Gorenstein, and P. S. Low, *Proc. Natl. Acad. Sci. USA* **92,** 3318 (1995).
[6] B. D. Ladbrooke and D. Chapman, *Chem. Phys. Lipids* **3,** 302 (1969).

Next, the antisense oligonucleotide, dissolved in sterile dimethyl sulfoxide (DMSO) to a stock concentration of 20–30 mg/ml, is added. Occasionally, a high salt concentration in the antisense preparation prevents complete solubilization of DNA; in this case, sterile water may also be added. We use P-ethoxy oligonucleotides as they are incorporated reproducibly and efficiently (95% efficiency for DOPC[7]) into lipid carriers. The antisense oligonucleotide sequences used in our laboratory include

> Grb2 antisense: 5′-ATATTTGGCGATGGCTTC-3′
> Bcl-2 antisense: 5′-CAGCGTGCGCCATCCTTCCC-3′

Antisense molecules are purchased from Oligos, Etc. (Wilsonville, OR). Both Grb2 and Bcl-2 antisense are targeted against the translation initiation site.

To facilitate the solubilization of DOPC and oligonucleotides, the detergent Tween 20, diluted 1:1000 in sterile water, is then added. Because DMSO prevents liquid crystalline lamellae formation required for good multilamellar vesicle (MLV) formation and hence for effective oligonucleotide encapsulation, sterile *tert*-butanol is added to at least a 95% volume of DMSO. To prepare 40 nmol of an antisense oligonucleotide consisting of 18 bases (estimating the molecular weight for the oligonucleotide at 325 g/base/mol), the formulation is as follows: 31.4 μl DOPC, 9.4 μl antisense oligonucleotide (25 mg/ml stock), 48.2 μl Tween 20 solution, and 3 ml *tert*-butanol.

Note that the ratio of lipid:oligonucleotide is 20:1 (mol:mol); our laboratory determined this to be the optimal ratio for the incorporation of oligonucleotides by the liposomes.[8] Also note that the volume of the stock oligonucleotide added depends on the concentration determined upon dissolving the stock oligonucleotide in DMSO and will vary between preparations. For example, an 18-mer oligonucleotide with stock concentrations of 20 and 30 mg/ml will require 11.7 and 7.8 μl, respectively. Finally, note that to prepare liposomal oligonucleotides containing greater or less than 40 nmol of oligonucleotide, DOPC and Tween 20 will be increased or decreased proportionately.

After the addition of *tert*-butanol, the vials are frozen in a dry ice–acetone bath until the frozen mixture is opaque and lyophilized overnight. Vials are then stored at −20°. Due to the low T_c of DOPC and the instability of the unsaturated bonds in the lipid chains, the recommended maximum storage time of lyophilized liposomes is 3 months.

[7] A. M. Tari and G. Lopez-Berestein, *J. Liposome. Res.* **7**, 19 (1997).
[8] A. M. Tari and G. Lopez-Berestein, unpublished results (1994).

To produce MLVs of liposomal antisense oligonucleotides, vials are warmed to room temperature, hydrated with 0.9% normal saline or phosphate-buffered saline (PBS), and vortexed. Upon hydration, liposomes are sonicated routinely to optimize oligonucleotide distribution in MLVs; however, we discovered that well-vortexed, but unsonicated, preparations are equally effective.[9] The working concentration of the liposomal antisense oligonucleotides is 100 μM (400 μl of solvent added to the 40-nmol preparation). The liposomes are then added to the cell growth medium. Typical final concentrations are 8–12 μM. We have demonstrated that liposomal antisense oligonucleotides are effective in medium containing 10% fetal bovine serum (FBS).[10]

Liposomal Grb2 (L-Grb2) Antisense Oligonucleotide Effects on Breast Cancer Cell Growth

Grb2 is an adaptor protein that can associate either directly[11] or indirectly[12] with ErbB2, an oncogene associated with poor survival in breast cancer patients.[13] Using the L-Grb2 antisense oligonucleotide, our laboratory has provided evidence that Grb2 plays a critical role in the growth regulation in ErbB2-activated breast cancer cells.[10]

ErbB2-overexpressing breast cancer cell lines MDA-MB-453 and SK-Br-3 are seeded at 1.0–2.5 × 10^3 cells/well in 96-well plates in 100 μl of DMEM/F12 + 10% FBS. After an overnight incubation, 0–12 μM of the L-Grb2 antisense oligonucleotide is added, and cells are incubated for an additional 5 days. Cell growth is measured by the incorporation of Alamar Blue dye (Accumed International Companies Westlake, OH). The percentage growth of treated cells is compared with untreated cells. The L-control oligonucleotide (Bcr-abl, an oncogene whose oligonucleotide sequence is undetected in breast cancer cells) is used as an additional control.

L-Grb2 antisense oligonucleotide treated ErbB2-overexpressing cells showed decreased cell growth (Fig. 3). In contrast, cells treated with the L-control oligonucleotide were not growth inhibited. In addition, breast cancer cell lines that do not overexpress ErbB2 (MCF-7 and MDA-MB-435) were also treated. These breast cancer cell lines were not growth inhibited upon L-Grb2 antisense oligonucleotide treatment (Fig. 4).

[9] A. M. Tari and G. Lopez-Berestein, unpublished results (1993).

[10] A. M. Tari, M. C. Hung, K. Li, and G. Lopez-Berestein, *Oncogene* **18,** 1325 (1999).

[11] P. W. Janes, R. J. Daly, A. deFazio, and R. L. Sutherland, *Oncogene* **9,** 3601 (1994).

[12] R. J. Fiddes, P. W. Janes, G. M. Sanderson, S. P. Sivertsen, R. L. Sutherland, and R. J. Daly, *Cell Growth Differ.* **6,** 1567 (1995).

[13] D. J. Slamon, G. M. Clark, S. G. Wong, W. J. Levin, A. Ullrich, and W. L. McGuire, *Science* **235,** 177 (1987).

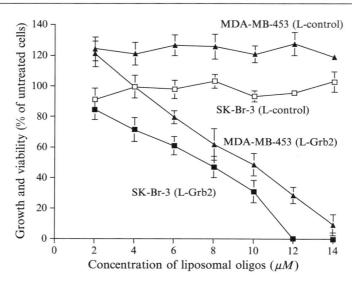

Fig. 3. The L-Grb2 antisense oligonucleotide inhibited the proliferation of ErbB2-overexpressing breast cancer cells. Cells were treated for 5 days with liposomes and growth was determined by Alamar Blue dye incorporation. Percentage growth was determined by comparing liposome-treated cells with respective untreated MDA-MB-453 or SK-Br-3 cells.

Analysis of Grb2-Mediated Signaling in Breast Cancer Cells Using the L-Grb2 Antisense Oligonucleotide

In addition to demonstrating that Grb2 can regulate cell growth, our laboratory has also used the L-Grb2 antisense oligonucleotide to dissect Grb2-mediated signaling pathways important for growth regulation in breast cancer cells. Analysis of two downstream signaling proteins, Akt and mitogen-activated protein kinase (MAPK), revealed that Grb2 preferentially regulates one protein rather than both, depending on the breast cancer cell line examined.[10,14]

To demonstrate these phenomena, ErbB2- or EGFR-overexpressing breast cancer cell lines are seeded in six-well plates in 1.5 ml of DMEM/ F12 + 5 or 10% FBS. In addition, a heregulin-transfected breast cancer cell line (MCF-7/T7) is examined. Heregulin is a growth factor that activates ErbB2[15–17] and hence its signaling pathways. After overnight incubation,

[14] S. J. Lim, G. Lopez-Berestein, M. C. Hung, R. Lupu, and A. M. Tari, *Oncogene* **19,** 6271 (2000).

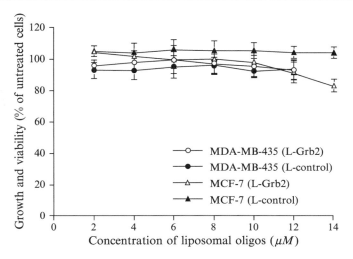

FIG. 4. The L-Grb2 antisense oligonucleotide did not inhibit the proliferation of breast cancer cells that express low levels of ErbB2. Cells were treated for 5 days with liposomes and growth was determined by Alamar Blue dye incorporation. Percentage growth was determined by comparing liposome-treated cells with respective untreated MDA-MB-435 or MCF-7 cells.

6–12 μM of the L-Grb2 antisense oligonucleotide is added to the cells and the cells are incubated for an additional 3 days. For EGFR-overexpressing cells, 80 ng/ml of EGF in DMEM/F12 + 0.5% FBS is added along with the L-Grb2 antisense oligonucleotide. The cells are then harvested and analyzed for Akt and MAPK activity by detection of their phosphorylation levels via Western blot. Antiphospho-Akt detects phosphorylation of the serine 473 residue, which is required for Akt activity.[18] The antiphospho-MAPK antibody detects phosphorylated threonine 202 and tyrosine 204 residues on MAPK, which are required for MAPK activation.[19]

Western blots are performed by lysing the harvested cells in 100 μl of lysis buffer (1% Triton X-100, 150 mM NaCl, 25 mM Tris, pH 7.4) and

[15] Y. Yarden and E. Peles, *Biochemistry* **30,** 3543 (1991).

[16] D. Wen, E. Peles, R. Cupples, S. V. Suggs, S. S. Bacus, Y. Luo, G. Trail, S. Hu, S. M. Silbiger, and R. B. Levy, *Cell* **69,** 559 (1992).

[17] W. E. Holmes, M. X. Sliwkowski, R. W. Akita, W. J. Henzel, J. Lee, J. W. Park, D. Yansura, N. Abadi, H. Raab, and G. D. Lewis, *Science* **256,** 1205 (1992).

[18] D. R. Alessi, M. Andjelkovic, B. Caudwell, P. Cron, N. Morrice, P. Cohen, and B. A. Hemmings, *EMBO J.* **15,** 6541 (1996).

[19] D. M. Payne, A. J. Rossomando, P. Martino, A. K. Erickson, J. H. Her, J. Shabanowitz, D. F. Hunt, M. J. Weber, and T. W. Sturgill, *EMBO J.* **10,** 885 (1991).

incubating on ice for 30 min, vortexing the cell suspension every 5–10 min. After centrifuging the lysates at 12,000g for 10 min at 4°, the concentration of total proteins in the supernatant is determined (DC Protein Assay, Bio-Rad, Hercules, CA) and normalized for the protein level in the samples. Thirty to 50 μg of protein is mixed with sample loading buffer containing 1% SDS and 1% 2-mercaptoethanol, boiled for 5 min, loaded on a 12% gel, and run by SDS–PAGE. The electrophoresed samples are then transferred from the gel to a nitrocellulose membrane (Schleicher and Schuell, Inc., Keene, NH). The membrane is blocked in Tris-buffered saline (TBS) containing 0.1% Tween 20 (TBS-T) and 5% nonfat dry milk (Bio-Rad Laboratories, Hercules, CA). The rabbit antiphospho-MAPK or phospho-Akt primary antibody (both from New England Biolabs, Beverly, MA) diluted 1:1000 in TBS-T + 1% milk is then added. After washing the membrane, the peroxidase-conjugated antirabbit antibody (Amersham Pharmacia Biotech, Piscataway, NJ) diluted 1:1000 in TBS-T + 2.5% milk is added. The membrane is then washed and developed using an enhanced chemiluminescence detection system (Kirkegaard & Perry Laboratories, Inc., Gaithersburg, MD).

MCF-7/T7 cells are treated with the L-Grb2 antisense oligonucleotide and analyzed for phospho-Akt and phospho-MAPK levels. Controls for this experiment are the Grb2 protein level and total MAPK and Akt levels, as well as the L-control (random) oligonucleotide. As shown in Fig. 5, MCF-7/T7 cells treated with 10 μM of the L-Grb2 antisense oligonucleotide had decreased Grb2 levels. Analysis of Akt revealed that the phospho-Akt level decreased upon L-Grb2 antisense, whereas L-control oligonucleotide treatment had a minimal effect. The total Akt level was unchanged between L-Grb2 antisense and control cells. Thus, decreased Akt activity upon Grb2 downregulation is not due to decreased Akt expression. Grb2 downregulation in two ErbB2-overexpressing breast cancer cell lines (MDA-MB-453 and BT-474) also resulted in decreased Akt activities (not shown), indicating that Grb2 regulation of Akt activity occurs in both heregulin-stimulated and ErbB2-overexpressing breast cancer cells.

For analysis of Grb2-mediated signaling in EGFR-overexpressing breast cancer cells, MDA-MB-468 cells are used. Treatment with 6 μM of the L-Grb2 antisense oligonucleotide showed a decreased Grb2 level as compared to untreated or L-control oligonucleotide controls (Fig. 6A). L-Grb2 antisense oligonucleotide-treated MDA-MB-468 cells also showed decreased phospho-MAPK levels compared to control treatments. Again, no decrease in total MAPK level was observed, indicating that decreased MAPK activity was not due to decreased MAPK protein expression. For comparison of MAPK activity in EGFR- vs ErbB2-overexpressing cells,

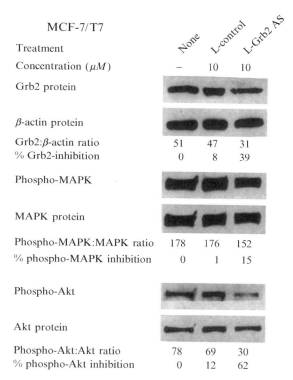

MCF-7/T7

Treatment	None	L-control	L-Grb2 AS
Concentration (μM)	–	10	10
Grb2 protein			
β-actin protein			
Grb2:β-actin ratio	51	47	31
% Grb2-inhibition	0	8	39
Phospho-MAPK			
MAPK protein			
Phospho-MAPK:MAPK ratio	178	176	152
% phospho-MAPK inhibition	0	1	15
Phospho-Akt			
Akt protein			
Phospho-Akt:Akt ratio	78	69	30
% phospho-Akt inhibition	0	12	62

FIG. 5. The L-Grb2 antisense oligonucleotide inhibited Akt but not MAPK activity in heregulin-transfected breast cancer cells. MCF-7/T7 cells were treated with liposomes for 3 days and analyzed for proteins indicated via Western blots. Ratios of Grb2:actin, phospho-MAPK:MAPK, and phospho-Akt:Akt were determined after the quantitation of bands via densitometric scanning. Percentage inhibition was calculated by $(1 -$ ratio in treated cells/ratio in untreated cells$) \times 100\%$.

the ErbB2-overexpressing breast cancer cell line SK-Br-3 was used. At 8 μM L-Grb2 antisense oligonucleotide treatment, a 5% decrease in MAPK activity was observed (Fig. 6B).

Because EGFR and ErbB2 are related proteins, and both are reported to activate Akt and MAPK,[20–24] these results were unexpected. In addition,

[20] R. Ben-Levy, H. F. Paterson, C. J. Marshall, and Y. Yarden, *EMBO J.* **13,** 3302 (1994).
[21] B. M. Marte, D. Graus-Porta, M. Jeschke, D. Fabbro, N. E. Hynes, and D. Taverna, *Oncogene* **10,** 167 (1995).
[22] W. Liu, J. Li, and R. A. Roth, *Biochem. Biophys. Res. Commun.* **261,** 897 (1999).
[23] N. G. Ahn, J. E. Weiel, C. P. Chan, and E. G. Krebs, *J. Biol. Chem.* **265,** 11487 (1990).
[24] B. M. Burgering and P. J. Coffer, *Nature* **376,** 599 (1995).

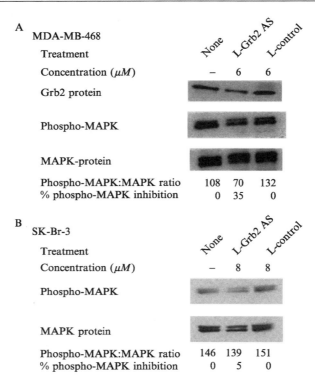

FIG. 6. The L-Grb2 antisense oligonucleotide inhibited MAPK activity in EGFR-overexpressing (A), but not in ErbB2-overexpressing (B) breast cancer cells. Cells were treated with liposomes for 3 days and analyzed for proteins indicated via Western blots. The ratio of phospho-MAPK:MAPK was determined after the quantitation of bands via densitometric scanning. Percentage inhibition was calculated by (1 − ratio in treated cells/ratio in untreated cells) × 100%.

L-Grb2 antisense oligonucleotide treatment decreases cell growth in EGFR-overexpressing breast cancer cells at concentrations similar to those that inhibited MAPK activity. Taken together, these results suggest that while several downstream proteins may be activated upon ErbB2 and EGFR stimulation, each protein has its preferred signaling pathway to regulate cell growth. These results also suggest that the therapeutic targets for ErbB2- and EGFR-overexpressing breast cancer cells would be most effective when aimed specifically toward the Akt and MAPK pathways, respectively.

Liposomal Bcl-2 Antisense Oligonucleotide in Acute Myeloid Leukemia Studies

Bcl-2 is a protein whose increased expression is associated with an increased resistance to apoptosis.[25–27] Using the L-Bcl-2 antisense oligonucleotide described earlier, Konopleva and colleagues[28] demonstrated that downregulating the expression of Bcl-2 in primary cultures of acute myeloid leukemia (AML) cells resulted in increased apoptosis. Samples of bone marrow or peripheral blood from newly diagnosed or recurrent AML patients are obtained and mononuclear cells are separated by Ficoll–Hypaque (Sigma Chemical, St. Louis, MO) density gradient centrifugation. The mononuclear cells are seeded at 5×10^5 cells/ml in RPMI + 10% FBS supplemented with 200 U/ml of granulocyte colony-stimulating factor (G-CSF). G-CSF enables AML blasts to proliferate and prevents spontaneous apoptosis without increasing Bcl-2 expression.[29] The primary AML cells are then treated with 8 μM of the L-Bcl-2 antisense oligonucleotide or L-control oligonucleotide (scrambled sequence) and incubated for 5 days.

To determine the proportion of apoptotic cells, the cells are harvested and suspended in PBS containing 2 mM MgCl$_2$ and 10% FBS. The cell membranes are then made permeable by mixing 80 μl of the cell suspension with 100 μl of an acridine orange solution containing 0.1% (v/v) Triton X-100, 0.05 N HCl, 0.15 N NaCl, and 8 μg/ml acridine orange (Polysciences, Inc., Warrington, PA). The cell fluorescence is measured within 5 min of staining. At least 3×10^4 cells are examined for the "sub-G$_1$" phase by flow cytometric analysis using the FACScan flow cytometer with a 488-nm wavelength excitation of a 15-mW argon laser. Filter settings are green (530 nm) for DNA and red (585 nm) for RNA fluorescence. Ten thousand events are stored in the list mode for analysis. The percentage of cells in the "sub-G$_1$" region is defined as the proportion of apoptotic cells in the tested populations. Statistical analyses of the results are performed using the two-tailed Student t test and the Spearman

[25] T. J. McDonnell, N. Deane, F. M. Platt, G. Nunez, U. Jaeger, J. P. McKearn, and S. J. Korsmeyer, *Cell* **57,** 79 (1989).
[26] Y. Tsujimoto, *Oncogene* **4,** 1331 (1989).
[27] D. Hockenbery, G. Nunez, C. Milliman, R. D. Schreiber, and S. J. Korsmeyer, *Nature* **348,** 334 (1990).
[28] M. Konopleva, A. M. Tari, Z. Estrov, D. Harris, Z. Xie, S. Zhao, G. Lopez-Berestein, and M. Andreeff, *Blood* **95,** 3929 (2000).
[29] M. Lisovsky, Z. Estrov, X. Zhang, U. Consoli, G. Sanchez-Williams, V. Snell, R. Munker, A. Goodacre, V. Savchenko, and M. Andreeff, *Blood* **88,** 3987 (1996).

rank correlation coefficient. P values of <0.05 are considered statistically significant.

L-Bcl-2 antisense oligonucleotide treatment resulted in a significant induction of apoptosis in 11 of 19 patient samples (57.9%). The overall proportion of apoptotic cells was $37.0 \pm 8.6\%$ ($P < 0.05$). When samples were divided into L-Bcl-2 antisense oligonucleotide responsive (group I) and nonresponsive (group II), the proportions for group I are $37.5 \pm 8.2\%$ apoptotic cells upon L-Bcl-2 antisense oligonucleotide treatment and 17.3 $\pm$ 5.2% for L-control treatment ($P < 0.005$). For group II, the frequencies were $25.1 \pm 7.4\%$ and $23.2 \pm 6.7\%$ ($P = 0.2$) apoptosis for L-Bcl-2 and L-control, respectively. No significant increase in apoptosis was observed in L-control oligonucleotide-treated primary AML cells compared to untreated cells ($18.6 \pm 4.5\%$ vs $15.1 \pm 4.3\%$, $P = 0.1$).

To determine whether L-Bcl-2 antisense oligonucleotide treatment could also sensitize AML cells to chemotherapeutic agents, Konopleva and colleagues.[28] also examined AML cell viability on the combined treatment of the L-Bcl-2 antisense oligonucleotide with cytosine-arabinoside (ara-C), the drug used most commonly for treating AML. Primary AML cells from 13 patients were treated simultaneously with 8 μM of the L-Bcl-2 antisense oligonucleotide and 1 μM ara-C and incubated for 72 h. Cells were then stained with acridine orange and examined for apoptosis.

The ara-C+L-Bcl-2 antisense oligonucleotide treatment increased apoptosis in L-Bcl-2 antisense oligonucleotide-responsive AML cells ($P < 0.05$, Table I). No significant difference was observed between untreated cells and cells treated with ara-C alone or ara-C + L-control ($P > 0.4$). In contrast, L-Bcl-2 antisense oligonucleotide-unresponsive cells had a low response rate when treated with the ara-C + L-Bcl-2 antisense oligonucleotide; only one of six samples showed increased apoptosis.

TABLE I

L-Bcl-2 Antisense Oligonucleotide Increases Apoptosis Induced by ara-C

Samples	Subdiploid apoptotic cells (%)		
	Control	Nonsense	Bcl-2 antisense
Responsive blasts ($N = 7$)			
No ara-C	18.5 ± 5.6	26.9 ± 6.6	43.8 ± 8.1
Ara-C	42.3 ± 5.8	42.6 ± 5.8	61.7 ± 7.0^a
Nonresponsive blasts ($N = 6$)			
No ara-C	20.5 ± 5.7	18.9 ± 5.4	20.5 ± 5.7
Ara-C	46.4 ± 11.2	42.5 ± 11.0	42.3 ± 11.6

$^a P < 0.05$ compared to ara-C-induced apoptosis alone.

Future Directions

Liposomal antisense oligonucleotides have been reported to treat xenograft tumors successfully, in mice.[30–32] While this is an encouraging step, its use as a therapeutic agent in cancer patients is not yet widely reported. Further pharmacologic and safety studies for human application will allow for the clinical development of liposomal antisense oligonucleotides. The selection of targets, lipids, and antisense composition will be vital to their broad application.

[30] O. Wilhelm, M. Schmitt, S. Hohl, R. Senekowitsch, and H. Graeff, *Clin. Exp. Metastasis* **13,** 296 (1995).

[31] Y. Kondo, S. Koga, T. Komata, and S. Kondo, *Oncogene* **19,** 2205 (2000).

[32] S. Mukai, Y. Kondo, S. Koga, T. Komata, B. P. Barna, and S. Kondo, *Cancer Res.* **60,** 4461 (2000).

Section IV

Liposomes *In Vivo*

[16]   Biodistribution and Uptake of Liposomes *In Vivo*

By JAN A. A. M. KAMPS and GERRIT L. SCHERPHOF

Introduction

In the development of liposomal drug delivery systems, pharmacokinetics, biodistribution, and cellular uptake of such systems are major issues to be dealt with. Depending on the aim of a particular study, a variety of experimental setups and methods are available to determine the *in vivo* behavior of liposomes or liposomal systems. This article touches briefly on some of these different methods, while discussing in more detail qualitative and quantitative methods concerning the role of the liver and spleen in the uptake and biodistribution of liposomes *in vivo*. *In vivo* studies with liposomes have been performed in a variety of animal models, including guinea pigs, dogs, hamsters, rabbits, and cats.[1–5] Most *in vivo* studies, however, have been performed in mice and rats. Interpretation of these *in vivo* studies deserves attention, especially because of differences among different animal species, which may hamper extrapolation of results to other species, e.g., humans. Experiments in rats and mice have shown that not only the clearance rate of liposomes varies between these species, but also the mechanism(s) by which this clearance rate is regulated.[6] The pharmacokinetics and tissue distribution of liposomes after parenteral administration are determined by a diversity of variables, including liposome size and composition, steric stabilization of the liposome, presence of surface-grafted molecules, such as targeting devices, and, obviously, the organism and the health status of that particular organism. Closely related to this last point is the accessibility of target tissues or cells for the liposomes. Upon injection, liposomes encounter anatomical barriers such as the endothelial lining of the vasculature and also of the blood–brain barrier, which will prevent access of the liposomes to extravascular sites. Only in organs such as the

[1] T. M. Huong, T. Ishida, H. Harashima, and H. Kiwada, *Biol. Pharm. Bull.* **24,** 439 (2001).

[2] I. Bekersky, G. W. Boswell, R. Hiles, R. M. Fielding, D. Buell, and T. J. Walsh, *Pharm. Res.* **16,** 1694 (1999).

[3] J. M. Devoisselle, S. Begu, C. Tourne-Peteilh, T. Desmettre, and S. Mordon, *Luminescence* **16,** 73 (2001).

[4] E. T. Dams, W. J. Oyen, O. C. Boerman, G. Storm, P. Laverman, E. B. Koenders, J. W. Van Der Meer, and F. H. Corstens, *J. Nuclear Med.* **39,** 2172 (1998).

[5] M. L. Matteucci, G. Anyarambhatla, G. Rosner, C. Azuma, P. E. Fisher, M. W. Dewhirst, D. Needham, and D. E. Thrall, *Clin. Cancer Res.* **6,** 3748 (2000).

[6] D. Liu, Q. Hu, and Y. K. Song, *Biochim. Biophys. Acta* **1240,** 277 (1995).

liver (see also later), spleen, and bone marrow and, under certain pathological conditions, such as occurring in solid tumors and at sites of infection or inflammation, the vascular endothelium may display enhanced permeability, allowing liposomes to extravasate.[7,8] Generally, small liposomes ($\leq$100 nm) are eliminated from the blood more slowly than large liposomes, whereas electrostatically charged liposomes disappear faster from the blood than uncharged liposomes. Under normal conditions, intravenously administered liposomes will be taken up from the blood circulation predominantly by cells of the mononuclear phagocyte system in the liver and spleen.[9–12] In addition, depending on size and composition of the liposomes, the parenchymal cells of the liver (hepatocytes) may also play a dominant role in the elimination of liposomes from the blood.[13] The ability of liposomes to pass the hepatic sinusoidal endothelial lining, allowing uptake by hepatocytes, relates to the presence of numerous open fenestrations in these cells. These fenestrations have a size of around 150 nm[14] and thus allow small liposomes to gain access to the hepatocytes. However, larger liposomes were also found to cross the endothelial lining, provided they contain the negatively charged phospholipid phosphatidylserine.[15]

Protein adsorption to the liposomal surface is an important determinant in liposome elimination from the blood compartment. In order to diminish protein adsorption and thus to prolong the circulation time of the liposomes, the concept of steric stabilization was developed.[16–19] Lipid-anchored poly(ethylene glycol) (PEG) has proven to be most valuable for this purpose. Coating of the liposomes with PEG increases the residence time in the blood circulation; in mice and rats, half-lives of as long as 20 h can be attained,[18,20] whereas in humans, half-lives of even up to 45 h have been

[7] O. Ishida, K. Maruyama, K. Sasaki, and M. Iwatsuru, *Int. J. Pharm.* **190,** 49 (1999).

[8] K. Maruyama, *Biol. Pharm. Bull.* **23,** 791 (2000).

[9] G. Poste, *Biol. Cell* **47,** 19 (1983).

[10] J. H. Senior, *Crit. Rev. Ther. Drug Carrier Syst.* **3,** 123 (1987).

[11] T. M. Allen, *Adv. Drug Deliv. Rev.* **2,** 55 (1988).

[12] M. C. Woodle and D. D. Lasic, *Biochim. Biophys. Acta* **1113,** 171 (1992).

[13] G. L. Scherphof and J. A. A. M. Kamps, *Prog. Lipid Res.* **40,** 149 (2001).

[14] E. Wisse, *J. Ultrastruct. Res.* **31,** 125 (1970).

[15] T. Daemen, M. Velinova, J. Regts, M. De Jager, R. Kalicharan, J. Donga, J. J. L. Van Der Want, and G. L. Scherphof, *Hepatology* **26,** 416 (1997).

[16] T. M. Allen and A. Cohn, *FEBS Lett.* **223,** 42 (1987).

[17] T. M. Allen, C. Hansen, and J. Rutledge, *Biochim. Biophys. Acta* **981,** 27 (1989).

[18] T. M. Allen, C. Hansen, F. Martin, C. Redemann, and A. Yau Yong, *Biochim. Biophys. Acta* **1066,** 29 (1991).

[19] D. Papahadjopoulos, T. M. Allen, A. Gabizon, E. Mayhew, K. Matthay, S. K. Huang, K. D. Lee, M. C. Woodle, D. D. Lasic, C. Redemann, and F. J. Martin, *Proc. Natl. Acad. Sci. USA* **88,** 11460 (1991).

[20] G. L. Scherphof, H. W. M. Morselt, and T. M. Allen, *J. Liposome Res.* **4,** 213 (1994).

reported.[21] It has been demonstrated that, because of their long-circulation properties, PEG–liposomes have a relatively high probability to extravasate and accumulate at sites that are characterized by increased vascular permeability.[19,22] As indicated earlier, this type of "leaky" endothelium can be found at various stages of tumor development and at sites of infection or inflammation.[23] Under nonpathological conditions, a major fraction of the PEG–liposomes ultimately end up, as conventional liposomes, in cells of the mononuclear phagocyte system in the liver and spleen. The contribution of the spleen to the uptake of PEG–liposomes from the blood is usually somewhat higher than in the case of conventional liposomes.[20] During the last decades, numerous efforts have been made to target liposomes to defined cells or tissues in order to improve biodistribution profiles of liposomes and their contents. The successes of this approach have been variable so far and are, as with nontargeted liposomes, restricted by physical accessibility of the target site and also by the circulation time.

To investigate the fate of liposomes *in vivo*, a large variety of liposomal markers are available. Radiolabels, either encapsulated water-soluble compounds or bilayer-incorporated lipid labels, provide a sensitive and powerful tool to determine liposome biodistribution. Most lipids can be purchased in one or more radiolabeled forms. The choice of the radiolabel depends primarily on the aim of the experiment. Radioactive lipid labels can be incorporated easily into the liposomal bilayer when mixed with other lipids during liposome preparation, irrespective of the method of liposome preparation. It is important that a lipid marker is stably incorporated in the liposomal bilayer to avoid transfer of the marker from the liposomal membrane to cellular membranes such as the plasma membrane or the endosomal membrane or to serum components such as lipoproteins. In addition to stability requirements, the radioactive marker also has to be metabolically inert on the timescale of the experiment to avoid unjustified interpretations of experimental data. Examples of radioactive markers that fulfill these characteristics are [³H]cholesterylhexadecyl ether or [³H]cholesteryloleyl ether. The ether bond in these markers ascertains that, following internalization, the marker is not metabolized significantly by cells within a period of days. These markers provide a convenient way for a rapid and quantitative assessment of tissue distribution after parenteral administration of liposomes and produce reliable and reproducible data on tissue distribution both shortly after injection (1–3 h) and at longer times after injection (24 h).

[21] T. M. Allen, *Drugs* **54,** 8 (1997).
[22] A. Gabizon, R. Catane, B. Uziely, B. Kaufman, T. Safra, R. Cohen, F. Martin, A. Huang, and Y. Barenholz, *Cancer Res.* **54,** 987 (1994).
[23] H. F. Dvorak, J. A. Nagy, J. T. Dvorak, and A. M. Dvorak, *Am. J. Pathol.* **133,** 95 (1988).

Several fluorescent lipid labels on the market have been shown to be valuable tools for studying the *in vivo* behavior of liposomes.[24] In addition to labeling of the liposome itself, there are a variety of possibilities to label liposome-associated compounds such as encapsulated material and homing devices. The usefulness of such labels as markers of the fate of liposomes depend largely on the specifications of the labeled compound and the nature of the label itself. To ensure that the fate of a liposomal marker is representative of the fate of the entire liposome, the use of double-labeled liposomes is strongly recommended. For morphological studies on the intracellular fate of liposomes and/or their components, fluorescent markers have been shown to be powerful tools at the light microscopic level, either with conventional fluorescence microscopy or with confocal laser-scanning microscopy.[25–27] For studies at the electron microscopic level, liposome-encapsulated colloidal gold or the reaction products of liposome-encapsulated horseradish peroxidase have been shown to be convenient and readily detectable markers.[28–31]

For determination of intraorgan distribution, a variety of methods are available that are generally qualitative or, at best, semiquantitative. Labeling of the liposome bilayer with a fluorophore such as 1,1'-dioctadecyl-3,3,3',3'-tetramethyl indocarbocyanine perchlorate (DiI) is, in many cases, a convenient method to obtain information on intraorgan distribution of the liposomes. The necessity to examine carefully intraorgan distribution is illustrated by a study in which we employed immunoliposomes, which *in vitro* interact highly specifically with a rat colon adenocarinoma cell line. Upon injection of these immunoliposomes in a rat bearing liver metastases originating from these tumor cells, we observed that the uptake of immunoliposomes by the metastatic tumor nodules in the liver was increased significantly compared to control liposomes. Although we had clearly achieved liposome accumulation in the tumor nodules, microscopic

[24] J. A. A. M. Kamps, G. A. Koning, M. J. Velinova, H. W. M. Morselt, M. Wilkens, A. Gorter, J. Donga, and G. L. Scherphof, *J. Drug Target* **8,** 235 (2000).

[25] D. L. Daleke, K. Hong, and D. Papahadjopoulos, *Biochim. Biophys. Acta* **1024,** 352 (1990).

[26] A. Cerletti, J. Drewe, G. Fricker, A. N. Eberle, and J. Huwyler, *J. Drug Target.* **8,** 435 (2000).

[27] E. Papadimitriou and S. G. Antimisiaris, *J. Drug Target.* **8,** 335 (2000).

[28] G. A. Koning, H. W. M. Morselt, M. J. Velinova, J. Donga, A. Gorter, T. M. Allen, S. Zalipsky, J. A. A. M. Kamps, and G. L. Scherphof, *Biochim. Biophys. Acta* **1420,** 153 (1999).

[29] K. Hong, D. S. Friend, C. G. Glabe, and D. Papahadjopoulos, *Biochim. Biophys. Acta* **732,** 320 (1983).

[30] S. K. Huang, K. Hong, D. Papahadjopoulos, and D. S. Friend, *Biochim. Biophys. Acta* **1069,** 117 (1991).

[31] H. Ellens, H. W. Morselt, B. H. Dontje, D. Kalicharan, C. E. Hulstaert, and G. L. Scherphof, *Cancer Res.* **43,** 2927 (1983).

examination of tumor slices, following an injection of fluorescently or gold-labeled immunoliposomes, revealed that the immunoliposomes were not associated with tumor cells but rather localized in other tumor-associated cells, probably macrophages.[24]

Intrahepatic distribution in rats can also be determined after iv administration of [^{3}H]cholesteryloleyl ether-labeled liposomes. The animal is treated as described for the quantitative determination of tissue distribution of liposomes in rats followed by collagenase perfusion of the liver and subsequent centrifugation and counterflow elutriation to isolate parenchymal, Kupffer, and liver endothelial cells from collagenase-perfused livers.[32–34] The collagenase treatment is relatively mild and thus leaves all cell types relatively unharmed. When only nonparenchymal liver cells, i.e., endothelial and Kupffer cells, are needed, digestion of the liver by pronase was found to be a more convenient isolation method.[35] Nonparenchymal cells are much more resistant to the relatively harsh pronase treatment than the more vulnerable hepatocytes.

Methodology

Quantitative Determination of Tissue Distribution of Liposomes in Rats

[^{3}H]Cholesteryloleyl ether-labeled liposomes, typically 1–2.5 μmol total lipid per 100 g of body weight (but more is acceptable), in an isotonic solution (e.g., phosphate-buffered saline or HEPES-buffered saline) are injected into male rats ($\sim$200 g) via the penile vein, using a 1-ml syringe with a 25-gauge needle under light halothane anesthesia. Blood samples can be taken from the tail vein under light anaesthesia when a limited number of samples ($<$6) of a relatively small volume ($<$300 μl) is needed. When more extensive blood sampling is required, one can consider the introduction of a permanent cannula in a vein or artery such as the jugularis or carotis. Blood samples are allowed to clot for 60 min at 4°. The samples are then centrifuged (5 min, 13,000g). Radioactivity is measured in the serum samples after addition of the appropriate amount of scintillation cocktail. The total amount of radioactivity in the serum can be calculated using the equation: serum volume (ml) = (0.0219 × body weight(g))

[32] E. Casteleijn, H. Van Rooij, Th. J. C. Van Berkel, and J. F. Koster, *FEBS Lett.* **201,** 193 (1986).

[33] T. Daemen, A. Veninga, F. H. Roerdink, and G. L. Scherphof, *Cancer Res.* **46,** 4330 (1986).

[34] J. Dijkstra, W. J. M. Van Galen, C. E. Hulstaert, D. Kalicharan, F. H. Roerdink, and G. L. Scherphof, *Exp. Cell Res.* **150,** 161 (1984).

[35] J. A. A. M. Kamps, H. W. M. Morselt, P. J. Swart, D. K. F. Meijer, and G. L. Scherphof, *Proc. Natl. Acad. Sci. USA* **94,** 11681 (1997).

+ 2.66.[36] At the last time point of blood sampling the rat is anesthetized either with halothane or by an intraperitoneal injection of nembutal (50 mg/kg body weight). The abdominal area is shaved and the anesthetized rat is placed on a warmed surface. Then the abdomen is opened with a V-shaped cut out of the skin in the middle of the abdomen. The abdominal flap is folded up onto the chest wall. The intestines are pushed to the right side of the animal and, if necessary, the lobes of the liver are displaced very gently upward to permit the identification of the portal vein. Forceps should not be used to handle the liver, and, in general, manipulation of the organ is kept to a minimum. No undue pressure should be applied on the diaphragm, as this will hinder breathing. Using a 1–in., 20-gauge hypodermic braunule, the portal vein is cannulated by gently introducing the needle into and along the vein (pointing toward the liver). The needle should not puncture the distal side of the vessel. The needle is tied in place with 4-0 braided silk using two ligatures; the ties should not be so tight as to cut into the wall of the vein. The posterior vena cava is ligated below the liver at the level of the renal vein. The animal is restrained by the limbs so that further movement does not disturb the needle. The vena cava is cut above the ligation to release the blood and the liver is perfused immediately, using a peristaltic pump (10–20 ml/min), with an isotonic buffer (37°) for 2–5 min to wash out the blood. The liver and other tissues of interest are removed gently. Blood, fat, and/or connective tissue is removed cautiously from the surface of the organ and the tissue is weighed. The organs are chopped using sharp scissors and the tissues are homogenized in a Potter Elvehjem tube in an appropriate, known volume of isotonic buffer. The radioactivity of the homogenized samples can be determined after solubilization of, typically, 400 μl homogenate in 100 μl 10% sodium dodecyl sulfate (SDS) and 4 ml scintillation cocktail. For calculation of the radioactivity per organ, the radioactivity in organs other than the liver has to be corrected for the amount of blood contents.[37]

When *in vivo* experiments of a short duration ($\leq$30 min) are performed, the entire procedure can be performed in anesthetized rats. After injection of the liposomes via the penile vein, the abdomen can be opened and blood samples can be taken from the vena cava inferior. This procedure also allows taking liver samples at given time points by excising liver lobules after ligation of these lobules to prevent blood leakage. Since in that case the tissue has not been perfused, a correction for blood content should be included (85 μl/g wet liver tissue).[37]

[36] M. K. Bijsterbosch, G. J. Ziere, and Th. J. C. Van Berkel, *Mol. Pharmacol.* **36,** 484 (1989).
[37] W. O. Caster, A. B. Simon, and W. D. Armstrong, *Am. J. Physiol.* **183,** 317 (1955).

Qualitative Determination of Intraorgan Distribution of Liposomes in Rats

One to 2.5 μmol (or more, if required) per 100 g of body weight of labeled (e.g., fluorescently) liposomes is injected into male rats via the penile vein under light halothane anesthesia. At the chosen time point the rat is anesthetized with halothane or with an intraperitoneal injection of nembutal (50 mg/kg body weight). The abdomen of the rat is opened as described earlier, and, after properly placing a cannula in the portal vein, the liver is perfused for 2–5 min with an isotonic buffer to rinse out the blood. The liver and other tissues of interest are removed gently and placed immediately on ice.

The blood, fat, and/or connective tissue is removed cautiously from the surface of the organ, the organ is cut into convenient pieces with a sharp surgical blade, and the pieces are placed on a buffer-drenched filter paper with a diameter of 2.5 cm. At this stage the different organs can already be compared visually with respect to differences in fluorescent appearance. The tissue pieces are snap-frozen in isopentane at $-80°$ and the frozen tissues are stored at $-80°$ until further processing. Four-micrometer cryostat sections are cut, and the air-dried sections are examined using a fluorescence microscope. Fixation of the tissue, for example, with acetone, may drastically diminish the fluorescent signal on the cryostat sections.

Isolation and Purification of Hepatocytes, Kupffer Cells, and Endothelial Cells by Collagenase Perfusion

Before starting liver cell isolation, the following solutions have to be prepared.

1. Preperfusion buffer [142 mM NaCl, 6.7 mM KCl, 10 mM N-2-hydroxyethylpiperazine-N'-2-ethanesulfonic acid (HEPES), pH 7.6]: The preperfusion buffer is oxygenated by carbogen bubbling for at least 20 min at 37°, adjust pH to 7.6 immediately before use of the buffer.

2. Collagenase buffer (66.7 mM NaCl, 6.7 mM KCl, 4.8 mM CaCl$_2$·2H$_2$O, 10 mM Hepes, pH 7.4): Collagenase buffer is oxygenated by carbogen bubbling for at least 20 min at 37°, bovine serum albumin (BSA) is dissolved in 10 ml of this buffer (2% BSA in final solution), collagenase (Boehringer Mannheim) is added (0.05% collagenase in final solution), and the solution is mixed gently with the rest of the buffer to avoid air bubbles and is adjusted pH to 7.4.

3. Postperfusion buffer: One hundred milliliters of Hanks' solution (137 mM NaCl, 5.4 mM KCl, 0.8 mM MgSO$_4$·7H$_2$O, 0.33 mM Na$_2$HPO$_4$·2H$_2$O, 0.44 mM KH$_2$PO$_4$, 10 mM HEPES, 5 mM glucose,

pH 7.4) are oxygenated by carbogen bubbling for at least 20 min at 37°, BSA is dissolved in 10 ml of this buffer (2% BSA in final solution), and the pH is adjusted to 7.4.

4. BSA is dissolved in 1.5 liters of Hanks' solution (0.3% BSA in final solution), the pH is adjusted to 7.4, and the solution is kept at 4°.

The perfusion system, consisting of a peristaltic pump, temperature controlled by a water bath, is first rinsed with 70% ethanol, followed by rinsing out the pump and tubing with water. The tubing is filled with the preperfusion buffer and the peristaltic pump is set to deliver 20 ml/min.

The rat injected with labeled liposomes, is anesthetized with nembutal (50 mg/kg body weight) administered intraperitoneally 20 min before the start of surgery. The abdominal area is shaved, and the anesthetized animal is placed on a warmed surface. The abdomen is opened, surgery is performed, and the portal vein is cannulated as described earlier. The chest is opened by cutting the rib cage up each side of the sternum. The diaphragm and the lowest ribs are left in place. The vena cava and the position where it penetrates the diaphragm are identified. Another 16-gauge braunule is inserted carefully into the vena cava between the heart and the diaphragm, pointing away from the heart. The braunule is tied into place with a 410 cotton or silk ligature. The vena cava is ligated (two ties) between the needle and the heart. The tubing of the peristaltic pump is connected from the perfusion medium to the portal vein braunule. Care is taken not to disturb the needle from its position in the hepatic portal vein. The vena cava braunule is connected to the tubing, which in turn is connected to a reservoir for waste perfusate.

The liver is perfused by allowing the peristaltic pump to pump 400 ml of preperfusion buffer at 20 ml/min. The temperature of the buffer is maintained at 37°. As the blood is displaced from the liver by the perfusate, the liver will become progressively paler (tan) in color. Then the pump speed is increased to 28 ml/min. After 300–400 ml of preperfusion medium has passed, the tubing is switched to the collagenase buffer reservoir. To prevent air bubbles from entering the liver, a small container (approximately 15 ml), often used in infusion systems, should be placed as a bubble trap between the pump and the portal vein cannula. The collagenase buffer is recirculated for 10 min at a pump speed of 28 ml/min. The surface of the liver is kept moist with warm buffer. The pH of the perfusate may drop after passage through the liver because of the metabolic activity of the cells. When necessary, the pH of the perfusate is adjusted with 0.1 M NAOH to pH 7.4. The tubing is switched to the postperfusion buffer and 100 ml is allowed to pass through the liver at a pump speed of 28 ml/min. The liver is removed by cutting around the diaphragm and subsequently cutting the liver free from the remnants of diaphragm. At this stage the liver should have changed into a fragile soft bag of cells, indicating a successful perfusion.

For the preparation of parenchymal cells, the perfused liver surface is washed with a small amount of cold Hanks' solution containing 0.3% BSA. The liver capsule is opened carefully with sharp scissors and the cell mixture is released. The cells are filtered gently through a Nylon mesh (100 μm). The cells which pass through the mesh are transferred into 50-ml plastic centrifuge tubes, which are then placed at 4° for 5–10 min so that the cells sediment by gravity. Alternatively, the tubes are centrifuged for 45 s at 50g (without braking). Cooling the cells too fast (e.g., directly on ice) leads to a loss of viability of the parenchymal cells. Note that calcium-free buffer must be used here, as calcium tends to cause macrophages to clump, which leads to contamination of the sediment with nonparenchymal cells. The supernatant from the thick pellet of parenchymal cells is removed and saved. The supernatant should contain only a few parenchymal cells and be rich in nonparenchymal cells. The parenchymal cell pellet is washed five times by centrifugation at 50g (with no brake) with Hanks' solution containing 0.3% BSA. The supernatants are removed and the first supernatant from the washing is pooled with the supernatant of the sedimented parenchymal cells. These supernatants are kept on ice until further processing for the isolation of nonparenchymal cells. The parenchymal cell pellet is resuspended in 10 ml of Hanks' solution or culture medium and the cells are counted. The viability of the cells can be assayed by a Trypan blue exclusion test (0.25% Trypan blue in phosphate-buffered saline). Cells with a stained nucleus are nonviable. The isolated and purified parenchymal cells can now be used for the determination of liposomal components.

For the preparation of liver endothelial and Kupffer cells, the pooled supernatants just described are pooled at 500g at 4° to sediment the nonparenchymal cells. The cells are washed once more with Hanks' solution containing 0.3% BSA by centrifugation for 10 min at 500g and 4°. The final pellet is resuspended in two plastic 14-ml tubes with 8.5 ml Hanks' BSA solution each. The nonparenchymal cell suspensions are mixed with 3.2 ml of Optiprep, a gradient solution used for the isolation of biological material. One milliliter of Hanks' solution is layered on the cell–Optiprep mixture. The tube is centrifuged for 15 min at 1350g, 4° (again, with no brake). The nonparenchymal cells are now separated from red blood cells and cell debris and can be collected easily as a cell layer in the buffer phase just on top of the Optiprep solution. The cells are resuspended in approximately 10 ml of Hanks' BSA solution and centrifuged for 10 min at 500g and 4°. The cell pellet is resuspended in 5 ml Hanks' BSA solution. The nonparenchymal cells are flushed into an elutriation rotor (Beckman, type

JE-6 elutriation rotor) at 4° at a flow rate of 13 ml/min and at a rotor speed of 2500 rpm (750g). At this flow rate, cell debris is flushed out in 200 ml of Hanks' solution containing 0.3% BSA. Liver endothelial cells are collected in 150 ml at a flow rate of 23 ml/min, an intermediate cell fraction containing large endothelial and small Kupffer cells is collected in 150 ml at a flow rate of 25 ml/min, and Kupffer cells are collected in 150 ml at 46 ml/min. The cells are concentrated by centrifugation for 10 min at 500g, 4°. The cell pellets are resuspended in 10 ml of Hanks' solution or culture medium, and the number of cells in each cell fraction is determined by microscopic examination. Cell viability can be assayed again by Trypan blue exclusion. Liver endothelial and Kupffer cells can now be used for the determination of liposomal components.

The contribution of each cell type to hepatic liposome uptake can be calculated as follows. First the specific radioactivity in each cell fraction is determined by dividing the amount of radioactivity by the number of cells in that fraction. Then the total number of cells of each cell population is determined from the body weight of the rat and the total number of cells of each individual cell population per gram body weight: parenchymal cells = 4.50×10^6; nonparenchymal cells = 1.94×10^6; sinusoidal endothelial cells = 1.45×10^6, and Kupffer cells = 0.49×10^6. The numbers thus obtained are multiplied by the measured cell-associated radioactivity in dps/cell. The radioactivities thus obtained are related to total hepatic uptake and/or to injected dose to obtain the relative contribution of each cell population to total hepatic uptake.[38]

Concluding Remarks

This article tried to provide the reader with some methodologies that have been shown over the past decades to give satisfactory data on liposome biodistribution and tissue and cell uptake after intravenous administration. However, different experimental approaches and various refinements are available and may also provide satisfactory results for a given experimental setup. An important requirement for the correct interpretation of experimental data from every *in vivo* distribution of liposomes is to ascertain the integrity of the liposome and its components, including the label, up to the moment of cellular uptake, and the firm retention of the label for an appreciable amount of time at the site of uptake. The references given in this article provide ample examples, including possible pitfalls, of the described methods.

[38] F. Roerdink, J. Dijkstra, G. Hartman, B. Bolscher, and G. L. Scherphof, *Biochim. Biophys. Acta* **677**, 79 (1981).

[17]   Transport of Liposome-Entrapped Substances into Skin as Measured by Electron Paramagnetic Resonance Oximetry *In Vivo*

By M. Šentjurc, J. Kristl, and Z. Abramović

Introduction

Over the past decade a number of efforts have been made to produce topical preparations that are easy to apply, enhance the local therapeutic index, and minimize unwanted systemic toxic effects. In pursuit of these goals, topical liposomes have attracted considerable interest for use in cosmetics and as drug delivery systems in dermatology.[1–4] They may serve as a solubilization matrix, as a local depot for sustained release of dermally active compounds, as penetration enhancers, or as rate-limiting membrane barriers for modulating systemic adsorption of drugs or delivery into the skin.[5] Faster and easier delivery of drugs through the skin lipid barrier by liposomal formulations has been established during the last decade.[6,7] For this purpose, liposomes should possess appropriate physical properties. With respect to their size and elastomechanical characteristics, they interact with skin in different ways.

1. They are held back by the barrier or disintegrate immediately upon contact with skin. This type of liposomes is less effective in topical application.

2. The other types of liposomes could make good contact with the skin surface. Some can penetrate intact into the dermis and are deposited there but most of them fuse with skin lipids and then diffuse in the form of small fragments or monomers into the deeper skin layers. In this way they enable better delivery of the entrapped substances. These are conventional liposomes, used most commonly in cosmetics or dermatology.[1–5]

[1] R. M. Fielding, *Clin. Pharmacokinet.* **21**, 155 (1991).

[2] J. Wepierre and G. Couarraze, *in* "Liposomes, New Systems and New Trends in Their Applications" (F. Puisieux, P. Couvreur, J. Delattre, and J. P. Devissaguet, eds.), p. 615. Edition de Sante, Paris, 1995.

[3] K. Stanzl, *in* "Novel Cosmetic Delivery Systems" (S. Magdassi and E. Touitou, eds.), p. 233. Dekker, New York, 1999.

[4] M. H. Schmid and H. C. Korting, *Crit. Rev. Ther. Drug Carrier Syst.* **11**, 97 (1994).

[5] H. Schreier and J. Bouwstra, *J. Control. Rel.* **30**, 1 (1994).

[6] G. Cevc, *Crit. Rev. Ther. Drug Carrier Syst.* **13**, 257 (1996).

[7] G. Betz, R. Imboden, and G. Imanidis, *Int. J. Pharm.* **229**, 117 (2001).

3. The third type possesses appropriate deformability to distort sufficiently on the skin surface, which allows them to pass through the narrow intercellular passages (<30 nm) in the outer skin layers. They can penetrate intact through the skin, carrying the entrapped drug into deeper skin layers. These ultradeformable lipid vesicles (transferosomes) may even reach the systemic blood circulation unfragmented. They provide a viable technological platform for the development of formulations that convey drugs—from low molecular weight molecules, through intermediate size polypeptides, to large macromolecular drug substances—into the skin or the body with high efficacy and good reproducibility.[6]

Evaluation of Liposomal Formulations for Topical Application

The influence of liposome composition on the physical characteristics of liposome membrane and its relation to the transport of encapsulated substances into the skin is a matter of many recent investigations.[7,8] One of the components that strongly influences the domain formation in the liposome membrane, its fluidity characteristics and consequently the transport of hydrophilic spin probes into the skin, is cholesterol. It was also found that the presence of 20–50% ethanol in liposomal formulations lowers the phase transition of the liposomal membrane, thereby promoting interactions with the stratum corneum and improving the permeation properties of drugs through the skin.[8] The role of other components and additives in the liposome membrane is less known.

The exact mechanisms responsible for the enhanced delivery of liposome-entrapped drugs into and through the skin are also still not clear.[9,10] Accordingly, there are many new techniques developed to follow these processes. One approach is to measure the release, penetration, and distribution in skin of substances entrapped in liposomes applied topically. The other possibility is to measure the biological responses of organism induced by compounds applied topically in liposomes.

Measurements of Release and Transport into Skin of Substances Entrapped in Liposomes Applied Topically

Radioactive- or fluorescence-labeled markers and drugs entrapped in liposomes in combination with various electron and (laser) light microscopic visualization techniques are widely used. Together with different

[8] E. Touitou, B. Godin, and C. Weiss, *Drug Dev. Res.* **50**, 406 (2000).
[9] D. D. Lasic, "Liposomes: from Physics to Applications." Elsevier, Amsterdam, 1993.
[10] M. E. M. J. van Kuijk-Meuwissen, L. Mougin, H. E. Junginger, and J. A. Bouwstra, *J. Control. Rel.* **56**, 189 (1998).

models they give some data about the interaction with and fate of vesicles in the skin.[5]

One of the methods by which it is possible to follow the time development of distribution profile and diffusion into the skin of paramagnetic substances delivered entrapped in liposomes is electron spin resonance (EPR).[11–13] This article describes this method in more detail.

Transport Measurements Based on EPR Spectroscopy In Vitro

The one-dimensional EPR imaging method, in combination with nitroxide reduction kinetic imaging, has been used to study the influence of liposome composition and size on the transport of a hydrophilic spin probe delivered to the skin entrapped in liposomes.[11–13] This method enables the differentiation of liposome decay in the stratum corneum from that in viable skin layers. The time development of distribution profiles in different skin layers of the total and entrapped spin probe delivered to the skin in liposomes can also be estimated. The method is based on the fact that nitroxide spin probes, when released from liposomes, are reduced to nonparamagnetic hydroxylamine by redox systems in the skin.[14] We have found that the composition of liposomes, which is reflected in the domain structure of liposome membranes and their fluidity characteristics, influences pronouncedly the transport into the skin.[15] The size of liposomes becomes important only when the diameter decreases below 200 nm, at which point transport decreases significantly.[12] However, these measurements were performed on conventional EPR spectrometers operating at microwave frequencies of 9 GHz. The measurements were therefore limited to small tissue slices (1 mm thick) only, and to a period of 30 min after application of the liposomal formulation to the skin. It is also difficult to prove that data obtained *ex vivo* are also relevant to *in vivo* systems.

Topical application of liposomal formulations on humans, however, makes it important to know if results obtained with *ex vivo* skin samples can be applied to conditions *in vivo*. For example, it is not known whether the transport of liposome-entrapped substances into the skin changes after sacrifice of the animal. Also, it is not known how the blood flow in the deeper skin layers influences the concentration profiles of drug molecules which penetrate into the skin. It is also necessary to know how prolonged

[11] V. Gabrijelčič, M. Šentjurc, and J. Kristl, *Int. J. Pharm.* **62,** 75 (1990).

[12] K. Vrhovnik, J. Kristl, M. Šentjurc, and J. Šmid–Korbar, *Pharm. Res.* **15,** 525 (1998).

[13] J. Fuchs, N. Groth, T. Herrling, and L. Packer, *Methods Enzymol.* **233,** 140 (1994).

[14] M. Šentjurc and V. Gabrijelčič, *in* "Nonmedical Applications of Liposomes" (D. D. Lasic and Y. Barenholz, eds.), p. 91. CRC Press, New York, 1996.

[15] M. Šentjurc, K. Vrhovnik, and J. Kristl, *J. Control. Rel.* **59,** 87 (1999).

and repeated applications of liposomal formulations on the skin surface affect penetration. To answer these questions, we have sought methods to measure penetration into the skin *in vivo*.

The recently developed technique of low-frequency EPR with surface coils allows the transport of spin-labeled substances delivered to the skin entrapped in liposomes to be studied *in vivo*. Two approaches have been used. One is based on the same principle as nitroxide reduction kinetic imaging *in vitro*, described in the previous paragraph, and the other is based on the biological response of the body on the action of liposome-entrapped vasodilators.

Nitroxide Reduction Kinetics In Vivo

We have used this method to study the enhancement of topical delivery of spin-labeled hydrophilic substances using multilamellar liposomes. The relative contributions of transepidermal and transfollicular routes of transport were investigated using hairless and normal mice.[16] Liposomes prepared from hydrogenated or nonhydrogenated soy lecithin with 30 wt% cholesterol (HSL and NSL, respectively) liposome dispersions that had been shown previously to have different transport characteristics on *ex vivo* skin are used. Using *ex vivo* EPR imaging methods on pig ear skin it was shown that HSL enhances the transport of the entrapped substance into deeper layers of the skin, whereas NSL remains in the epidermis.[15] The kinetics of the reduction of the hydrophilic spin probe GluSL [N-(1-oxyl-2,2,6,6-tetramethyl-4-piperidinyl)-2,3,4,5,6-pentahydroxyhexane-amide] applied to the skin encapsulated in liposomes are measured. To distinguish the reduction of GluSL on the skin surface from its reduction inside the skin, the oxidizing agent potassium ferricyanide (KFeCN) is used. KFeCN does not penetrate into the skin and therefore it oxidizes hydroxylamines back to nitroxide only on the surface of the skin. We observed significant differences in the properties of different types of liposomes with respect to their stability when in contact with skin and to their penetration into the skin. Results measured *in vivo* are in agreement with those obtained *ex vivo*, demonstrating that L-band EPR *in vivo* is a powerful technique for following pharmacokinetics in the skin of living animals. Results show that the blood flow clearance and possible alterations of skin after sacrifice of animals do not influence the results of penetration of liposome-entrapped substances into the skin during the time of our experiment (typically around 60 min). It was also shown that transfollicular penetration was not of major importance *in vivo* in this experimental model.

[16] L. Honzak, M. Šenjurc, and H. M. Swartz, *J. Control. Rel.* **66,** 221 (2000).

Another method that seems promising in studying the release, penetration, and distribution of spin-labeled substances in skin after topical application in liposomal formulations is EPR imaging *in vivo*.[13] It is a novel approach that could stimulate progress in skin pharmacology. However, according to recent developments, the method is not sensitive enough for reliable images of the limited amounts of spin probes delivered to the skin in liposomes.

Evaluation of Biological Responses Induced by Entrapped Compounds in Liposomes

The *in vivo* study of drug action after topical application in liposomes has often involved the detection of some local biological response that the drug elicits upon penetration into the skin.[17–19] Examples of responses that have been utilized include vasodilation, vasoconstriction, keratinization, epidermal proliferation, and changes in blood pressure.

For example, the skin color changes arising from a vasodilatory response to a local application of nicotinates are measured by different methods. The erythematous reaction to nicotinate application can be followed by photographic and micrometric techniques, which provide elegant ways for visual documentation.[20] However, they do not yield quantitative results. Skin reflectance spectrophotometry enables the induced erythema to be monitored directly. Laser Doppler velocimetry (LDV) and photopulse plethysmography (PPG) have been used for monitoring blood perfusion through the cutaneous microcirculation. These methods employ different optical principles to generate an output related to either velocity (LDV) or amount (PPG) of cutaneous blood vessel perfusion. However, they are not able to correlate directly the blood flow characteristics with skin oxygen levels.[21,22]

A powerful method by which it is possible to follow directly the oxygen level in skin *in vivo* with time after topical application of the drug, and with repeated application over longer periods of time, is EPR oximetry.[23] If a vasodilator (e.g., benzyl nicotinate) is introduced into liposomes (or

[17] H. C. Korting, P. Blecher, and M. Schafer-Korting, *Eur. J. Clin. Pharmacol.* **39,** 349 (1990).

[18] M. Foldvari, A. Geszetes, and M. Mezei, *J. Microencaptulation* **7,** 479 (1990).

[19] M. Kržič, M. Šentjurc, and J. Kristl, *J. Control. Rel.* **70,** 203 (2001).

[20] S. Y. Chan and A. Li Wan Po, *Skin Pharmacol* **6,** 298 (1993).

[21] R. H. Guy, R. C. Wester, E. Tur, and H. I. Maibach, *J. Pharm. Sci.* **72,** 1077 (1983).

[22] R. Kohli, W. I. Archer, and A. Li Wan Po, *Int. J. Pharm.* **36,** 91 (1987).

[23] H. M. Swartz, S. Boyer, D. Brown, K. Chang, P. Gast, J. F. Glockner, H. Hu, J. Liu, M. Moussavi, M. Nilges, S. W. Norby, A. Smirnov, N. Vahidi, T. Walzcak, M. Wu, and R. B. Clarkson, *in* "Oxygen Transport to Tissue XIV" (W. Erdmann and D. F. Bruley, eds.), p. 221. Plenum Press, New York, 1992.

any other carrier) and the oxygen level in skin is measured with time following topical application, the role of different carriers in the physiological response of the organism to the action of the vasodilator can be investigated. The oxygen level will change as a consequence of the vasodilator action on the blood–vascular system in the skin, which depends on the efficacy and rate of transport of the vasodilator through the stratum corneum to the dermis.[19] The rate of transport and time of action depend on the type of delivery system.

The remainder of this article describes the use of EPR oximetry *in vivo*, together with its advantages and pitfalls in characterizing the roles of different liposomal preparations in the topical delivery of drugs into the skin. Some examples are given that show how the liposome composition influences the transport of an entrapped vasodilator into the skin and, consequently, the oxygenation of the latter.

Basic Principles of EPR Oximetry

Electron paramagnetic resonance oximetry is based on the fact that molecular oxygen, having two unpaired electrons, is paramagnetic and causes fast relaxation of other paramagnetic species by Heisenberg spin exchange interaction. To measure oxygen concentration in a biological sample, it is necessary to introduce an oxygen-sensitive paramagnetic material into the tissue and to measure its EPR spectrum. In the presence of oxygen, the exchange interaction between unpaired spins of oxygen and a paramagnetic probe can shorten the spin–lattice and spin–spin relaxation times (T_1 and T_2) of the paramagnetic probe. Therefore, an increase in oxygen concentration broadens the EPR spectral line width and decreases the microwave power at which saturation occurs.[24] The line width of the EPR spectrum can be calibrated with known concentrations of oxygen and then used to measure the amount of oxygen in tissue, expressed as oxygen concentration $[O_2]$ or as partial pressure of oxygen pO_2.

$$[O_2] = pO_2 \times \text{solubility} \tag{1}$$

For heterogeneous biological tissues, it is usually more appropriate to express the amount of oxygen as partial pressure, as its solubility in different parts of tissues could be different. The development of low-frequency L-band EPR spectrometers (about 1.2 GHz) makes it possible to use this method in living animals.

[24] H. M. Swartz and R. B. Clarkson, *Phys. Med. Biol.* **43**, 1957 (1998).

Types of Oxygen-Sensitive Paramagnetic Probes

A variety of paramagnetic materials have been developed and used for pO_2 measurements in living animals. For successful clinical application, they should have a high degree of chemical and physical stability and should be nontoxic and completely inert in tissues. They must be highly sensitive to oxygen, with retention of their response to oxygen for periods of years.[25] Basically two types of oxygen-sensitive paramagnetic materials have been developed.

Soluble Radicals

Nitroxides in Solution. Most typically the soluble radicals are nitroxides and their derivatives. They have the advantage of diffusing through the tissue and are distributed over larger areas of the tissue, or even over the whole body. Therefore, they provide a potentially uniform sampling of oxygen concentrations over large areas of tissue. They may be synthesized with properties that affect their distribution in the tissue. However, water-soluble nitroxides diffuse through the tissues and are finally secreted from the body. In addition, they are metabolically sensitive and are reduced quickly to nonparamagnetic hydroxylamines. Therefore, these compounds can monitor the distribution of oxygen in the body only over a period of several minutes.

Nitroxides Entrapped in Liposomes or Microspheres. The mentioned disadvantage can be overcome by encapsulating nitroxides in liposomes, nanoparticles, or microspheres.[26,27] Cationic nitroxide entrapped inside liposomes cannot penetrate through the liposome membrane due to its charge, and measurements can be made *in vivo* for several hours after intramuscular administration. The liposomes protect the nitroxides from bioreduction and, with appropriate design and composition, can be delivered to specific areas of the body. Further, by encapsulating lipophilic nitroxides in the liposome membrane or in solid lipid nanoparticles or microspheres, the sensitivity to oxygen can be increased considerably because oxygen is several fold more soluble in lipophilic solvents than in water. Although the liposomes and microspheres are stable *in vivo* for several hours, they are metabolized over an extended period of time.[26] Therefore, they are intended for short-term measurements of oxygen

[25] N. Vahidi, R. B. Clarkson, K. J. Liu, S. W. Norby, M. Wu, and H. M. Swartz, *Magn. Res. Med.* **31,** 139 (1994).

[26] J. F. Glockner, H. C. Chan, and H. M. Swartz, *Magn. Res. Med.* **20,** 123 (1991).

[27] K. J. Liu, M. Grinstaff, J. Jiang, K. Suslick, H. M. Swartz, and W. Wang, *Biophys. J.* **67,** 896 (1994).

concentration only. For longer times, such as weeks or months, particulate probes are more appropriate.

Insoluble Solid Particles

These are paramagnetic materials with strong, simple spectral lines, which broaden reproducibly in the presence of oxygen. They are carbon-based materials such as different types of coals or chars of carbohydrates, usually in the form of fine powder (which can be as small as 5 μm in diameter[28]), lithium phthalocyanine-[29,30] or naphthalocyanine[31]-based probes in the form of small crystals (<0.1 μm diameter), and Indian ink,[32] which has already been used widely in human subjects and therefore provides the most feasible paramagnetic substance for application of EPR oximetry in humans. These materials are stable physicochemically, have little or no toxicity *in vitro* and *in vivo*, and reflect the partial pressure of oxygen rather than oxygen concentration. They can be introduced into the body in the form of dry particles or in suspension. In the former state, the measurement is limited to their immediate surroundings, whereas in suspension the particles are distributed uniformly over a larger volume in the tissue and will give information about the average level of oxygen in this region. After the initial placement of the particles in the selected place, measurements can be made noninvasively, as frequently as desired, over longer periods of time. They provide sensitive and accurate measurements of pO_2 in the immediate surroundings of the particle. At low pO_2 it is possible with some of these materials to resolve differences of the order of 0.1 mm Hg.

It should be mentioned, however, that some paramagnetic markers can lose their responsiveness to oxygen with time after insertion into the tissue, depending on the particulate system and on the tissue. This effect may be due to chemical changes on the surface of the particles or, more probably, to the response of tissues to the foreign substance, leading to the formation of a capsule surrounding the particles, which makes a barrier to oxygen. By coating the paramagnetic particles with a film of biocompatible, oxygen transparent and nonbiodegradable polymers, long-term studies can be

[28] B. F. Jordan, C. Baudelet, and B. Gallez, *MAGMA* **7,** 121 (1998).
[29] K. J. Liu, P. Gast, M. Moussavi, S. W. Norby, N. Vahidi, T. Walczak, M. Wu, and H. M. Swartz, *Proc. Natl. Acad. Sci. USA* **90,** 5438 (1993).
[30] S. W. Norby, H. M. Swartz, and R. B. Clarkson, *J. Microsc.* **192,** 172 (1998).
[31] G. Ilangovan, A. Manivannan, H. Li, H. Yanagi, J. L. Zweier, and P. Kuppusamy, *Free Radic. Biol. Med.* **32,** 139 (2002).
[32] H. M. Swartz, K. J. Liu, F. Goda, and T. Walczak, *Magn. Res. Med.* **31,** 229 (1994).

achieved. The materials used for coating should have good tolerance in tissues, and preferably be already accepted for human use.[33]

With particulate paramagnetic materials, spectra can be obtained from several sites simultaneously if the particles are located at discrete positions and if an appropriate magnetic field gradient is applied to separate the absorption lines of paramagnetic material at different sites.

Instrumentation

For *in vivo* EPR oximetry, low-frequency EPR spectrometers, operating at frequencies below 2 GHz (L-band EPR spectrometers), are used. With recent advances in this field, the sensitivity of L-band spectrometers now makes it feasible to use *in vivo* EPR for measuring pO$_2$ in tissues of living animals, with the sensitivity, accuracy, and repeatability that is required for most purposes. An especially important part of *in vivo* measurement is the design of the resonator for detecting EPR signals. These can be whole body resonators, where the whole animal can be put into the cavity. At this time the dimensions are limited to the size of a mouse. The other types are surface coil resonators.[34] Here there is no limitation on the size of animal that can be studied, but without specially designed resonators or accessories that allow the insertion of a needle-catheter probe into the body, the measurements are limited to a depth of 5–10 mm. For measurements in skin this is quite convenient.

Assay for Measuring the Efficacy of a Topically Applied, Liposome-Entrapped Vasodilator

Materials

Vasodilator. Benzyl nicotinate (BN, Lek, Ljubljana, Slovenia) is used as a model lipophilic drug. It acts as a pro-drug, which crosses the skin rapidly and, on enzymatic hydrolysis, releases nicotinic acid. This agent frequently provokes an increase in cutaneous blood flow, at least partly via the formation of vasodilator prostaglandins (PGD2).[35] As a consequence of the dilatation of small arterioles, the level of oxygen in skin increases. The amount of BN (free or entrapped) in all samples was 0.83% (w/w).

Preparation of Liposomes with BN. Multilamellar liposomes (MLV) are prepared from hydrogenated soy lecithin (HSL; Emulmetic 320, Lucas

---

[33] B. Gallez and K. Mäder, *Free Radic. Biol. Med.* **29,** 1078 (2000).

[34] H. M. Swartz and T. Walczak, *Phys. Med.* **9,** 41 (1993).

[35] J. K. Wilkin, G. Fortner, L. A. Reinhardt, O. V. Flowers, S. J. Klipatrick, and W. C. Steeter, *Clin. Pharmacol. Ther.* **38,** 273 (1985).

Mayer, Hamburg, Germany) or nonhydrogenated soy lecithin (NSL; Phospholipon 80, Natterman Phospholipid GmbH, Cologne, Germany) and cholesterol (Ch, Sigma, St. Louis, MO) at a weight ratio of 7:3. The lipids (17.5 mg/ml phospholipids and 7.5 mg/ml cholesterol) and 1.25% (w/w) benzyl nicotinate (12.5 mg/ml) are dissolved in dichloromethane for NSL and chloroform:methanol (1:1) for HSL and are dried in a rotary evaporator in a round-bottom flask to obtain a thin film on the wall. The remaining solvent is removed by a vacuum pump (10–15 min at 40° and a pressure of 100 Pa). The dry film is hydrated with distilled water above the phase transition temperature of pure phospholipids (45° for HSL and room temperature for NSL). The flask is shaken until the film is removed completely from the walls. The liposome dispersion is stabilized by stirring for 2 h on a magnetic stirrer (300 turns/min) at room temperature. It is not possible to form liposomes at concentrations of BN higher than 1.25% (w/w).

Liposomes in Hydrogel. Hydrogel is composed of 2.0 g hydroxyethylcellulose, 20.0 g glycerol, and 66.0 g of liposome dispersion with BN and 12.0 g of distilled water. Hydroxyethylcellulose is dispersed in water and one-third of the liposome dispersion at room temperature. After 1 h, glycerol and the remaining liposome dispersion are added and mixed to yield a homogeneous hydrogel.

Benzyl Nicotinate in Hydrogel. Two grams of hydroxyethylcellulose (Natrosol 250 HHX, Aqualon, Great Britain) is dispersed in two-thirds of the total mass of distilled water (78.0 g) at 70° and is then cooled to room temperature with continuous stirring to give a homogeneous mixture. BN dispersed in 20.0 g glycerol is mixed with the hydroxyethylcellulose dispersion and the remaining water to give homogeneously dispersed BN in the hydrogel.

Paramagnetic Probes. Most experiments are performed with lithium phthalocyanine (LiPc), and in some experiments the results are compared with those obtained with carbon-based paramagnetic probe, a char of the bubinga tree—Bubinga. Both probes are generous gifts from the EPR Center for Viable Tissues, Dartmouth Medical School, Hanover, New Hampshire.

EPR Measurements. EPR spectra are recorded on a Varian E-9 EPR spectrometer with a custom-made, low-frequency microwave bridge operating at 1.1 GHz and an extended loop resonator (11 mm diameter), both designed by Dr. T. Walczak (Dartmouth Medical School, Hanover, NH).

Calibration Curves. To obtain the relationship between the line width of EPR spectra and pO_2, calibration curves for both paramagnetic probes are measured. For this purpose, a suspension of the paramagnetic probe in phosphate-buffered saline (PBS, pH 7.4, 320 mmol/kg) is placed in a gas-permeable Teflon tube for EPR measurements. Gases with known concentrations of oxygen and nitrogen, controlled with an oxygen analyzer (MK200, IJS, Ljubljana, Slovenia), are flushed over the samples (gas flow was 6 liters/min), and spectra are recorded after the EPR line width becomes constant, typically 20–30 min. Typical EPR spectra of Bubinga for three concentrations of oxygen are presented in Fig. 1. Calibration curves for Bubinga and LiPc are given in Fig. 2. In the physiological range of pO_2 (5–40 mm Hg), both curves are linear. The relation between pO_2 and line width is calculated from Eqs. (2) and (3) for Bubinga and LiPc, respectively,

$$\text{Bubinga: } \Delta B \text{ (mT)} = 0.039 + 4.4 \times 10^{-3} \times pO_2 \text{(mm Hg)} \tag{2}$$

$$\text{LiPc: } \Delta B \text{ (mT)} = 0.0059 + 3.9 \times 10^{-4} \times pO_2 \text{(mm Hg)} \tag{3}$$

The lines for LiPc are about 10 times more narrow than those for Bubinga and broaden to about 1000 μT in air.[29] Due to the narrow lines, the signal to noise is appropriate for measurements over a larger range of oxygen concentration, from very low to physiological levels of pO_2. Bubinga and other chars are readily available and, in some cases, more stable in tissues. Therefore, they have some advantages for measurements at low oxygen levels from 0 to 20 mm Hg.

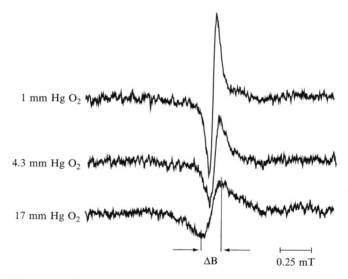

1 mm Hg O_2

4.3 mm Hg O_2

17 mm Hg O_2

ΔB 0.25 mT

FIG. 1. EPR spectra of the paramagnetic probe Bubinga as a function of pO_2. ΔB is the peak-to-peak line width, the parameter being measured.

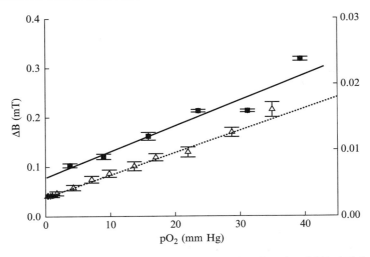

FIG. 2. Calibration curves for Bubinga (Δ) (scale on left ordinate) and LiPc ($\blacksquare$) (scale on right ordinate) in saline give the relation between line width (ΔB) of EPR spectra and partial oxygen pressure (pO$_2$). Each point represents the average of three experiments $\pm$ SD.

Experimental Animals. Female mice (BalbC, 20–25 g) are used, and guidelines and legislative regulations on the use of animals for scientific purposes are followed. The mice are anesthetized by an intraperitoneal injection of a mixture containing 16 mg/kg xylazine and 100 mg/kg ketamine. Because most experiments last more than 1 h, it is necessary to add additional anesthetic (half of the initial dose) during the experiment when the animal starts to wake up (about 40 min after the application of anesthetic).

The hair from their back is shaved and the remainder is removed with a depilatory cream (Vitaskin, Krka, Slovenia). Since depilation causes a complete removal of stratum corneum which is restored after 4 days,[16] measurements are performed not earlier than 4 days after depilation.

Small crystals of LiPc (15–40 μm diameter; the volume is estimated to be 0.5 μm^3) or amorphous particles of Bubinga (5–15 μm diameter, volume 1 μm^3) are put into an injection needle (23 gauge) and introduced at different depths into the mouse skin by pushing the particles from the needle with thin stainless steel wire.

Spectra are recorded continuously for at least 90 min after topical application of the formulation. During the measurements, the animals are kept at 36.5 $\pm$ 0.5° by blowing hot air around them. The body temperature is measured rectally with a thermocouple inserted into a glass capillary.

In control experiments, the time necessary for the paramagnetic probe to equilibrate with its surroundings is measured, as well as the influence of anesthesia on body temperature and on pO_2. How the sensitivity of Bubinga and LiPc to oxygen changes with time after insertion of the probe into the body is also checked.

Selection of Appropriate Experimental Conditions for pO_2 Measurements in Skin

Spectrometer Settings. Because the signal-to-noise ratio in L-band spectrometers is lower than in conventional X-band spectrometers and because the intensity of EPR spectral lines decreases with the square of the distance from the surface coil, there are frequently problems to reach appropriate signal-to-noise ratios for reliable measurements of line widths. To overcome this problem, it is necessary to introduce sufficient amounts of paramagnetic probe into the skin and to find the optimal spectrometer settings. The latter depend on the paramagnetic probe used. Because the line widths of the probes are usually very narrow and increase sharply with increasing oxygen concentration, special attention should be paid not to overmodulate the EPR signal. The modulation amplitude of the alternating magnetic field should not exceed one-third of the peak-to-peak line width. For pO_2 in the range of physiological conditions in skin (5–30 mm Hg), the line width varies from 0.06 to 0.17 mT for Bubinga and from 0.0078 to 0.0176 mT for LiPc. Consequently, the modulation amplitude for LiPc should be about 10 times smaller than for Bubinga. It is also important to take care to keep the microwave power below saturation of the energy levels, which depend on spin–lattice relaxation time and could vary with the amount of oxygen in the sample. In our experiments, the microwave power is 30 mW. Other spectrometer settings are modulation frequency, 100 kHz; microwave frequency, 1.1 GHz; and scan range, 2 mT for Bubinga and 0.4 mT for LiPc.

Effect of Site of Placement and Amount of Paramagnetic Probe. The site and amount of the paramagnetic probe are very important, as the line width of the spectrum depends on pO_2 in the immediate surroundings of the probe (e.g., in the vicinity of blood vessels). The site and the amount of the introduced paramagnetic probe were checked by histological examination at the Medical School, Institute of Histology, Ljubljana, Slovenia, at the end of some experiments performed with Bubinga.[36] It was found that the baseline of pO_2 is slightly higher in subcutis than in dermis and

[36] M. Šentjurc, M. Kržič, J. Kristl, O. Grinberg, and H. M. Swartz, *Polish J. Med. Phys. Eng.* **7,** 165 (2001).

that the response to the application of BN is faster and more pronounced in subcutis.

When the probe was concentrated in a small volume, no response to the drug action was observed. In such cases the probe is not sensitive to oxygen, probably because only the particles that are on the surface of the probe, in intimate contact with the tissue, are exposed to pO_2 changes, and the sensitivity of the paramagnetic probe and the time of response depend on the surface of the probe.[30]

It is important to stress that the intensity of EPR spectra is proportional to the amount of the spin probe in the tissue. The amplitude of the EPR signal increases with decreasing line width. Because the line width of LiPc is narrower than that of Bubinga, much less LiPc is necessary in the tissue to get the same signal-to-noise ratio if the spin densities of the probes are comparable. From this point of view LiPc is the better probe.

Equilibration of the Paramagnetic Probe with Its Surroundings. When the probe is inserted into the skin it needs some time to equilibrate with its surroundings. The particulate paramagnetic markers used are very porous[30] and it takes some time before the oxygen level in the pores is equilibrated with that in the surrounding tissue. Thus, soon after introduction of the markers into the skin, the lines are very broad; with time they become narrower until equilibrium is reached. Typically it takes about 12 h for LiPc or Bubinga to equilibrate with their surroundings. Therefore, it is advisable to start the measurements 1 day after the introduction of the probe into the skin.

The Sensitivity of Paramagnetic Probe to Oxygen. For experiments lasting for longer periods, for example, for prolonged or sustained release and measurements of repeated application of formulations, one should be aware of the fact that some paramagnetic probes become less sensitive or insensitive to oxygen.[33] The histological examination, performed about 1 month after the insertion of paramagnetic markers into the skin, in some cases showed the presence of fibrous tissue around the particles, which prevents the adequate response of the probe to changes in pO_2. In our experiments it was found that the sensitivity of Bubinga to oxygen remained unchanged for more than 2 months, whereas LiPc lost its sensitivity to oxygen after about 21 days.

Influence of Anesthesia on pO_2 in Skin. pO_2 is measured at constant body temperature as a function of time after anesthesia. There is a decrease in pO_2 in skin after anesthesia of mice with the xylazine/ketamine mixture used in our experiments. A typical dependence is presented in Fig. 3, but it varies from mouse to mouse, and depends also on the location of the paramagnetic probe in the skin. Typically pO_2 decreases by about 2 mm Hg within the first 30 to 40 min following the application of anesthetic, and

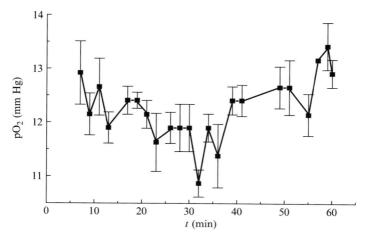

FIG. 3. Changes of pO_2 in dermis of mice with time after the beginning of action of the anesthetic drug (mixture containing 16 mg/kg xylazine and 100 mg/kg ketamine) at constant body temperature ($36.5 \pm 0.5°$). A typical curve is presented. Each point is the mean of three measurements $\pm$ SD.

starts to increase when the mouse starts to wake up. If at that time more anesthetic is introduced, the pO_2 does not change significantly, corroborating the experiments of other authors[37] who demonstrated that, after the first 40 min, pO_2 does not change significantly over the next 60 min. In future experiments a better alternative would be to apply isoflurane gas instead of ketamine/xylazine anesthetic. This would enable the animal to be kept asleep for several hours, allowing longer measurements without changes in the baseline due to anesthesia.

Influence of Body Temperature on pO_2. pO_2 in skin was shown to change with body temperature following anesthesia, as measured from the line width of the paramagnetic probe LiPc (Fig. 4). The results are in very good agreement with previous measurements, where a similar response of pO_2 in subcutis was obtained with Bubinga.[36] There is a sharp decrease of more than 5 mm Hg in pO_2 as the body temperature decreases from 36 to 33°, whereas in the region of physiological temperatures (around 37°), the decrease in pO_2 is in the same range as the decrease due to anesthesia. Because the body temperature changes significantly

[37] F. Goda, J. A. O'Hara, K. J. Liu, E. S. Rhodes, J. F. Dunn, and H. M. Swartz, *in* "Oxygen Transport to Tissue XVIII" (E. M. Nemoto and J. C. LaManna, eds.), p. 543. Plenum Press, New York, 1997.

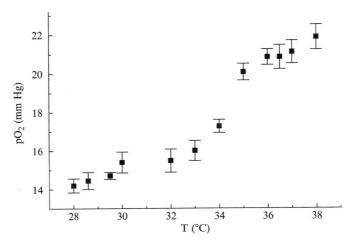

F<small>IG</small>. 4. Influence of body temperature on pO$_2$ in skin. The decrease in body temperature after anesthesia was measured. pO$_2$ was obtained from the line width of EPR spectra of inserted LiPc using the calibration curve in Fig. 2.

after anesthesia,[36] it is of prime importance to keep the mice at constant temperature during the measurements.

Influence of Liposome Composition on the Action of the Entrapped Drug

Topical Application of Formulations with BN on Mouse Skin

To obtain the physiological pO$_2$ in mouse skin (baseline), the animal is anesthetized and the EPR spectral line width is measured with time (for about 15 min). Formulations are then applied to the skin using a plastic ring, 1 mm thick with a 18-mm-diameter hole. The ring is placed on the mouse skin and the hole is filled with the formulation. The ring is then removed from the skin but the formulation remains on the site of application during the measurements. EPR spectra are recorded every 2 min for at least 90 min to get the pO$_2$ changes in skin as a response of the organism to the topical application of vasodilator BN in different formulations.

Effect of Different Formulations on pO$_2$ in Skin. The time dependence of pO$_2$ after the application of BN in HSL or NSL liposomes is presented in Fig. 5 for both paramagnetic probes used. As the baseline varies from animal to animal from 6 to 20 mm Hg due to different positions of the paramagnetic probe in the skin, the difference in pO_2 between the baseline and the measured pO_2 is shown (ΔpO_2). Curves in Fig. 5 represent mean values from at least five animals. From the individual time dependence curves, several parameters can be defined, which characterize the time response of individual animals to the topical application of BN in different formulations. These are maximal relative increase of pO_2 due to the action of BN (ΔpO_{2max}), the time expired from the application of the formulation and the time when pO_2 starts to increase (lag time t_{lag}), the time when

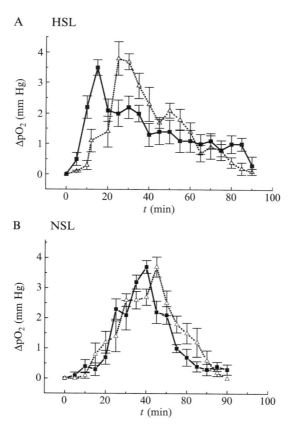

FIG. 5. Variation of pO_2 in skin with time after topical application of BN in liposomes: (A) hydrogenated (HSL) and (B) nonhydrogenated (NSL) soybean lecithin as measured with Bubinga ($\triangle$) or LiPc ($\blacksquare$). The difference in pO_2 (ΔpO_2) between the baseline and the measured pO_2 is presented. Each measurement is a mean value from at least five animals $\pm$ SE. As a starting point ($t = 0$), lag time is taken, as this value changes from animal to animal.

maximal increase in pO_2 is achieved (t_{max}), and the area under the curve (AUC), which measures the efficacy of BN action (Table I):

$$AUC = 1/2 \ \Sigma \ (t_{i+1} - t_i) \ (pO_{2i} + pO_{2i+1}) \tag{4}$$

where t_i is the time of measurement and pO_{2i} is the corresponding partial pressure for $i = 0$–80.

Figure 5 and Table I show that there is a marked difference in pharmacokinetics between BN delivered at the same concentration in different formulations. It is important to note that the lag times for the liposomal formulations are at least half those for the hydrogel. They differ significantly also for both liposomal preparations. The pronounced difference between the two liposomal formulations is observed in t_{max}, indicating that the permeation of BN from HSL liposomes is faster than from NSL. Results corroborate previous findings obtained with EPR imaging methods in *in vitro* experiments. For HSL liposomes, the transport of the entrapped hydrophilic probe into deeper skin layers was observed in the first 30 min, whereas for NSL only a small amount of the spin probe was observed in the stratum corneum in this time.[12] Similarly, in the reduction kinetics experiment *in vivo*, faster transport was observed for HSL than for NSL.[16]

TABLE I

INFLUENCE OF DIFFERENT VEHICLES ON OXYGENATION OF MOUSE SKIN AFTER
TOPICAL APPLICATION OF BENZYL NICOTINATE (BN) FREE IN HIDROGEL (HG),
ENTRAPPED IN NONHYDROGENATED (NSL), AND HYDROGENATED (HSL)
SOYBEAN LECITHIN LIPOSOMES[d]

	HG		NSL		HSL	
	Bubinga	LiPc	Bubinga	LiPc	Bubinga	LiPc
ΔpO_{2max}[b] (mm Hg)	3.4 ± 0.3	3.6 ± 0.5	3.7 ± 0.3	3.7 ± 0.2	3.8 ± 0.6	3.5 ± 0.3
t_{lag}[c] (min)	44 ± 2	39 ± 2	22 ± 2	17 ± 1	18 ± 1	9 ± 1
t_{max}[d] (min)	54 ± 3	55 ± 2	47 ± 1	40 ± 1	27 ± 1	16 ± 2
AUC_{0-80}[e] (mm Hg × min)	89 ± 4	90 ± 24	118 ± 18	113 ± 5	145 ± 14	132 ± 17

[a] Each value represents the mean ± SE of measurements on at least five mice.
[b] The maximal relative increase of pO_2 after application of BN in formulation.
[c] The time between application of the formulation and when pO_2 starts to increase.
[d] The time when the maximal increase in pO_2 is achieved.
[e] The area under the curve.

Results confirm the statement that the vehicle is very important for percutaneous absorption.[38] We assume that the delivery system influences the skin permeability and BN flux. Accordingly, during the permeation of BN out of the vehicle, the stratum corneum becomes the major rate-limiting component for transport through the skin. The relatively shorter t_{lag} and t_{max} values obtained with liposomal formulations can be explained by different effects of the vehicles on the skin barrier. BN, as a lipophilic substance, passes preferentially through the double-layer lipid matrix between corneocytes of the stratum corneum. The faster pharmacodynamic effect of BN observed in liposome vehicles could be the result of changed permeability of the stratum corneum caused by the liposomal lipids and different elastomechanical properties of the two liposomal formulation, which define the rate of decay of liposomes as well as their penetration into the skin.

Measured maximal pO_2 is within experimental error for all three formulations, showing that liposomes do not influence the vasodilatory action of BN, only the kinetics. The results of AUC of BN are informative because they clearly show the basic kinetics of the drug action in different formulations. They indicate that the total action of BN is slightly greater when applied in liposomes, especially in HSL, as compared to the single vehicle, hydrogel.

Comparison of Results Obtained with Two Different Paramagnetic Probes, LiPc and Bubinga. Figure 5 and Table I present results for the paramagnetic probes Bubinga and LiPc. It is evident that the time dependence for both types of liposomes and both probes is the same, as is the maximal increase in pO_2. The values of AUC are within experimental error. A significant difference is observed in the response of the probes to pO_2 changes. It could be due to the different amount of the probe inserted into the tissue and, connected with this, a different surface exposed to the oxygen. As already mentioned, because the line widths for LiPc are narrower than for Bubinga, less material is used to achieve the same signal-to-noise ratio. In addition, the small crystals of LiPc could be distributed more uniformly in tissue, and the surface of each crystal exposed to oxygen, while in Bubinga, due to the larger amount of powder, the particles are likely to be associated, and therefore Bubinga needs more time to equilibrate with the surroundings. The time needed to reach the equilibrium for a paramagnetic probe also varies with the type of paramagnetic material and its physical configuration, especially the size of the particles and the surface area of the material.[30,33]

---

[38] G. L. Flynn, *in* "Modern Pharmaceutics" (G. S. Banker and C. T. Rhodes, eds.), p. 239. Dekker, New York, 1996.

Effect of Repeated Application. One of the big advantages of this method is the possibility of following the response of the body to repeated applications of a drug. Usually the vasodilators that are used for topical application are applied to the skin repeatedly over a longer period of time. It is important to measure how the body responds to such treatment. EPR oximetry, unlike any other method, can be used to measure oxygen concentration changes directly at the same position for longer periods of time.

The action of BN incorporated in HSL on pO_2 is followed during daily application for a period of 5 days. The baseline of pO_2 before application of HSL and the relative increase of pO_2 after successive applications of HSL, as well as t_{lag} and t_{max}, are presented in Table II. A systematic increase in pO_2 in the skin tissue baseline, measured before each day's application, is observed with successive applications. The second day pO_2 baseline is 1 mm Hg higher than on the first day and, on the fifth day, it is 2.5 mm Hg higher than on the fourth day. The total increase of the baseline is 6.9 mm Hg in comparison to the initial baseline. The increase in pO_2 after the application of HSL is greater in the first 3 days than in the following days. This can be explained by the dilation of vessels, which remain dilated after several applications. When the dilation of vessels, and consequently the level of pO_2 in the skin, reaches some maximal value it cannot increase any more. It is also interesting to note that, with repeated application of HSL, the response of the organism is faster and that t_{lag} and t_{max} decrease with time. The third day after stopping the application, pO_2 in the skin is the same as before the first application. A similar response to repeated application is also observed when BN is applied free in the hydrogel.[19]

TABLE II

EFFECT OF DAILY REPEATED TOPICAL APPLICATION OF BN ENTRAPPED IN HSL LIPOSOMES OVER A PERIOD OF 5 DAYS ON OXYGENATION OF MOUSE SKIN

Time (days)	Daily baseline (mm Hg)	ΔpO_{2max} (mm Hg)	t_{lag} (min)	t_{max} (min)
1	13.9 ± 0.1	3.1 ± 0.1	10 ± 2	19 ± 1
2	14.9 ± 0.4	3.2 ± 0.2	7 ± 2	18 ± 2
3	16.4 ± 0.6	5.1 ± 0.2	6 ± 1	10 ± 1
4	18.3 ± 0.5	4.1 ± 0.3	6 ± 2	8 ± 1
5	20.8 ± 0.3	2.1 ± 0.3	6 ± 1	8 ± 2

Conclusions

The good agreement obtained for all three methods based on EPR provides additional confidence in the results obtained *in vitro* with hydrophilic probes on the skin. Many studies performed *in vitro* have measured the influence of cholesterol, the fluidity characteristics of liposome membranes, and the size of liposomes involved in the transport. The agreement of some of these experiments with the new EPR oximetry method, which measures directly the response of an organism to the liposome-entrapped drug, gives additional validity to the results. In future work it would be interesting to find some hydrophilic vasodilators, as hydrophilic probes cannot penetrate into the skin if not applied in liposomes. The influence of liposomes on pO_2 would be expected to be much greater with a hydrophilic vasodilator.

A big advantage of EPR oximetry is the possibility of measuring the time profile of a biological response during and after drug application, i.e., the pharmacodynamics of the drug. This is important in studying the influence of different liposome systems and any other vehicles or final formulations on the pharmacokinetics of the administered vasodilator. After initial insertion of the paramagnetic probe into the tissue, measurements can be made noninvasively over longer periods of time of several weeks. There are some limitations with respect to long-term measurements, which are connected with the long-term sensitivity of the paramagnetic probes to oxygen. Oxygen probes that are biocompatible and remain sensitive to oxygen are under development.[33]

Results obtained in these experiments at least qualitatively relate to the kinetics of release of the drug from liposomes into the skin in a clinical situation. Thus, it is necessary to know the influence of experimental conditions on results. Proper interpretation and integration of the results from such studies as this one should lead to efficient and safe drug formulations.

[18]   Liposomal Gels for Vaginal Drug Delivery

By Željka Pavelić, Nataša Škalko-Basnet,
Rolf Schubert, and Ivan Jalšenjak

Introduction

The intravaginal route of drug administration is used commonly for the local therapy of specific gynecological diseases (bacterial vaginosis, candidiasis, sexually transmitted diseases). Due to the large surface area of the vaginal wall, its rich blood supply, and permeability, it offers great potential for systemic drug delivery to a wide range of compounds, including peptides and proteins. Delivery via the vagina has some additional advantages over oral delivery, such as avoiding hepatic first-pass metabolism, reducing or eliminating the incidence and severity of side effects, and gastrointestinal incompatibility. Administration via this route is relatively easy and allows complete privacy of therapy.[1-3] However, the effectiveness of currently available vaginal dosage forms (solutions, foams, creams, gels, and vaginal tablets) is often limited by their low residence time at the target site. In order to overcome these limitations, attempts are being made to develop novel vaginal delivery systems with desirable distribution, bioadhesion, and release characteristics.[4,5]

Liposomes, i.e., phospholipid vesicles, are widely applied for the topical treatments of diseases in dermatology. Many drugs encapsulated into liposomes show enhanced skin penetration.[6] Because of their ability to provide a sustained and controlled release of the incorporated material, liposomes also have a potential for being applied vaginally.[3,7,8] The major disadvantage of using liposomes topically and vaginally lies in the liquid nature of the preparation. To achieve the viscosity desirable for application, liposomes should be incorporated into a suitable vehicle. It has been well established that liposomes are fairly compatible with viscosity increasing agents (methylcellulose) and polyacrylic acid (Carbopol).[9,10] In addition, several Carbopol resins have good bioadhesive properties and the ability to prolong the retention of the formulation on the mucosal surface.[3,11] Therefore, it seems reasonable to use hydrogels prepared from these polymers as vehicles for the incorporation of liposomes destined for vaginal delivery.

This article reports on the development of a liposomal carrier system, which would be able to provide the sustained and controlled release of entrapped compounds in vaginal therapy. Because potential, new drugs are often hydrophilic and of higher molecular weight, we have chosen

[1] L. Brannon-Peppas, *Adv. Drug Deliv. Rev.* **24,** 161 (1993).
[2] J. Brown, G. Hooper, C. J. Kenyon, S. Haines, J. Burt, J. M. Humphries, S. P. Newman, S. S. Davis, R. A. Sparrow, and I. R. Wilding, *Pharm. Res.* **14,** 1073 (1997).
[3] S. K. Jain, R. Singh, and B. Sahu, *Drug Dev. Ind. Pharm.* **23,** 827 (1997).
[4] J. R. Robinson and W. J. Bologna, *J. Control. Release* **28,** 87 (1994).
[5] K. Vermani and S. Garg, *PSTT* **3,** 359 (2000).
[6] J. Lasch and J. Bouwstra, *J. Liposome Res.* **5,** 543 (1995).
[7] M. Foldvari and A. Moreland, *J. Liposome Res.* **7,** 115 (1997).
[8] Ž. Pavelić, N. Škalko-Basnet, and I. Jalšenjak, *Eur. J. Pharm. Sci.* **8,** 345 (1999).
[9] M. Foldvari, *J. Microencapsul.* **13,** 589 (1996).
[10] N. Škalko, M. Čajkovac, and I. Jalšenjak, *J. Liposome Res.* **8,** 283 (1998).
[11] K. Knuth, M. Amiji, and J. R. Robinson, *Adv. Drug Deliv. Rev.* **11,** 137 (1993).

two commonly used model substances (markers), namely fluorescein-isothiocyanate dextran (FITC-dextran, MW 4400) and FITC-dextran (MW 21,200), and have encapsulated them into liposomes. Liposomal preparations are tested for their encapsulation efficiency and *in vitro* stability in media simulating human vaginal conditions. After the incorporation of liposomes into Carbopol gels, studies on the *in vitro* release of encapsulated compounds are performed.

Experimental Procedures

Chemicals

Phospholipids, egg phosphatidylcholine (EPC), and egg phosphatidyl-glycerol sodium (EPG-Na), as well as Carbopol 974P NF, were generous gifts from Lipoid GmbH (Ludwigshafen, Germany) and BFGoodrich (Brussels, Belgium), respectively. FITC-dextrans (MW 4400 and 21,200) are from Sigma Chemicals (Deisenhofen, Germany). Phosphate buffer, pH 7.4, is composed of 8 g/liter (137 mM) NaCl, 0.19 g/liter (1.4 mM) KH_2PO_4, and 2.38 g/liter Na_2HPO_4. Phosphate buffer, pH 4.5, is made of 13.61 g/liter (100 mM) KH_2PO_4. These and all other chemicals used in experiments are of analytical grade.

Preparation of Liposomes

To achieve high trapping efficiency, liposomes composed of EPC and EPG-Na in the molar ratio 9:1 are prepared by a modified proliposome[12] or polyol dilution[13] method. The phospholipid concentration of all liposome dispersions is 12.8 mM, the concentrations of the markers are 100 μM (FITC-dextran MW 4400) or 50 μM (FITC-dextran MW 21,200) in phosphate buffer.

Proliposome Method

1. EPC (88 mg), EPG-Na (12 mg), 80 mg of warm ethanol, and 200 mg of FITC-dextran buffer solution (pH 7.4) are mixed.
2. The mixture is stirred for a few minutes at approximately 40° to form the initial proliposome mixture.
3. The mixture is cooled to room temperature and 10 ml of phosphate buffer, pH 7.4 is added dropwise, with stirring at 600 rpm.
4. Stirring is continued for 60 min.

[12] S. Perrett, M. Golding, and W. P. Williams, *J. Pharm. Pharmacol.* **43**, 154 (1991).
[13] H. Kikuchi, H. Yamauchi, and S. Hirota, *J. Liposome Res.* **4**, 71 (1994).

Polyol Dilution Method

1. EPC (88 mg) and 12 mg of EPG-Na are dissolved in 500 mg of propylene glycol.
2. FITC-dextran (500 mg) buffer solution (100 μM for FITC-dextran 4400 or 50 μM for FITC-dextran 21,200) is added to the lipid–polyol mixture.
3. The liposomal dispersion is diluted with 10 ml of phosphate buffer, pH 7.4 (warmed previously to 60°) and stirring is continued for 45 min.

Note: The entire procedure is carried out at 60° with stirring at 600 rpm.

All liposomal preparations are extruded once through polycarbonate membrane filters with 0.4 μm pores (LiposoFast, Avestin, Ottawa, Canada).

Vesicle Size Measurements

A microscopic image analysis technique for the determination of liposomal size distributions is applied. Morphology and particle size distribution (based on the number of particles) are determined in an Olympus BH-2 microscope equipped with a computer-controlled image analysis system (Optomax V, Cambridge). A microscopic field is scanned by a television camera and is digitized or broken into electronically defined individual picture elements. In all measurements, approximately 10,000 liposomes are examined.[14]

Trapping Efficiency Determination

In order to compare liposome preparation methods, the trapping efficiencies are determined. The separation of free markers from those trapped in liposomes is performed by gel chromatography on a Sepharose CL-4B (Pharmacia, Uppsala, Sweden) column using phosphate buffer, pH 7.4, for elution. The concentrations of markers (both in liposomes and free) are determined in all collected fractions by measuring their fluorescence intensities on a Perkin-Elmer luminescence spectrometer (494 nm excitation wavelength and 524 nm emission wavelength). To determine the amount of trapped marker, liposomal fractions are treated with a detergent (10% Triton X-100 solution). After dissolving phospholipids, the samples are diluted with buffer, pH 7.4, whereupon the final concentration of detergent

---

[14] N. Škalko-Basnet, Ž. Pavelić, and M. Bećirević-Laćan, *Drug Dev. Ind. Pharm.* **26**, 1279 (2000).

is 1%. In order to avoid the influence of the detergent on the fluorescence intensity, blank tests (1% Triton X-100 buffer solution, pH 7.4) are performed prior to the measurements.[15]

The recovery of the model substance determined for all preparations is between 92 and 96% of the starting material.

Preparation of Hydrogel[15]

1. One gram of Carbopol 974 P NF is dispersed in 88 g of distilled water with stirring at 800 rpm for 60 min.
2. The stirring rate is reduced to 400–600 rpm and 10 g of propylene glycol is added.
3. The mixture is neutralized with dropwise addition of 10% NaOH and mixing is continued until a transparent gel appears (quantity of neutralizing agent is adjusted to achieve a gel with pH 5.5).

Incorporation of Liposomes into Mucoadhesive Gel[10,15]

Liposomes, free from unentrapped model substance, are mixed into the 1% (w/w) Carbopol 974P NF gel by an electrical mixer (25 rpm, 5 min) with the concentration of liposomes in the gel being 10% (w/w, liposomal dispersion/total). Control gels (10%, w/w) are prepared under the same conditions. Instead of liposomes, these samples contain FITC-dextran (MW 4400 or 21,200) solutions in buffer, mixed in the gel.

In Vitro Stability Studies

Liposomal preparations, both dispersions and gels, are tested for *in vitro* stability in phosphate buffer, pH 4.5 (corresponding to normal human vaginal pH).

Liposomal Dispersions (Procedure)[8]

1. Five milliliters of testing media is incubated in a sealed tube at 37° using a water bath.
2. One milliliter of liposomes (separated previously from the unentrapped substance) is added and mixed well, and the incubation is continued for 24 h.
3. One milliliter of liposomal dispersion is removed at certain time intervals (1, 2, 4, 6, and 24 h).
4. The substance released is separated from liposomes and the concentration is determined fluorimetrically (as described earlier).

[15] Ž. Pavelić, N. Škalko-Basnet, and R. Schubert, *Int. J. Pharm.* **219,** 139 (2001).

Liposomal Gels (Modified Procedure by Peschka et al.[16])

1. Three grams of liposomal gel is weighed in a glass vial (10 ml).
2. One milliliter of 2% (w/v) liquefied agarose is poured over the gel.
3. Vials are kept at room temperature for 5 min to let the agarose (covering) layer harden.
4. Five milliliters of receptor solution (buffer, pH 4.5) is added and the vials are incubated at $37°$ (water bath).
5. At certain time points (1, 2, 4, 6, and 24 h) the receptor solution is replaced completely.
6. The amount of released substance is determined fluorimetrically.

Control gels are examined simultaneously and under exactly the same conditions.

Rheological Measurements

Carbopol 974P NF gel with incorporated liposomes containing FITC-dextran 4400 is tested for flow properties using a CS-rheometer (RheoStress RS 100 1 Ncm, Peltier TC81, Haake, Germany). Measurements are performed at $20°$, whereby a cone/plate C 35/1° (0.05 mm) measuring system is applied. The flow properties of the blank gel and of the gel-containing buffer instead of liposomes are examined under the same conditions.

Storage Stability Study

The physical stability of liposomes incorporated in the gel is determined by image analysis microscopy (for details, see vesicle size measurements). Samples are stored at 20 and $40°$.[10]

Characterization of Liposomal Gels

The potential of the intravaginal route for the local, as well as for the systemic delivery of drugs, has been well known. In general, hydrophilic substances, especially those of higher molecular weight, show a low bioavailability from the mucosal site because of their limited absorption and lability to mucosal enzymes. Several novel carrier systems were proposed to be suitable for vaginal delivery, such as polycarbophilic gel,[2,4,11] bioadhesive tablets,[17-19] and liposomes.[3,7,8]

---

[16] R. Peschka, C. Dennehy, and F. C. Szoka, *J. Control. Release* **56**, 41 (1998).

[17] A. Gürsoy, I. Sohtorik, N. Uyanik, and N. A. Peppas, *S. T. P. Pharma* **5**, 886 (1989).

[18] S. Bouckaert, M. Temmerman, J. Voorspoels, H. van Kets, J. P. Remon, and M. Dhont, *J. Pharm. Pharmacol.* **47**, 970 (1995).

To achieve the desirable therapeutic effects of liposomes as drug carriers, they must be loaded with a sufficient quantity of active compounds; therefore, high trapping efficiency is required. This can be achieved easily with lipophilic and amphiphilic molecules, which have the tendency to incorporate themselves in the liposomal membrane, while only a small amount of hydrophilic substances ends up inside the liposomes.[20]

We have reported on the polyol dilution and the proliposome method as optimal for the encapsulation of water-soluble compounds of small molecular weight. Comparisons with other preparation techniques, such as the film method, detergent removal, and high-pressure homogenization, showed an extremely high trapping efficiency of calcein in polyol dilution liposomes (63%). Those prepared by the proliposome method could trap about 32%, whereas liposomes prepared by other techniques could encapsulate only up to 7.7% of the starting material.[15] Furthermore, both methods have already been proven as appropriate for the incorporation of lipophilic antimicrobial drugs used commonly in the treatment of bacterial vaginosis.[8] This article describes the preparation of liposomes containing hydrophilic substances of higher molecular weight (FITC-dextrans 4400 or 21,200) using the proliposome and polyol dilution methods.

As shown in Table I, liposomes prepared by both methods have mean diameters between 320 and 370 nm. Regardless of the encapsulated substance, liposomes are of spherical shape, and the image analysis technique revealed no morphological differences between the vesicles prepared by the two methods.

As in studies with calcein,[15] better trapping is achieved with liposomes prepared by polyol dilution than with proliposomes. The encapsulation efficiency is the highest for liposomes containing FITC-dextran MW 21,200 (58%). Those containing FITC-dextran MW 4400, prepared by the same method, could entrap 49% of the starting material (see Table I).

Liposomes destined for vaginal delivery should be stable in conditions chosen to simulate the vaginal environment. In *in vitro* stability studies, we have checked the stability of liposomes containing hydrophilic substances of different molecular weights and prepared by the different procedures. As can be seen in Fig. 1, liposomes containing the hydrophilic model substances and prepared by the polyol dilution method and stable in media chosen to simulate normal vaginal pH. Regardless of the encapsulated substance, even after 24 h of incubation at 37°, approximately 80% of the originally entrapped substance is still retained in the liposomes.

[19] L. Genç, C. Oğuzlar, and E. Güler, *Pharmazie* **55**, 297 (2000).

[20] G. Gregoriadis, "Liposome Technology," 2nd Ed. CRC Press, Boca Raton, FL, 1993.

TABLE I

MEAN DIAMETERS (nm) AND ENCAPSULATION EFFICIENCIES (%) OF LIPOSOMES[a]

FITC-dextran	Proliposome method		Polyol dilution method	
	Mean diameter (nm)	Encapsulation (%)	Mean diameter (nm)	Encapsulation (%)
MW 4400	354.9 ± 15.9	35.89 ± 2.34	367.3 ± 11.4	49.04 ± 5.02
MW 21,200	365.3 ± 9.2	46.25 ± 4.89	320.9 ± 8.9	57.86 ± 6.73

[a] Liposomes containing FITC-dextrans of different molecular weights were prepared by the proliposome and polyol dilution methods. Values denote the mean of three preparations ±SD.

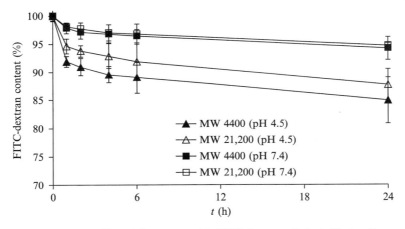

FIG. 1. *In vitro* stability of liposomes with FITC-dextrans. Polyol dilution liposomes, separated from an unentrapped marker, were incubated in the pH 4.5 buffer at 37°. Samples were taken at certain time intervals, and the concentration of marker still present in the liposomes was determined fluorimetrically. Values denote the mean of three preparations ±SD.

In previous experiments performed with calcein, liposomes retained a lower amount of originally entrapped marker (65% after 24 h).[15] A better stability of liposomes containing FITC-dextrans could be a consequence of steric effects due to the higher molecular weights of encapsulated compounds. When we examined the stability of lecithin liposomes containing lipophilic drugs, prepared by the same methods, a similar stability of liposomes containing metronidazole and clotimazole was observed. The only exception were liposomes containing chloramphenicol, where those prepared by the polyol dilution method retained less of the originally incorporated drug than proliposomes (30 and 46%, respectively).[8] Comparison of those findings with results obtained with hydrophilic substances, both calcein[15] and FITC-dextrans, demonstrate a better stability of liposomes containing water-soluble compounds. This could be explained by the hydrophilic nature of calcein and FITC-dextrans at neutral pH inside the liposomes. Therefore, membrane penetration is very slow and the liposomes retain a higher amount of originally entrapped substances.

An appropriate viscosity of liposomal preparations with regard to administration in humans can be achieved by liposome incorporation in a suitable vehicle. One of the major disadvantages of commercially available vaginal dosage forms is the short residence time of drug at the site of application. Because an improved retention on the mucous is often required for the desired therapeutic effect, research has focused on using bioadhesive polymers (such as polyoxiethylene and polyacrylic acid) in vaginal delivery.[1] Also, hydrogels of low viscosity prepared from these polymers alter the hydration level of the vaginal tissue and can be applied as moisturizers for the treatment of dry vagina.[4] Since the compatibility of liposomes with polymers derived from acrylic acid (Carbopol) has already been proven,[9] it seemed worthwhile to use these hydrogels as vehicles for liposomes.

Polyol dilution liposomes are mixed with 1% Carbopol 974P NF hydrogel and are tested for the *in vitro* release of encapsulated material. The results are compared with those obtained when liposomes are dispersed in the pH 4.5 buffer. A simple and reproducible *in vitro* model introduced by Peschka *et al.*[16] is modified and applied to follow the release of FITC-dextrans from liposomes incorporated in gel. The porosity of the agarose matrix permits intact liposomes and released (free) FITC-dextran to diffuse through the matrix into the receptor solution. The amount of substance released from the gel is determined fluorimetrically before and after the addition of Triton X-100 to the supernatant over the gel.

Results (Table II, Fig. 2) confirm the prolonged release of both FITC-dextrans from liposomes incorporated in the gel. Even after 72 h of incubation at 37°, only 26% of released FITC-dextran MW 4400 is detected (absence of detergent). Upon the addition of detergent, a small increase in fluorescence intensity is found, indicating that only a very small amount of intact liposomes is released into the receptor solution (buffer, pH 4.5). A better retention of encapsulated marker in the gel is observed with FITC-dextran MW 21,200 (only about 19% is released from liposomes incorporated in the gel after 72 h). In a study with calcein,[15] after only 24 h of incubation, 16% of the originally entrapped substance is released from liposomes incorporated in the gel. These findings are expected

TABLE II

In Vitro Release of Encapsulated FITC-Dextrans from Carbopol 974P NF Gel[a]

Time (h)	Free FITC-dextran released (%)	Total FITC-dextran released (%)	Control (%)
1	0.70 ± 0.15^b	0.71 ± 0.16^b	2.87 ± 0.53^b
	0.16 ± 0.07^c	0.16 ± 0.07^c	1.00 ± 0.53^c
2	1.39 ± 0.87^b	1.44 ± 0.73^b	6.78 ± 0.98^b
	0.47 ± 0.08^c	0.49 ± 0.10^c	3.50 ± 0.87^c
4	2.79 ± 0.45^b	2.91 ± 0.39^b	13.01 ± 1.45^b
	2.16 ± 0.41^c	2.65 ± 0.38^c	6.98 ± 1.02^c
6	4.31 ± 0.85^b	4.54 ± 0.68^b	19.23 ± 2.01^b
	2.84 ± 0.80^c	3.46 ± 0.76^c	11.32 ± 3.07^c
24	12.56 ± 1.05^b	13.30 ± 1.23^b	58.00 ± 4.02^b
	8.44 ± 1.01^c	9.42 ± 0.08^c	38.56 ± 3.21^c
48	18.32 ± 1.09^b	23.30 ± 1.13^b	96.53 ± 9.53^b
	14.53 ± 1.48^c	17.11 ± 1.87^c	68.00 ± 9.33^c
72	26.32 ± 4.08^b	36.43 ± 4.23^b	99.54 ± 13.56^b
	19.78 ± 2.68^c	24.97 ± 2.94^c	88.56 ± 10.21^c

[a] Samples of gels containing liposomes with encapsulated FITC-dextrans were incubated in phosphate buffer, pH 4.5, at 37°. The amount of FITC-dextran was determined in release media before and after the disruption of liposomes by the addition of detergent. Control experiments (release of free substance from gel) were performed simultaneously. Values denote the mean of three preparations ±SD.
[b] FITC-dextran MW 4400.
[c] FITC-dextran MW 21,200.

because of an inverse correlation between the release rate and the molecular weight of the encapsulated marker.

When the amount of retained material in gel is plotted against the square root of time, a linear correlation according to the Higuchi equation is obtained for both preparations (Fig. 2), showing matrix-controlled diffusion of the released model substances.

A comparison with the results of *in vitro* stability of liposomes dispersed in buffer (Fig. 1) confirms the protective effect of a hydrogel matrix on liposomes. Thus, considering the stability of liposomes during vaginal administration in humans, as well as the prolonged and controlled release of encapsulated drugs for systemic delivery, the Carbopol 974P NF gel would be a right choice of a vehicle.

Regarding the physical properties, a gel for the incorporation of liposomes should have an adequate pH value, as well as improved rheological characteristics and stability. Empty Carbopol 974P NF hydrogel (blank) and those containing liposomes are examined for basic rheological measurements. As can be seen in Fig. 3, a pronounced loss of viscosity is

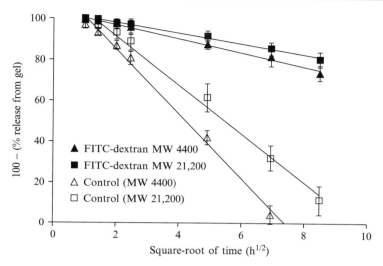

FIG. 2. Release of FITC-dextrans (MW 4400 and 21,200) fitted against the square root of the time. Liposomes containing FITC-dextrans and incorporated in Carbopol 974P NF hydrogels were examined for *in vitro* release (pH 4.5). The amounts of released markers were detected in receptor media before the addition of a detergent. Control gels were tested under the same conditions. Values denote the mean of three preparations ±SD.

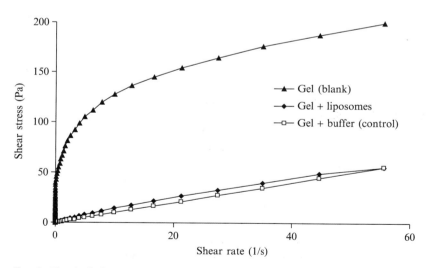

FIG. 3. Rheological properties of liposomes incorporated in a hydrogel. Polyol dilution liposomes containing FITC-dextran MW 4400 were incorporated in preformed 1% Carbopol 974P NF hydrogel and tested for flow behavior on a CS-rheometer at 20° (cone/plate measuring system). Control and blank gels were examined under the same conditions.

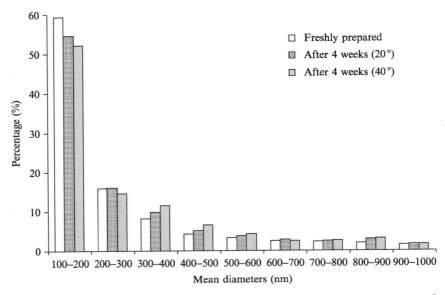

Fig. 4. Size distributions of liposomes stored in a hydrogel for 4 weeks at 20 and 40°. Polyol dilution liposomes containing FITC-dextran MW 4400 were incorporated in 1% Carbopol 974P NF and tested for storage stability by image analysis microscopy (for details, see the text). Mean diameters of liposomes were 350 nm (immediately after incorporation in a vehicle), 391 nm (after 4 weeks storage at 20°), and 416 nm (after 4 weeks at 40°), respectively.

observed in samples with incorporated liposomes and in those containing buffer instead of liposomes. The changed viscosity is affected by the cationic ingredients (sodium ions) from the buffer that are incompatible with the anionic Carbopol resins.[21] Therefore, instead of the commonly used 0.5%,[11] we prepare 1% Carbopol 974P NF hydrogel and, after addition of 10% liposomes, a suitable application viscosity is achieved.

To provide a stable vehicle suitable for vaginal application in which liposomes are distributed uniformly and their structure is preserved, a storage stability study is performed. Liposomal gels are kept for 4 weeks at 40° (stress testing) and at 20°. During those tests, the size distribution and mean diameter of the incorporated liposomes are determined. Results presented in Fig. 4 demonstrate the ability of the Cabopol 974P NF hydrogel to preserve the original size distributions of liposomes. After 4 weeks of storage at stress conditions (40°), the mean diameter changed from 350 to

[21] M. Dittgen, M. Durrani, and K. Lehmann, *S.T.P. Pharma* **7,** 403 (1997).

415 nm, but the distribution stayed statistically similar. The same behavior was observed with samples of gel containing FITC-dextran MW 21,200.

Concluding Remarks

All experiments performed in this and previous studies[8,11] confirm the possibility of using liposomes as a novel vaginal delivery system. Polyol dilution and proliposome methods have been proven to be appropriate not only for the incorporation of lipophilic drugs, but also for hydrophilic substances, regardless of their molecular weight. Both methods are simple, reproducible, and acceptable for the upscaled production of liposomes that are stable in media chosen to simulate vaginal pH. The incorporation of those liposomes in the Carbopol 974P NF hydrogel further improves their stability and confirms the applicability of liposomal gels as a novel vaginal delivery system that is able to provide the controlled and sustained release of an entrapped drug in local and systemic treatments.

Acknowledgments

This work was supported partially by a Deutscher Akademischer Austauschdienst (DAAD) grant to Ž. Pavelić, and some experiments were performed at Albert-Ludwidge-University Freiburg, Germany. We thank W. Michaelis for his valuable help in performing the rheological measurements. The generosity of Lipoid GmbH (Ludwigshafen, Germany) and BFGoodrich (Brussels, Belgium) is greatly appreciated.

[19] Use of Nitroxides to Protect Liposomes Against Oxidative Damage

By AYELET M. SAMUNI and YECHEZKEL BARENHOLZ

Introduction

Liposomes, formed upon exposure of amphiphiles to aqueous media, are spherical, self-closed structures composed of one or several concentric curved lipid bilayers, entrapping part of the aqueous medium in which they are dispersed.[1–3] The major amphiphiles of liposomes used for drug

[1] D. D. Lasic, "Liposomes: From Physics to Applications." Elsevier Science, Amsterdam, 1993.

[2] Y. Barenholz and D. J. A. Crommelin, in "Encyclopedia of Pharmaceutical Technology" (J. Swarbrick and J. C. Boylan, eds.), Vol. 9, p. 1. Dekker, New York, 1994.

delivery, including 10 formulations approved for clinical use,[4] are diacyl phospholipids (PL), mainly phosphatidylcholines (PC). In many liposomes, cholesterol is an additional major lipid component.[2–5] The phospholipids form the liposome lipid bilayer, and cholesterol, when a major (>30 mol%) bilayer component, increases membrane stability and decreases its permeability.[4,6,7] Cholesterol transforms both phases, the liquid disordered (LD) and the solid ordered (SO) phases in the membrane, to a liquid ordered (LO) phase. At the molecular level (short range), cholesterol reduces the "kinks" formed due to trans–gauche isomerization that run along the hydrocarbon chains. At the lateral organization level (long range), cholesterol above 33 mol% may eliminate membrane defects formed at phase boundaries present when LD and SO phases coexist. Both effects involve reduction in free volume.[4,8] Cholesterol has also been shown to reduce the hydration of the bilayer, the rate of water diffusion, and the depth of water penetration into the bilayer, while reducing free volume in LD and increasing the bilayer packing density.[9–12]

The unique properties of liposomes have triggered numerous applications in various fields of science and technology, from basic studies to drug delivery and transfection vectors.[1–3,13–15] Liposome stability, which can be divided into physical, chemical, and biological stability, is one of the most important factors in liposome applications. All three aspects are interrelated. Chemical degradation reduces the biological and physical stability of liposomes. Reduction of physical stability due to aggregation or drug leakage reduces liposome utility. The major chemical reactions involved are acyl ester bond hydrolysis and oxidative damage to polyunsaturated

[3] Y. Barenholz and D. J. A. Crommelin, *in* "Encyclopedia of Pharmaceutical Technology" (J. Swarbrick and J. C. Boylan, eds.), 2nd Ed. Dekker, New York.

[4] Y. Barenholz, *Prog. Lipid Res.* **41,** 1 (2002).

[5] J. K. Lang and C. Vigo-Pelfrey, *Chem. Phys. Lipids* **64,** 19 (1993).

[6] T. H. Hains, *Prog. Lipid Res.* **40,** 229 (2001).

[7] P. L. Yeagle, "The Membrane of Cells," 2nd Ed., p. 139. Academic Press, New York, 1993.

[8] Y. Barenholz and G. Cevc, *in* "Physical Chemistry of Biological Surfaces" (A. Baszkin and W. Norde, eds.), p. 171. Dekker, New York, 2000.

[9] T. Parasassi, M. DiStefano, M. Loiero, G. Ravagnan, and E. Gratton, *Biophys. J.* **66,** 763 (1994).

[10] T. Parasassi, A. M. Giusti, M. Raimondi, and E. Gratton, *Biophys. J.* **68,** 1895 (1995).

[11] S. A. Simon, T. J. McIntosh, and R. Latorre, *Science* **216,** 65 (1982).

[12] A. M. Samuni, A. Lipman, and Y. Barenholz, *Chem. Phys. Lipids* **105,** 121 (2000).

[13] D. Lichtenberg and Y. Barenholz, *in* "Methods of Biochemical Analysis" (D. Glick, ed.), Vol. 33, p. 337. Wiley, New York, 1988.

[14] Y. Barenholz and D. D. Lasic, eds., "Handbook of Nonmedical Applications of Liposomes," Vols. I–IV. CRC Press, Boca Raton, FL, 1996.

[15] R. R. C. New, "Liposomes: A Practical Approach." IRL Press, Oxford, 1990.

acyl chains, cholesterol, and (primary) amino groups.[2,3,13–15] As for physical stability, the most important parameters in quality control and characterization of liposomal formulations are liposome size distribution and liposome physical integrity.

Lipid Peroxidation

This article focuses on oxidative damage to lipids in liposomes and on the use of nitroxides as antioxidants capable of decreasing such damage. Lipids, like most biomolecules, undergo degradation reactions such as oxidation and hydrolysis.[1–3,13,15–18] Lipid oxidation, also called lipid peroxidation (LPO) or lipid autooxidation, is mediated by free radicals, leads to the formation of a broad spectrum of intermediates and products, and has long been a problem in the preparation and preservation of foods and of lipid-based formulations.

In liposomal formulations, the two main types of lipid components, phospholipids and cholesterol, are susceptible to peroxidation reactions.[1–3,13–16] Oxidation of phospholipids takes place in their unsaturated, mainly polyunsaturated fatty acid (PUFA) chains.[1,5,12,19] Examples of phospholipid acyl chain peroxidation are described in Fig. 1. Like phospholipids, cholesterol, although less sensitive, has been shown to decompose by a peroxidation mechanism.[5,12,20] Acyl chain and cholesterol peroxidation are interrelated. Cholesterol seems to inhibit phospholipid acyl chain peroxidation in the lipid bilayer, and the PUFA level influences cholesterol peroxidation, with higher PUFA levels decreasing cholesterol peroxidation.[5,12,19]

Detection and Measurement of LPO

Several techniques are available for measuring and quantitating the rate of LPO in membranes. Each technique measures something different, some measuring products that appear only transiently and undergo fast changes, as in the case of conjugated dienes,[16,17] other lipid radicals, lipid

[16] Y. Barenholz and S. Amselem, in "Liposome Technology" (G. Gregoriadis, ed.), 2nd Ed., Vol. I. CRC Press, Boca Raton, FL, 1993.

[17] D. Pinchuk and D. Lichtenberg, *Prog. Lipid Res.* **41,** 279 (2002).

[18] E. Niki, *Chem. Phys. Lipids* **44,** 227 (1987).

[19] B. Halliwell and J. M. C. Gutteridge, "Free Radicals in Biology and Medicine." Clarendon Press, Oxford, 1989.

[20] L. L. Smith, J. I. Teng, Y. Y. Lin, P. K. Seitz, and M. F. McGehee, in "Lipid Peroxides in Biology and Medicine" (K. Yagi, ed.), p. 89. Academic Press, New York, 1982.

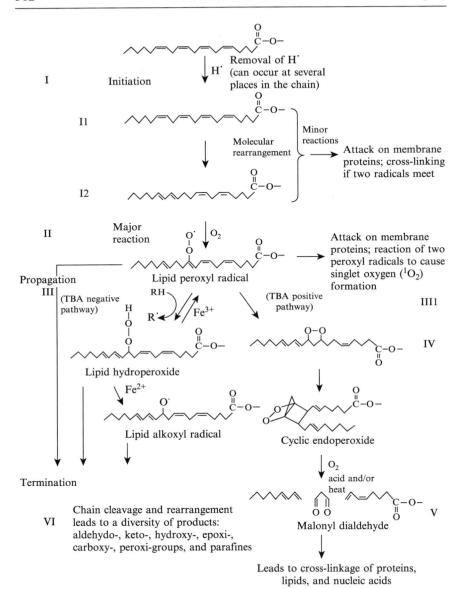

Fig. 1. Lipid peroxidation pathways.

hydroperoxide (LOOH), and malonyl dialdehyde (MDA) (Fig. 1). In contrast, a quantitative determination of individual fatty acids (using GC) or cholesterol (using HPLC) enables an accurate follow-up of liposome lipid degradation.[5,16]

Inhibiting Lipid Peroxidation

Preventive and protective measures are generally taken to minimize oxidative damage. Preventive measures include the efficient chelating of ions of transition metals such as Fe and Cu and protection from light and from exposure to air.[2,3,13,15] These preventive measures by themselves are not sufficient to prevent LPO completely. Therefore, in many cases there is a need to use additional preventative measures.[13] A common strategy of protection against LPO employs reducing agents (conventional antioxidants) that act as preventive and chain-breaking antioxidants. However, their efficacy is limited because they (a) are being consumed and therefore depleted, (b) give rise to secondary radicals that may be deleterious themselves, and (c) may act as prooxidants.[21] This article focuses on a nonconventional group of antioxidants—nitroxides.

Nitroxides

Nitroxides represent a new and alternative strategy in combating oxidative damage inflicted by deleterious species. Nitroxides are stable cyclic radicals, differing in size, charge, and lipophilicity (and therefore permeability through biological membranes).[22–24] Their use as biophysical probes or contrast agents for nuclear magnetic resonance imaging has been investigated extensively and reviewed.[25] Nitroxides, which lack any prooxidative effect, unlike commonly used antioxidants, act catalytically and are self-replenished. They have a protective effect from radiation in cells in culture and in the whole animal[26–29] and serve as a new class of nonthiol aerobic

[21] Y. Yoshida, J. Tsuchiya, and E. Niki, *Biochim. Biophys. Acta* **1200,** 85 (1994).

[22] V. Afzal, R. C. Brasch, D. E. Nitecki, and S. Wolff, *Invest. Radiol.* **19,** 549 (1984).

[23] E. G. Ankel, C. S. Lai, L. E. Hopwood, and Z. Zivkovic, *Life Sci.* **40,** 495 (1987).

[24] W. G. DeGraff, M. C. Krishna, A. Russo, and J. B. Mitchell, *Environ. Mol. Mutagen.* **19,** 21 (1992).

[25] N. Kocherginsky and H. M. Swartz, "Nitroxide Spin Labels: Reactions in Biology and Chemistry." CRC Press, Boca Raton, FL, 1995.

[26] T. Goffman, D. Cuscela, J. Glass, S. Hahn, C. M. Krishna, G. Lupton, and J. B. Mitchell, *Int. J. Radiat. Oncol. Biol. Phys.* **22,** 803 (1992).

[27] S. M. Hahn, L. Wilson, C. M. Krishna, J. Liebmann, W. DeGraff, J. Gamson, A. Samuni, D. Venzon, and J. B. Mitchell, *Radiat. Res.* **132,** 87 (1992).

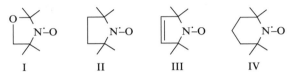

FIG. 2. Types of commonly used nitroxide rings.

radioprotectors. The reactions of nitroxides, being radicals themselves, as antioxidants and radioprotectors do not yield secondary radicals, but instead terminate radical chain reactions. The variety of chemical structures of nitroxides, including the types of rings on which they are based (Fig. 2), makes them suitable for a large number of applications in biological systems.

Nitroxides as Antioxidants

The role of nitroxides as antioxidants has been established in recent years, and their protective activity has been investigated using diverse means of insult, various nitroxides, and a variety of experimental models. Physical, chemical, biochemical, and cellular means are used to initiate injurious processes mediated by reactive oxygen species (ROS). H_2O_2,[30–32] organic peroxides [such as *t*-butyl hydroperoxide (*t*-BuOOH)],[33,34] and O_2^-, generated enzymatically by hypoxanthine/xanthine oxidase (HX/ XO),[35] are the more common means of insult used. Nevertheless, other insults, such as ionizing radiation,[28] ADP-FeII,[36] azo initiators such as

[28] S. M. Hahn, Z. Tochner, C. M. Krishna, J. Glass, L. Wilson, A. Samuni, M. Sprague, D. Venzon, E. Glatstein, and J. B. Mitchell, *Cancer Res.* **52,** 1750 (1992).

[29] J. B. Mitchell, W. DeGraff, D. Kaufman, M. C. Krishna, A. Samuni, E. Finkelstein, M. S. Ahn, S. M. Hahn, J. Gamson, and A. Russo, *Arch. Biochem. Biophys.* **289,** 62 (1991).

[30] D. Gelvan, V. Moreno, D. A. Clopton, Q. Chen, and P. Saltman, *Biochem. Biophys. Res. Commun.* **206,** 421 (1995).

[31] J. B. Mitchell, A. Samuni, M. C. Krishna, W. G. DeGraff, M. S. Ahn, U. Samuni, and A. Russo, *Biochemistry* **29,** 2802 (1990).

[32] A. Samuni, D. Godinger, J. Aronovitch, A. Russo, and J. B. Mitchell, *Biochemistry* **30,** 555 (1991).

[33] J. Antosiewicz, J. Popinigis, M. Wozniak, E. Damiani, P. Carloni, and L. Greci, *Free Radic. Biol. Med.* **18,** 913 (1995).

[34] F. Tanfani, P. Carloni, E. Damiani, L. Greci, M. Wozniak, D. Kulawiak, K. Jankowski, J. Kaczor, and A. Matuszkiewics, *Free Radic. Res. Commun.* **21,** 309 (1994).

[35] V. Gadzheva, K. Ichimori, H. Nakazawa, and Z. Raikov, *Free Radic. Res. Commun.* **21,** 177 (1994).

[36] U. A. Nilsson, L. I. Olsson, G. Carlin, and F. A. Bylund, *J. Biol. Chem.* **264,** 11131 (1989).

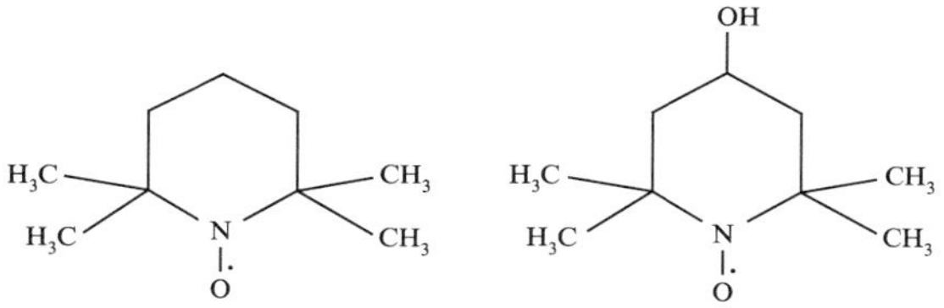

Fig. 3. Structure of the nitroxides Tempo and Tempol.

2,2'-azobis(2-amidinopropane)dihydrochloride   (AAPH),   2,2'-azo(2,4-dimethylvaleronitrile) (AMVN), and 2,2'-azobis(2-methylpropionamidine) dihydrochloride (AMPH),[36,37] and mechanical trauma[38] are used as well.

Most studies are performed using commercially available nitroxides, particularly 2,2,6,6-tetramethylpiperidine-1-oxyl (Tempo) and 4-hydroxy-2,2,6,6-tetramethylpiperidine-1-oxyl (Tempol) (Fig. 3). In some studies, new nitroxide derivatives were synthesized in order to improve biological protective activity. The test systems used range from isolated compounds, membranes, and cells to organs and whole animals.[31,38–42]

Nitroxides in the Protection of Liposomes

Radioprotection

γ-Irradiation, being also a natural factor causing LPO, serves as a convenient means for inducing oxidative damage. It is well defined, allowing good control of the insult, and, due to the fact that the radical species formed are known, can help clarify the mechanism of LPO and its inhibition.

Egg phosphatidylcholine (EPC) small unilamellar vesicles (SUV) are γ-irradiated with a dose of 10–12 kGy, using a ^{60}Co γ-source, in the absence and presence of Tempo and Tempol (Aldrich, Milwaukee, WI).

[37] M. Takahashi, J. Tsuchiya, and E. Niki, *J. Am. Chem. Soc.* **111,** 6350 (1989).

[38] E. Beit-Yannai, R. Zhang, V. Trembovler, A. Samuni, and E. Shohami, *Brain Res.* **717,** 22 (1996).

[39] G. Cighetti, P. Allevi, S. Debiasi, and R. Paroni, *Chem. Phys. Lipids* **88,** 97 (1997).

[40] D. Gelvan, P. Saltman, and S. R. Powell, *Proc. Natl. Acad. Sci. USA* **88,** 4680 (1991).

[41] D. Rachmilewitz, F. Karmeli, E. Okon, and A. Samuni, *Gut* **35,** 1181 (1994).

[42] H. Sasaki, L. R. Lin, T. Yokoyama, M. D. Sevilla, V. N. Reddy, and F. J. Giblin, *Invest. Opthalmol. Vis. Sci.* **39,** 544 (1998).

Egg PC and phospholipids derived from it by transphosphatidylation [such as egg phosphatidylglycerol (PG)], which contain more than 10% PUFA, are good examples for investigating the sensitivity to oxidative damage as well as a means to protect against such damage. The acyl chain composition of the phospholipids before and after irradiation is determined using gas chromatography (GC) (Fig. 4). Tempo and Tempol protect EPC acyl chains against irradiation damage in a concentration-dependent manner, with 5 mM nitroxide providing full protection (Fig. 5). Almost complete protection is also found in the case of EPC:EPG (10:1 mol/mol) and EPC:EPG:cholesterol (10:1:4 mol/mol) liposomes. Both nitroxides similarly prevent radiation-induced degradation of both the liposomal phospholipids and the liposomal cholesterol. The addition

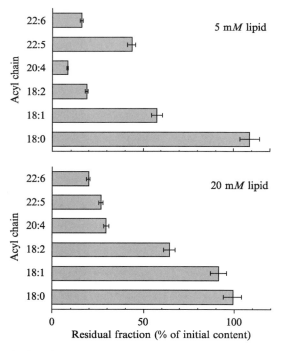

FIG. 4. Effect of γ-irradiation on the composition of fatty acids in EPC SUV. Liposomal dispersion of EPC SUV in saline was γ-irradiated with 10 kGy at room temperature. Fatty acid composition was determined after methyl esterification followed by GC separation. The new levels of fatty acids, related to palmitic acid, which served as internal standards, are presented as a fraction of their original levels. *Top*—5 mM phospholipid; *bottom*—20 mM phospholipid.

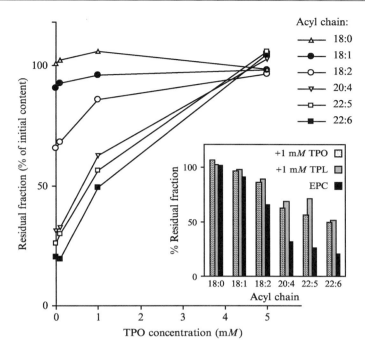

FIG. 5. Tempo inhibits radiation damage induced in PUFA. Liposomal dispersion of 20 mM EPC SUV in saline was γ-irradiated at room temperature with 10 kGy in the presence of concentrations of 0.1–5.0 mM Tempo. Fatty acid composition was determined after methyl esterification followed by GC separation. The residual levels of six unsaturated fatty acids, related to palmitic acid, which served as internal standard, are presented as a fraction of their original levels. (*Inset*) Comparison of the inhibitory effect provided by Tempo (TPO) and Tempol (TPL) against radiation damage induced in PUFA. Liposomal dispersion of 20 mM EPC SUV in saline was γ-irradiated at room temperature with 10 kGy in the presence of 1 mM Tempo or Tempol.

of 5 mM nitroxides does not modify the physical or chemical properties of the liposomal dispersions.

Degradative Damage Upon Long-Term Storage

We have studied the effect of long-term storage under defined conditions (temperature, daylight) on the chemical and physical stability of liposomes and evaluated the protection against liposome degradation provided by cholesterol and antioxidants. Cholesterol in lipid bilayers has a pronounced effect on bilayer characteristics and has been suggested to also play a role as an antioxidant in biological membranes.[43] Liposomes composed of EPC and EPC/cholesterol (10:1 mol/mol) are stored at room temperature, either exposed to or protected from daylight, in the presence or absence of antioxidants. Chemical and physical changes are monitored at several time points to assess the oxidative and hydrolytic degradation of liposomal lipids (Figs. 6 and 7).

No cholesterol degradation occurred in any of the cholesterol-containing dispersions, even after 6 months of storage, whereas PUFA showed $\geq 70\%$ degradation (Fig. 7). These findings point to the fact that phospholipid PUFA are the components that "take the heat," thereby protecting cholesterol, as suggested previously by Lang and Vigo-Pelfrey.[5] However, in the presence of cholesterol, PUFA showed less degradation than in the EPC liposomes (lacking cholesterol), which points to the stabilizing and protective effect of cholesterol on membrane lipids.

Polyunsaturated fatty acids were shown to be the most sensitive part of the liposome to oxidative degradation during long-term storage, and the protective effect against the oxidation of PUFA was most pronounced.

In EPC liposomes, which were clearly more sensitive to degradation than cholesterol-containing liposomes, Tempol (1 mM) was a better antioxidant than vitamin E (1 mM), providing significantly greater protection to PUFA (Fig. 6), a considerable amount persisting after 6 months storage at room temperature, while vitamin E was consumed completely.

Selecting the Desired Nitroxide

To select the nitroxide of choice for the radioprotection of liposomal dispersions, we have studied two nitroxides, Tempo and Tempol, which differ in their lipophilicity. We have also studied the correlation between the radioprotection of the nitroxides and their concentration in the lipid bilayer and in the aqueous phase of the liposome. Coefficients of nitroxide partition between the liposomal lipid bilayer and the aqueous phase are determined using a two-compartment cell specially designed for equilibrium dialysis ($K_{lipid/aq}$ values: Tempo, ~ 31.1; Tempol, ~ 3.5). This method obviates the need to separate the lipid bilayer from the aqueous phase in a liposomal dispersion.

We have demonstrated that the protective effect is dependent on the concentration of the nitroxide(s) in the aqueous phase, with both Tempo and Tempol achieving complete protection at 5 mM (for 20 mM EPC). The protective effect is related neither to the lipophilicity of the nitroxide

[43] L. L. Smith, *Free Radic Biol. Med.* **11**, 47 (1991).

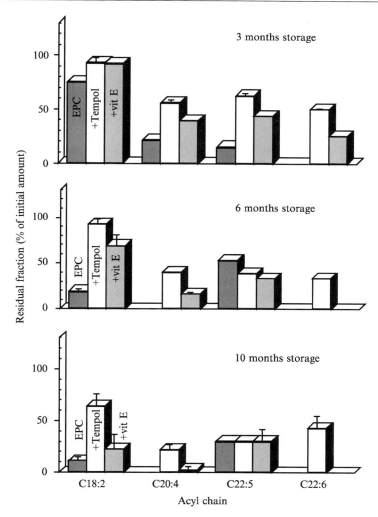

Fig. 6. Effect of Tempol and vitamin E on damage caused to acyl chain residues of EPC liposomes upon long-term storage. Liposomal dispersions of EPC SUVs in 10 mM HEPES buffer (pH 7.4) were stored at room temperature for a period of 10 months, with and without 1 mM Tempol or vitamin E. Acyl chain composition was determined after methyl esterification followed by GC separation. The levels of acyl chains at 3, 6, and 10 months of storage, related to palmitic acid, which served as internal standard, are presented as a fraction of the original levels.

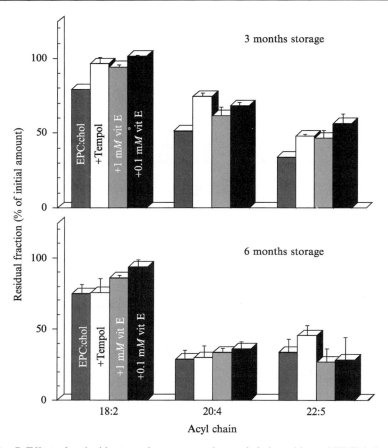

Fig. 7. Effect of antioxidants on damage caused to acyl chain residues of EPC/cholesterol (10:1) liposomes upon long-term storage. Liposomal dispersions of EPC/cholesterol (10:1) SUVs in 50 mM HEPES buffer (pH 7.4) were stored at room temperature for a period of 6 months, with and without 1 mM Tempol, 1 mM vitamin E, or 0.1 mM vitamin E. Acyl chain composition was determined after methyl esterification followed by GC separation. The levels of acyl chains at 3 and 6 months of storage, related to palmitic acid, which served as internal standard, are presented as a fraction of the original levels.

(Tempo $\gg$ Tempol) as assessed by octanol/buffer partition coefficient ($K_{oct/aq}$) nor to its concentration in the egg PC lipid bilayer (Tempo $>$ Tempol). Results suggest that for antioxidants that lack a prooxidant effect and that can be regenerated during the oxidation, a low concentration in the membrane is not an obstacle to the protection against γ-irradiation oxidative damage, as long as sufficient concentrations in the aqueous phase can be achieved.[44]

Methods Used to Evaluate Liposome Physical and Chemical Stability

Chemical and physical changes in the liposomes are studied to evaluate the effect of oxidative damage on the liposomal dispersions. For this purpose, liposome size distribution and pH are measured. Cholesterol degradation is quantified using HPLC, phospholipid hydrolysis by nonesterified (or free) fatty acid (NEFA) levels,[45] and acyl chain oxidation by analyzing phospholipid acyl chain composition using GC.[16]

Liposome Preparation

EPC small unilamellar vesicles (SUV) are prepared by extrusion using the LiposoFast-Basic device (Avestin, Inc., Ottawa, ON, Canada).[46] EPC: cholesterol (10:1) (mol/mol) SUV liposomes are prepared by dissolving the two lipids in *t*-butanol, lyophilizing the lipid mixture, resuspending in buffer, and extruding the suspension using the LiposoFast-Basic device.[44] The final phospholipid concentration is determined using a modification of the Bartlett procedure.[16,47]

Particle Size Determination

Liposome size distribution is determined by photon correlation spectroscopy using a Coulter Model N4 SD apparatus.[16]

Acyl Chain Oxidation

Oxidative damage to each of the individual EPC acyl chains is determined by following changes in acyl chain composition using GC analysis, as described previously by Barenholz and Amselem.[16] In brief, after a Bligh and Dyer extraction,[48] the lower (chloroform-rich) phase is transferred to a small glass bottle, evaporated under N_2 to complete dryness, and dissolved in 50 μl of toluene. Transmethylation is performed by a 30-min incubation at room temperature in the presence of 20 μl of

[44] A. M. Samuni and Y. Barenholz, *Free Radic. Biol. Med.* **22,** 1165 (1997).

[45] H. Shmeeda, S. Even-Chen, R. Nissim, R. Cohen, C. Weintraub, and Y. Barenholz, *Methods Enzymol.* **367,** 272 (2003).

[46] R. C. MacDonald, R. I. MacDonald, B. P. M. Menco, K. Takeshita, N. K. Subbarao, and L. Hu, *Biochim. Biophys. Acta* **1061,** 297 (1991).

[47] N. Düzgüneş, *Methods Enzymol.* **367,** 23 (2003).

[48] E. G. Bligh and W. J. Dyer, *Can. J. Biochem. Physiol.* **37,** 911 (1959).

Meth-Prep II (Alltech, Deerfield, IL). A volume of 2 μl of this mixture is injected into a Perkin Elmer AutoSystem GC and autosampler using a 6-ft. 10% Silar 10C column (Alltech), dry N_2 as the carrier gas, and flame ionization detection. The initial temperature of the run is 140° for 5 min and then the oven temperature is raised, at a rate of 5°/min, to 240°, and kept there for 5 min. Methyl esters are identified by comparing their retention times with those of known standards. Methyl palmitate (C_{16}), a saturated acyl chain that had been found previously to be highly resistant to peroxidation by γ-irradiation[44,49] [confirmed by using methyl pentadecanoate (C_{15}) as an external standard], is selected as an internal reference for the determination of the extent of degradation of acyl chains of EPC.

Determination of Phospholipid Hydrolysis by Quantification of Nonesterified Fatty Acids (NEFA)

The NEFA concentration in the samples is determined using the NEFA C kit (Wako Chemicals, GmbH, Neuss, Germany). The details are described elsewhere.[45] In brief, samples are diluted to adjust the sensitivity range of the kit and 60-μl aliquots of the diluted dispersion are drawn into Nunclon microwell plates. To all samples, 20 μl of 20% Triton X-100 in water (fresh solution) is added to solubilize the liposomal dispersion. Following the addition of color reagents A and B and proper incubation times at 37°, the optical density of the samples at 540 nm is measured using a Biochromatic ELISA reader (Labsystems Multiskan, Finland). The NEFA concentration in the samples is calculated using the kit standard curve.

Quantification of Cholesterol

The amount of cholesterol degradation is quantified by HPLC, from the reduction in the cholesterol level and the appearance of the major degradation product of cholesterol, 7-keto-cholesterol, following extraction of the aqueous liposomal dispersion using the procedure of Ansari and Smith.[50] Extraction is performed using the Dole extraction procedure,[16] and the heptane-enriched upper phase, containing >98% of the cholesterol and 7-keto-cholesterol, is concentrated and analyzed. The analysis is performed at ambient temperature on an Econosphere silica column (10 × 0.46 mm i.d.) using a silica precolumn (Alltech). The mobile phase consists of hexane:isopropanol (500:6, v/v) at a flow rate of 1 ml/min. HPLC is carried out using a Kontron (Switzerland) HPLC system: 425 pump, 430 detector,

[49] N. J. Zuidam, C. Versluis, E. A. A. M. Vernooy, and D. J. A. Crommelin, *Biochim. Biophys. Acta* **1280,** 135 (1996).

[50] G. A. S. Ansari and L. L. Smith, *J. Chromatogr.* **175,** 307 (1979).

460 automatic injector, and 450 data analysis system. Spectrophotometric detection is carried out at 212 nm.

Methods Used for the Evaluation and Quantification of Nitroxides

Electron Paramagnetic Resonance (EPR) Measurements

EPR spectrometry is employed to detect and follow the nitroxide-free radicals using a JES-RE3X ESR spectrometer (JEOL, Tokyo, Japan). Samples are drawn by a micropipette into a gas-permeable Teflon capillary of 0.81 mm i.d., 0.05 mm wall thickness, and 15 cm length (Zeus Industrial Products, Raritan, NJ). Each capillary is folded twice, inserted into a 2.5-mm-i.d. quartz tube open at both ends, and placed in the EPR cavity. EPR spectra are recorded with the center field set at 3361 G, 100-kHz modulation frequency, 1-G modulation amplitude, and nonsaturating microwave power.

Nitroxides decay in biological systems predominantly through a one-electron reduction, yielding the respective cyclic hydroxylamines.[51] For determination of the total concentration of nitroxide + hydroxylamine, the hydroxylamine is oxidized by 1 mM ferricyanide or H_2O_2 + NaOH, final concentration 0.3% and 0.01 N, respectively.[52]

Partition of Nitroxide Between Lipid Bilayer and Saline

Partitioning of the nitroxides between the liposome lipid bilayer and the aqueous phase is measured using a two-compartment Lucite cuvette specially designed for equilibrium dialysis. The two compartments, each of volume 0.3 ml, are separated by a dialysis membrane with a molecular weight cutoff of 12–14 kDa. One compartment contains the liposomal dispersion and the other compartment contains saline. The samples are incubated under continuous shaking at room temperature to allow the dialysis membrane-permeable nitroxide to equilibrate between the two compartments. Following a 24-h incubation, samples from the liposome-free and liposome-containing compartments are taken, scanned in the EPR spectrometer, and the respective intensities $C_{liposome}$ and C_{aq} of the EPR signal of nitroxide in each compartment are compared. Because nitroxide EPR signals in the lipid and aqueous phases are different, and the differences are dependent on the specific nitroxide, the intensities of the EPR signal of Tempo and Tempol at various lipid concentrations are measured, and

[51] H. M. Swartz, M. Sentjurc, and P. Morse, *Biochim. Biophys. Acta* **888,** 82 (1986).
[52] E. G. Rozantsev and V. D. Sholle, *Synthesis* **190,** (1971).

calibration curves are constructed by means of which the actual nitroxide concentrations in each compartment are calculated.

Conclusions

PUFA are the components of the liposome bilayer most sensitive to radiation-induced damage and to oxidative degradation during long-term storage. Both Tempo and Tempol provide similar radioprotection to liposomal lipids. Their protective effect cannot be correlated with their lipid-bilayer/aqueous partition coefficient. It can best be correlated with their concentration in the aqueous phase. Lipophilic nitroxides show no advantage over hydrophilic nitroxides for protection against oxidative damage in liposomal preparations. Nitroxides themselves cause neither physical nor chemical changes in liposomal dispersions. EPC liposomes are more sensitive to degradation during storage than EPC/cholesterol liposomes. Cholesterol in the lipid bilayer has a stabilizing and protective effect, mainly due to decreasing lipid-bilayer hydration. Tempol provides significantly greater protection than vitamin E to EPC liposomal PUFA against degradation during long-term storage.

Acknowledgments

The support of the U.S.–Israel Binational Science Foundation (BSF) for the development of the choline-phospholipid determination and the Israel Science Foundation (ISF) for the development of the nonesterified cholesterol assay is gratefully acknowledged.

Author Index

Numbers in parentheses are footnote reference numbers and indicate that an author's work is referred to although the name is not cited in the text.

A

Abadi, N., 246(17), 247
Abernethy, D. R., 231
Abra, R. M., 19
Abramovic, Z., 267
Adamis, A. P., 209
Afzal, V., 303
Agrawal, S., 171, 172, 186, 203, 205(53), 210, 221(1), 231
Ahmad, A., 230, 231
Ahmad, I., 230, 232
Ahmad, L., 231
Ahn, M. S., 303(29), 304, 305(31)
Ahn, N. G., 249
Akita, R. W., 246(17), 247
Aksentijevich, I., 199
Alakoskela, J.-M., 85, 86(16a)
Albericio, F., 54, 63(37)
Alessi, D. R., 247
Alford, D. R., 146
Alila, H., 213, 214(17)
Allen, T. M., 3, 14, 50, 51, 51(4), 52, 53, 54(33), 56(33), 62, 62(12), 63(33), 65(15; 33), 66(15), 68, 69, 69(12; 51), 134, 137, 171, 174, 175, 177(39), 178(39), 179, 179(39), 181, 181(38), 185, 185(40), 187(38), 188(37–39), 190(7), 191, 198, 198(7), 231, 258, 259, 259(19; 20), 260
Allevi, P., 304, 305(36)
Altendorf, K., 139
Alton, E. W. F. W., 191
Amiji, M., 288, 292(11), 298(11), 299(11)
Amiot, M., 16
Amselem, S., 301, 311(16), 312(16)
Anderson, K. C., 230(8), 231
Anderson, P. M., 185
Anderson, R., 34(9), 35
Anderson, R. G., 34(8), 35
Anderson, R. U., 93
Anderson, V. C., 154, 155, 156(25; 35)
Anderson, W. B., 231
Andjelkovic, M., 247
Andreeff, M., 251
Ankel, E. G., 303
Ansari, G. A. S., 312
Anscher, M. S., 83
Ansell, S. M., 3, 114, 115(7), 117(7), 123(7), 124(7), 126(7), 127(7), 128(7), 183, 188(47)
Antao, V. P., 145(32), 146, 212
Antimisiaris, S. G., 260
Antony, A. C., 33
Antopolsky, M., 223
Antosiewicz, J., 304
Anyarambhatla, G. R., 75, 84, 87(7), 93(6; 7), 106(7), 112, 257
Aoki, M., 199
Archer, W. I., 271
Armstrong, W. D., 262
Aronovitch, J., 304
Arora, V., 172
Arsenault, A. L., 115
Aruffo, A., 16, 20
Arveiler, B., 214
Asselbergs, F., 223
Auriola, S., 223
August, J. T., 16
Auriola, S., 223
Avigan, M., 242
Azhayeva, E., 223
Azuma, C., 257

B

Bacus, S. S., 246(16), 247
Bae, S. K., 73
Bae, Y. H., 76

M

Subject Index

A